Promoting pollination and pollinators in farming

It is widely recognised that agriculture is a significant contributor to global warming and climate change. Agriculture needs to reduce its environmental impact and adapt to current climate change whilst still feeding a growing population, i.e. become more 'climate-smart'. Burleigh Dodds Science Publishing is playing its part in achieving this by bringing together key research on making the production of the world's most important crops and livestock products more sustainable.

Based on extensive research, our publications specifically target the challenge of climate-smart agriculture. In this way we are using 'smart publishing' to help achieve climate-smart agriculture.

Burleigh Dodds Science Publishing is an independent and innovative publisher delivering high quality customer-focused agricultural science content in both print and online formats for the academic and research communities. Our aim is to build a foundation of knowledge on which researchers can build to meet the challenge of climate-smart agriculture.

For more information about Burleigh Dodds Science Publishing simply call us on +44 (0) 1223 839365, email info@bdspublishing.com or alternatively please visit our website at www.bdspublishing.com.

Related titles:

Pesticides and agriculture
Print (ISBN 978-1-78676-276-4); Online (ISBN 978-1-78676-279-5, 978-1-78676-278-8)

Integrated management of insect pests: Current and future developments
Print (ISBN 978-1-78676-260-3); Online (ISBN 978-1-78676-262-7, 978-1-78676-263-4)

Climate change and agriculture
Print (ISBN 978-1-78676-320-4); Online (ISBN 978-1-78676-322-8, 978-1-78676-323-5)

Biopesticides for sustainable agriculture
Print (ISBN 978-1-78676-356-3); Online (ISBN 978-1-78676-358-7, 978-1-78676-359-4)

Reconciling agricultural production with biodiversity conservation
Print (ISBN 978-1-78676-348-8); Online (ISBN 978-1-78676-350-1, 978-1-78676-351-8)

Chapters are available individually from our online bookshop:
https://shop.bdspublishing.com

climate•SMART•publishing
IN AGRICULTURAL SCIENCE

BURLEIGH DODDS SERIES IN AGRICULTURAL SCIENCE
NUMBER 126

Promoting pollination and pollinators in farming

Edited by Emeritus Professor Peter Kevan and Dr D. Susan Willis Chan, University of Guelph, Canada

burleigh dodds
SCIENCE PUBLISHING

Published by Burleigh Dodds Science Publishing Limited
82 High Street, Sawston, Cambridge CB22 3HJ, UK
www.bdspublishing.com

Burleigh Dodds Science Publishing, 1518 Walnut Street, Suite 900, Philadelphia, PA 19102-3406, USA

First published 2023 by Burleigh Dodds Science Publishing Limited

Library of Congress Control Number: 2022949278

British Library Cataloguing in Publication Data
A catalogue record for this book is available from the British Library

ISBN 978-1-80146-098-9 (print)
ISBN 978-1-80146-101-6 (PDF)
ISBN 978-1-80146-100-9 (ePub)
ISSN 2059-6936 (print)
ISSN 2059-6944 (online)

DOI 10.19103/AS.2022.0111

Typeset by Deanta Global Publishing Services, Dublin, Ireland

Contents

Series list

Title	Series number
Achieving sustainable cultivation of maize - Vol 1 From improved varieties to local applications Edited by: Dr Dave Watson, CGIAR Maize Research Program Manager, CIMMYT, Mexico	001
Achieving sustainable cultivation of maize - Vol 2 Cultivation techniques, pest and disease control Edited by: Dr Dave Watson, CGIAR Maize Research Program Manager, CIMMYT, Mexico	002
Achieving sustainable cultivation of rice - Vol 1 Breeding for higher yield and quality Edited by: Prof. Takuji Sasaki, Tokyo University of Agriculture, Japan	003
Achieving sustainable cultivation of rice - Vol 2 Cultivation, pest and disease management Edited by: Prof. Takuji Sasaki, Tokyo University of Agriculture, Japan	004
Achieving sustainable cultivation of wheat - Vol 1 Breeding, quality traits, pests and diseases Edited by: Prof. Peter Langridge, The University of Adelaide, Australia	005
Achieving sustainable cultivation of wheat - Vol 2 Cultivation techniques Edited by: Prof. Peter Langridge, The University of Adelaide, Australia	006
Achieving sustainable cultivation of tomatoes Edited by: Dr Autar Mattoo, USDA-ARS, USA and Prof. Avtar Handa, Purdue University, USA	007
Achieving sustainable production of milk - Vol 1 Milk composition, genetics and breeding Edited by: Dr Nico van Belzen, International Dairy Federation (IDF), Belgium	008
Achieving sustainable production of milk - Vol 2 Safety, quality and sustainability Edited by: Dr Nico van Belzen, International Dairy Federation (IDF), Belgium	009
Achieving sustainable production of milk - Vol 3 Dairy herd management and welfare Edited by: Prof. John Webster, University of Bristol, UK	010
Ensuring safety and quality in the production of beef - Vol 1 Safety Edited by: Prof. Gary Acuff, Texas A&M University, USA and Prof. James Dickson, Iowa State University, USA	011
Ensuring safety and quality in the production of beef - Vol 2 Quality Edited by: Prof. Michael Dikeman, Kansas State University, USA	012
Achieving sustainable production of poultry meat - Vol 1 Safety, quality and sustainability Edited by: Prof. Steven C. Ricke, University of Arkansas, USA	013
Achieving sustainable production of poultry meat - Vol 2 Breeding and nutrition Edited by: Prof. Todd Applegate, University of Georgia, USA	014

Introduction

It has been reported that up to 95% of all flowering plants require the services of other organisms to move pollen from male to female flower parts during the pollination process. These organisms, including bees, are collectively known as pollinators. However, in light of the growing evidence of global declines in pollinator species, the management, ecology and conservation of wild and managed pollinators is a subject of growing importance and research activity.

This volume reviews the wealth of research on our current understanding of existing pollination processes and their importance to our global ecosystems. The book considers how pollinators interact with plants, as well as the major threats to pollinator species, including climate change, diseases and pesticide exposure. The book is split into three parts: Part 1 chapters focus on understanding pollinators and pollination, focusing specifically on their role in agriculture, the role and application of olfaction and the use of wind pollination in crop plants. Chapters in Part 2 examine the various threats to pollinators, such as the impact of climate change, disease, the use of neonicotinoid insecticides and pesticides as well as the impact of alien bees on native ones. Part 3 chapters review promoting pollinators and pollination, focusing specifically on the role of habitat conservation and restoration, the alteration of crop management practices and the use of ecological network approaches. Chapters also highlight best management practices for pollinator protection, focusing specifically on US apple production and the use of entomovectoring technology.

Part 1 Understanding pollinators and pollination

The first chapter of this volume provides an overview of what pollination is and the role of pollinators in agriculture. Chapter 1 begins by discussing the basics of pollination and goes on to discuss pollinators and their diversity. The chapter moves on to review the ecology and evolution of floral traits, focusing specifically on the origins of flowering plants and their relationship with pollinators, floral traits and the ecology of pollination as well as the co-evolution of plants and pollinators. A section on domestication and its impact on plant-pollinator relationships is also provided, which is then followed by an analysis of how pollinators impact agriculture and how pollinators are present in modern agriculture practices.

The next chapter examines the role and application of olfaction in crop plant-pollinator interactions. Chapter 2 introduces the background and theory underpinning the use of odours by insects in pollinators. It discusses how flowers produce odours and highlights issues specific to crops such as selective breeding. It then explores current technologies and case studies in which natural or synthetic odours on or near the crop, and the interaction with insects, influences visitation, pollination success and yield.

Chapter 3 reviews the role of wind pollination in crop plants. This chapter reflects on the early transition from wind-pollinated species (especially Poaceae) into crop species and the current state of wind-pollination in agricultural pollination networks. It then provides a short review of the evolution and biomechanics of wind pollination to establish why humans moved away from open pollination. A more deep exploration of the relative importance (or lack thereof) in crops such Zea, Triticum, Oryza, and Secale follows, as does a focus on the extant risk of heterospecific pollen transfer to wind-pollinated weeds. The chapter also discusses the role of anthropogenic climate change on wind-pollinated crops and the future and relative importance of wind-pollination in widespread and niche crops. The chapter ends with a case study of how one might create a 'win-win' in terms of conservation and restoration of natural habitats to promote wind-pollination in farming.

Part 2 Threats to pollinators

Part 2 opens with a chapter that focuses on assessing the impacts of climate change on pollinators. Chapter 4 first presents an overview of observed and predicted impacts of climate change on pollinators, focusing on elements such as rising temperatures, loss of habitat, temporal and spatial mismatches between plants and pollinators and extreme events. The chapter also highlights the effect climate change has on pollinator cues, factors that may mediate species' responses, including synergisms with other threatening processes. A section emphasising the importance of understanding climate change impacts on pollinators is also provided, which is then followed by a section presenting a number of ways in which we can help pollinators face the challenge of climate change.

Chapter 5 looks at the impact of diseases on pollinators. The chapter first provides an analysis of various honey bee diseases, focusing specifically on ectoparasitic mites, viruses, fungi, bacteria, trypanosomes and common pests such as the small hive beetle and wax moths. It also discusses how disease impact on honey bee colonies and pollination can be measured. A section on the poorly known wild bee diseases is also provided, discussing the major known pests and pathogens of both bumble bees and other solitary bee species. The chapter reviews disease transmission, spillover and spillback in all three pollinator taxa as well as some of the current methods and approaches for controlling pests and pathogens.

The next chapter focuses on how neonicotinoid insecticides affect bees and other pollinators. Chapter 6 discusses the wide range of research on how neonicotinoids affect both managed and wild bee populations in particular. It reviews levels of environmental contamination and how this affects topical and oral exposure routes. The chapter also addresses sublethal effects and briefly discusses the interaction of different stressors. It includes a number of case studies showing the negative effects of neonicotinoids on a range of bee species.

Moving on from Chapter 6, Chapter 7 examines the impact of pesticides on pollinators. The chapter begins by providing an overview of the origins of bee testing and risk assessment, then moves on to discuss the testing of pesticide effects in pollinators, drawing attention to species testing, testing methodologies and designs and guidelines and risk assessment guidance documents. The chapter also examines test method development and validation and highlights the principles of Good Laboratory Practice. A section on the ecotoxicological risk assessment for pollinators is included, which is then followed by an analysis of both the indirect and sublethal effects of pesticides. Risk mitigation and pesticide incident monitoring are also discussed.

The final chapter of Part 2 reviews the impact of alien bees on native ones. Chapter 8 explores the main impacts of alien bees on native bees through competition for food or nesting resources, interference, pathogen spillover and genetic contamination. Implications for native bee conservation are also discussed. In addition, in this chapter, a particular focus on the first alien bee that colonised Europe, *Megachile sculpturalis* is also emphasised. The chapter also highlights the main knowledge gaps and important trends for future research. Finally, avenues for managing alien bee species and preventing their introduction are provided.

Part 3 Promoting pollinators and pollination

Chapter 9 focuses on challenges and options in habitat restoration for solitary bees which account for 90% of bee species. As well as being important parts of local ecosystems, these species are pollinators of a wide range of crops such as alfalfa, tomato, eggplant and blueberries, cucurbit crops (such as pumpkin, squash and watermelon) as well as orchard crops such as almonds and cherries. The chapter looks at challenges facing habitat restoration and research on the effectiveness of existing restoration projects. The chapter also includes case studies on commercial management of Alkali bees, promoting stem or cavity-nesting bees and providing underground nesting boxes for bumblebees.

The next chapter of Part 2 draws attention to the alteration of crop management practices to promote pollinators. Chapter 10 begins by introducing the current approaches that are currently used when it comes to managing production space for pollinators within the context of various cropping systems, such as row crops, specialty crops, perennial orchards and perennial forage and pasture systems. The chapter then goes on to provide several case studies demonstrating these approaches, drawing specific attention to annual row and specialty crop systems as well as perennial orchard and forage systems. Assessing the efficacy of alternative agronomic practices is also discussed.

The subject of Chapter 11 is the use of ecological network approaches for promoting pollinators in agriculture. The chapter begins by describing the important and agriculturally relevant characteristics and structures in plant-pollinator networks. It then moves on to examine the use of ecological networks at the field, farm, landscape and national scale. The chapter also highlights the importance of filling in major gaps in pollinator networks. A section on embedding pollination within wider agroecosystem networks is also provided, which is followed by an analysis of the potential next steps for pollination network research.

Chapter 12 describes a set of pollinator best management practices (BMPs) for US apple production. It first provides an overview of apple orchard pollinators, drawing specific attention to ground nesting bees, mason bees, small carpenter bees and wild bumble bee queens. The chapter moves on to provide an overview of the various pests and diseases that can affect apple production. This is then followed by a section that highlights the potential hazards to pollinators from apple pest management practices. The chapter also examines best management practices to promote pollinators as well as the use of integrated pest and pollinator management and pesticide application to manage apple pests and diseases.

The final chapter of the book focuses on using pollinators to spread biocontrol agents via the use of entomovectoring technology. Chapter 13 first highlights the factors that are important for successful entomovectoring, such as the type of vector, the dispenser, the biocontrol agent product and biosafety. The chapter moves on to provide successful and failed attempts at using bumble bees as vectors for entomovectoring. This section is then followed by a case study that focuses on entomovectoring (also known as apivectoring) of microbial biological control agents by *Bombus* spp., especially *B. terrestris*, against grey mould *Botrytis cinerea* and other pests in strawberry production, drawing specific attention to the importance of tackling and controlling *Botrytis cinerea* infection in strawberries.

Acknowledgement

Cover image copyright

- Cover image: A male hoary squash bee on staminate pumpkin flower. 2011. © Sheila Macleod Potter.

Part 1

Understanding pollinators and pollination

Chapter 1

What is pollination and what are pollinators in agriculture?

Seanne Clemente and Lynn Adler, University of Massachusetts, USA

1 Introduction

The Altamira cave system in northern Spain is renowned for the contemporary depictions of Paleolithic life that adorn its walls in vivid blacks, reds, and ochres. But tucked away from the main chamber's iconic herds of bison is an inconspicuous painting in an unassuming side chamber. It depicts a gangly human figure climbing a winding structure - most likely, a ladder. At the upper end of the ladder are four thick, overlapping arches nested within each other, surrounded by thin winged figures - the nest of a wild honey bee colony, and the bees themselves. This painting at Altamira, dated 13 500 years old, is a prehistoric account of beekeeping - the oldest one known to this day (Crane, 1999). Similar paintings depicting ladders, bees' nests, and bees are found in prehistoric art across the world: primarily in South Africa, but also in Australia, India, France, and Spain (Crane, 1986). They are a testament that even in the earliest chapters of human history, humankind has relied on the activities of pollinators.

Honey bees, their flower-visiting habits, and their industrious honey-making skills have undoubtedly been appreciated by the humans of the ancient world, appearing in Aristotle's *History of Animals*, Virgil's *Georgics*,

http://dx.doi.org/10.19103/AS.2022.0111.01

and the Old Testament. The Romans and Greeks were well aware of the importance of pollination in the production of date palms (*Ficus carica*); the polymath Theophrastus (c 371–287 BC) described the plant's 'male and female' flowers and the necessity of a 'union' between two flower types for the trees to bear fruit. But it was only much later, in the late seventeenth century, that Rudolf Jakob Camerarius (1665–1721) observed that male anthers and female stigmas were prerequisites to plant reproduction. His work was built upon by the investigations of several botanists, including Bradley (1731), Miller (1724), and Vaillant (1718); by the early eighteenth century, it was widely accepted that pollen, known as the *Farina Fecundens*, was the mode of sexual transmission of plants.

The work of Kölreuter (1733–1806) was fundamental in establishing the importance of insect vectors in the transport of pollen, the structure and mechanism of pollen, and the role of nectar in attracting pollinators. Sprengel (1750–1816) further examined the structure and function of flower parts, describing in detail a diversity of floral adaptations involved in attracting pollinators. Darwin's (1809–1882) publication of *The Origin of Species* and his studies on pollination mechanisms (notably his work with orchids) were followed by a surge of interest in pollination; this included Hermann Müller's (1829–1883) and Federico Delpino's (1868–1875) investigations into a myriad of plant-pollinator systems and Knuth's seminal Handbook of Flower Pollination series (Knuth 1906–1909). With the twentieth century came the advent of genetics, the discovery of DNA, and an explosion of interest in plant ecology. Such discoveries shifted the study of pollination toward understanding the molecular mechanisms, evolutionary history, and interspecies relationships that govern pollination. There are countless contributors to this recent era of pollination studies, in addition to others that were crucial in the discovery of pollination and its basic mechanisms. Their work is beyond the scope of this brief introduction; for further reading, Proctor (1996a) provides a comprehensive summary.

2 The basics of pollination

Pollination is a key first step in plant reproduction. But to understand pollination, one must first be familiar with how pollen is produced. Pertinent floral structures are shown in Fig. 1. The *corolla*, which is the shape and structure formed by a flower's petals, serves chiefly to attract pollinators. The *calyx* (or sepals) protects the flower in the bud and serves as a structural support for the corolla. The *androecium* (or *stamens*) are the male reproductive structures of flowers consisting of *anthers* that each rest atop a structural *filament*. The *gynoecium* is the collective female reproductive structure comprised of individual *stigmas* resting upon a *style* that connects to the *ovary*, where the flower's *ovules* reside.

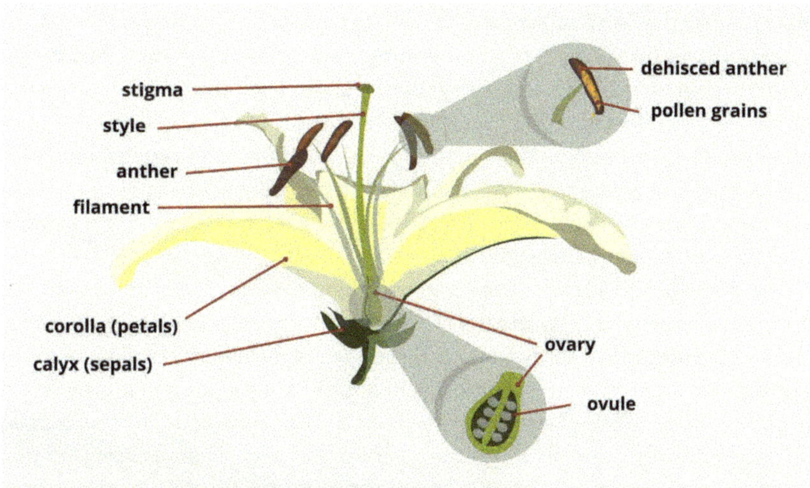

Figure 1 Cross-section of a mature flower and its components.

The anthers are the site of pollen production. At maturity, the inner cells of the anther undergo meiosis to produce individual pollen grains. The outer layers of the anthers then split open, or *dehisce*, to release pollen. Appearing as a powdery, fine substance to the naked eye, pollen is made up of microscopic grains, each surrounded by an outer wall, or *exine*, that gives the grains their distinct shape and rigidity. The aid of a microscope will reveal that the pollen exines are covered in pores. An inner wall of cellulose and pectin, known as an *intine*, further protects the interior of the grain, which contains the gametes in the form of a *generative nucleus* and a supplemental *vegetative nucleus*.

Pollination is simply the transfer of pollen from the anthers of the androecium to the stigmas of the gynoecium. Following pollination is *fertilization*, which begins as the pollen grain attaches to the stigma, imbibes water, and *germinates* to form a pollen tube that penetrates the stigma's tissue and carves a path through the *style* to ovules in the *ovary*. The grain's vegetative nucleus moves through the newly formed tube, mediating the movement of the generative nucleus (Zhou and Meier, 2014), which has now divided into two separate male *gametes*. Upon arrival at the ovule, the vegetative nucleus dissolves, and one of the male gametes fuses with the egg to form a diploid embryo. The second male gamete fuses with a *diploid fusion nucleus* in the ovary, forming the endosperm that provides nutrients for the newly formed seed.

In angiosperms (flowering plants), most species bear hermaphrodite *perfect* flowers, containing both pollen-producing and ovule-bearing structures. An obvious consequence of this hermaphroditism is *autogamy*, the self-fertilization of a flower by its own pollen. Plants may also self-fertilize through *geitonogamy*, the fertilization of a flower by the pollen of another flower on the same plant. About 20% of angiosperms reproduce predominantly by self-fertilization (Barrett, 2003), and many plants reproduce through self-fertilization to some degree (Barrett, 2003; Goodwillie et al., 2005; Charlesworth, 2006; but see Lande and Schemske, 1985; Igić and Kohn, 2006). Self-fertilization grants the advantage of guaranteed reproduction, especially in environments where mates or pollinators are scarce (Kalisz et al., 2004; Aizen and Harder, 2007; reviewed in Busch and Delph, 2012).

The alternative reproductive strategy is *outcrossing*, fertilization with non-self-pollen. Darwin (1877) wrote that '[Nature] abhors perpetual self-fertilization'. Although inbreeding depression – the reduced survival and fertility of inbred offspring – is surprisingly uncommon in predominantly selfing plants (Husband and Schemske, 1996), self-fertilization is disadvantageous to outcrossing when considering gene flow, genetic variability, and diversification of plant lineages (Goldberg et al., 2010). Outcrossing plants employ mechanisms to minimize self-fertilization (Barrett, 2003). A common anti-selfing mechanism in plants is the spatial (*herkogamy*) and temporal (*dichogamy*) separation of pollen and stigma. In herkogamy, the anthers and stigma are physically separate, such that pollen cannot be passively transferred to the stigma. In dichogamy, the anthers mature asynchronously with the receptive period of the stigmas. Rather than relying on spatio-temporal separation, some plants separate anthers and stigmas into unisexual *imperfect flowers*, which can be a more effective mechanism to avoid self-fertilization. *Monoecious* plants bear both male and female imperfect flowers on the same plant (Fig. 2b); the plants are still hermaphroditic and may self-fertilize. Monoecious crops include corn, oil palms, and members of the Cucurbitaceae family (e.g. melons, pumpkins, cucumbers, and zucchini). *Dioecious* plants only bear flowers of one sex and must rely on a plant with opposite-sexed flowers to reproduce (Fig. 2c) – in this way, dioecy is the only mechanism that fully prevents self-fertilization. Dioecious crops are uncommon; some examples are asparagus, kiwi fruit, and hops. Finally, plants may prevent self-fertilization through *self-incompatibility*, a variety of biochemical barriers to the fertilization process. The mechanisms governing self-incompatibility vary widely across taxa; Glover (2014) provides a review. One mechanism of self-incompatibility is evident to many humans across the globe: the allergic response to pollen, commonly known as hay fever, is caused by self-incompatibility proteins that coat the pollen grain surfaces of some plant species (Proctor, 1996b).

Pollination vectors transport pollen from anthers to stigmas. Pollen transfer can be achieved abiotically, through wind (anemophily) or water (hydrophily). The vast majority of abiotically pollinated plants are anemophilous (Ackerman, 2000), with nearly all gymnosperms (Faegri and van der Pijl, 1979) and 10% of all flowering plants transporting their pollen through the wind (Friedman

Figure 2 Floral traits and reproductive systems vary widely across plant taxa, including crop plants. (A) *Brassica napus*. The oilseed rape has bright, four-petalled flowers with four prominent anthers and two shorter anthers. Most other *Brassica* crops (e.g. kale, turnips, and bok choy) display similar four-petalled, yellow flowers. (B) *Cucurbita pepo*. Cucurbit crops, such as squash, pumpkin, melons, and cucumber, are monoecious. Individual plants produce both pollen-producing male flowers and stigma-bearing female flowers. On female flowers, the stigma and style connect to the developing fruit at the flower's base. (C) *Actinidia deliciosa*. Kiwifruit vines are dioecious, with male and female flowers occurring on separate plants. The male flowers bear numerous showy anthers. Prominently displayed in the centers of female flowers are fertile stigmas surrounded by a crown of sterile 'false anthers'. (D) The floral head, or *capitulum*, of a sunflower, *Helianthus annus*, is made up of many small ray florets (*rf*), with yellow petals surrounding a head of disk florets (*df*). Composite flower heads are characteristic in Asteraceae plants. (E) *Medicago sativa*. The complex flowers of alfalfa are wholly reliant on pollinators. The flowers' stigmas and anthers are sprung up inside two 'keel petals.' When a bee lands on the keel petal, it 'trips' the flower and the reproductive structures are explosively unfurled, releasing pollen. In the untripped flower (*un*), the intact keel petals enclose the reproductive structures. The exposed stigma and a remaining keel petal are visible on the tripped flower (*tr*). (F) *Oryza sativa*. Most major cereal crops, like rice shown here, are either self- or wind-pollinated. Characteristic of Poaceae plants: the inflorescence of rice bears dull, inconspicuous florets with many anthers.

and Barrett, 2009). Several key cereal crops, including maize (*Zea mays*), rye (*Secale cereale*), and foxtail millet (*Setaria italica*), are strictly anemophilous (Saunders, 2018). Anemophilous flowers follow an archetypal structure or *syndrome*: small, inconspicuous floral displays and copious production of dry, powdery pollen. Wind-pollinated plants maximize pollination success by producing large amounts of pollen. A single inflorescence of rye, for example, produces over 67 000 pollen grains (Piotrowska, 2008). Furthermore, anemophilous flowers commonly bear stigmas with large surface areas, increasing the 'target area' for pollen to land on (Proctor, 1996c). There are no crops that are pollinated via hydrophily - hydrophily is relatively rare, occurring in about 2.7% of angiosperm families (Ackerman, 2000) and 2% of plants (Faegri and van der Pijl, 1979). Unlike anemophilous plants, hydrophilous flowers do not follow a syndrome; this reflects the scattered and diverse origins, occurrence, and modes of hydrophily across plant taxa (Proctor, 1996c). Generally, hydrophilous pollen floats across or under water to reach mature stigmas. Hydrophilous plants are more diverse in their adaptations but generally rely on elongated stigmas and pollen modifications such as larger size or stickiness that allow grains to accumulate in 'rafts' (Cox, 1988). Wind and water are commonly used but often unreliable modes of transferring pollen to another flower of the same plant species: relying on a fluid pollination vector is a game of chance, dependent on the direction and flow of the fluid. As a protein-rich substance, pollen is costly to make, and in abiotic pollination a very low proportion of pollen ends up fertilizing ovules. A vast majority of flowering plants have instead evolved to rely on animals as pollination vectors.

3 Pollinators and their diversity

Pollinators are biotic pollen vectors: animals that visit flowers and transfer pollen. The importance of animal pollination cannot be overstated: 87.5% of angiosperms are pollinated by animals in some form (Ollerton et al., 2011). Many animals visit flowers incidentally or occasionally to consume nectar or pollen. Although these visitors may sometimes transfer pollen - and are in a sense still 'pollinators' - their efficiency and significance as pollinators are negligible compared to those that regularly visit flowers and have co-evolved specialized structures and behaviors for collecting floral rewards and facilitating pollination. Pollinators' visitations to plants can be described as *generalist* - involving a wide array of flower taxa, or *specialist* - sharing pollination relationships with one or a few species of plants. But pollinators themselves are not dichotomously generalist nor specialist, instead falling within a continuum of specialization ranging from extreme generalization to obligate specialization (Waser et al., 1996; Waser, 2006).

The pollination services provided by insects are undoubtedly the most consequential form of biotic pollination. Most insect pollinators are in the orders: Coleoptera (beetles), Lepidoptera (butterflies and moths), Hymenoptera (bees, ants, wasps,) and Diptera (flies), but significant pollinators are also in the insect orders: Plecoptera (stoneflies), Hemiptera (bugs), Mecoptera (scorpionflies), and Thysanoptera (thrips). Incidental insect visitors have also been observed in nine other orders of insects (Yeo, 1996; Ollerton, 2017). However, by far the most prolific flower visitors and important insect pollinators are the bees (Anthophila), which have co-evolved a suite of specialized pollen-collecting structures and behaviors with a suite of flowering plant families (Neff and Simpson, 1996; Danforth, 2007). There are 20000 bee species worldwide, and with few exceptions, all are pollinators (Ascher and Pickering, 2018). Eusocial bees live in populous colonies with reproductive queens and males and sterile 'workers.' Eusociality can be found in many species in the family Apidae (honey bees, bumble bees, carpenter bees, and stingless bees) and the family Halictidae (sweat bees). But over 90% of bee species are instead *solitary* - females are fertile, build their own nests, and forage alone for floral resources for their offspring (Danforth et al., 2019a).

Significant pollinators are also found in the insect families Lepidoptera, Diptera, and Coleoptera. Dipterans are the second most frequent flower visitors after bees (Larson et al., 2001), with visitation observed in at least 86 families of Diptera, especially in the families Syrphidae (hover flies), Bombyliidae (bee flies), Empididae (dance flies), Tabanidae (horse and deer flies), and Muscidae (muscid flies) (Winfree et al., 2011; Inouye et al., 2015). Dipterans are primary pollinators of cacao, mango, cashew, cherry, and tea; they have also been used to substitute hand pollination in small-flowered plants such as onions, parsnips, carrots, celery, and brassica crops (Free, 1970; Rader et al., 2012, 2016; Inouye et al., 2015). Nectar-feeding Lepidopterans are also prevalent pollinators, with over 140000 flower-visiting species; they are the most diverse group of pollinators (Ollerton, 2017). Flower visitors are prevalent in the butterfly families Hesperiidae (skippers) and Papilionidae (common butterflies). Moths are major nocturnal pollinators (Willmer, 2011), particularly those in the families Sphingidae (hawkmoths), Noctuidae (owlet moths), and Geometridae (geometer moths) (Winfree et al., 2011). In North America and Europe, Lepidopterans are likely not crucial to crop pollination but are essential in maintaining the biodiversity of agroecosystems by serving as pollinators of non-crop plants (Hahn and Brühl, 2016). Coleopterans are particularly important pollinators in tropical ecosystems, but compared to the insect groups discussed so far, their role as pollinators is least understood (Wardhaugh, 2015; Sayers et al., 2019). Flower visitors have been observed in 60 beetle families, with pollinators found in the families Cantharidae (e.g. soldier beetles), Cerambycidae (longhorn beetles), Chrysomelidae (leaf beetles), Curculionidae (weevils), Elateridae (click beetles), Melyridae (e.g. soft-winged flower beetles), Mordellidae (tumbling flower beetles), Nitidulidae (sap beetles), Scarabaeidae (scarabs), and Staphylinidae (rove beetles) (Sayers et al., 2019). They are primary pollinators of oil palms, sugar and custard apples, soursop, cherimoya, and nutmeg (Irvine and Armstrong, 1990; Westerkamp and Gottsberger, 2000; Rader et al., 2016).

Pollinators can also be found in various non-insect taxa including birds, mammals, reptiles, crustaceans, and marine worms (Ollerton, 2017). Of these, birds and bats are of particular importance for flowering plants. Bird pollinators contribute to the pollination of 65 flowering plant families and are important pollinators in situations with limited insect density, such as high-altitude ecosystems, dry environments, winter months, and islands where insect colonization is poor (Cronk and Ojeda, 2008; Regan et al., 2015). An array of birds can be pollinators, but the specialized nectar-collecting tongues and flower-seeking behaviors typical in families Trochilidae (hummingbirds), Meliphagidae (honeyeaters), and Nectariniidae (sunbirds and spiderhunters) make them especially effective as pollinators. Independent radiations of these nectar-eaters across the world have established a dominant group in each region: hummingbirds in the Americas, honeyeaters in Australasia and the Pacific, and the sunbirds and spiderhunters in Africa, southern Asia, and Australasia (Cronk and Ojeda, 2008). Birds are primary pollinators of pineapple guava (*Acca sellowiana*) and manila hemp (*Musa textilis*) (Itino et al., 1991; Macfarlane, 1995). Bat pollinators occur primarily in the Old World Pteropidae and the New World Phyllostomatidae (Kunz et al., 2011), collectively contributing to the pollination of 528 species across 67 plant families (Fleming and Sosa, 1994). As with birds, nectarivorous bats possess specialized nectar-feeding morphologies, such as elongated snouts and long tongues (Kunz et al., 2011). Bats are primary pollinators for wild bananas and durian (Vogel and Westerkamp, 1991; Westerkamp and Gottsberger, 2000).

4 The ecology and evolution of floral traits

4.1 The origins of flowering plants and their relationships with pollinators

> The rapid development as far as we can judge of all the higher plants within recent geological times is an abominable mystery.
>
> – *Charles Darwin in a letter to James Hooker (Darwin, 1879)*

Angiosperms, or flowering plants, comprise 79% of the estimated 374 000 listed plant species (Christenhusz and Byng, 2016) – a remarkable diversity that is crucial in the structure and ecology of terrestrial ecosystems (Friis et al.,

1987). Equally impressive is the rate at which angiosperms have emerged: over half of angiosperm taxa are thought to have emerged no later than 77 million years ago (Mya), with all angiosperms descending from lineages that appeared no later than 130 Mya (Magallón and Castillo, 2009; reviewed in Doyle, 2012; Coiro et al., 2019). Darwin's 'abominable mystery' is popularly used to refer to questions referring to this 'angiosperm radiation', as the seemingly sudden appearance of flowering plants in the fossil record must have been perplexing to his gradualist views toward evolution (Friedman, 2009; but see Buggs, 2021). What was the catalyst of such extraordinary radiation in angiosperms? Though many other factors such as climate and superior growth traits over gymnosperms have been implicated (reviewed in Augusto et al., 2014), it is widely accepted that insect pollinators were a substantial driver in the angiosperm radiation of the mid-Cretaceous (Grimaldi, 1999; Ollerton et al., 2011; van der Niet et al., 2014).

The origins of the intimate relationships between flowering plants and their pollinators are equally unresolved. It is likely that insects had been pollinating gymnosperms before angiosperms were prevalent; this is bolstered by numerous amber specimens of insects with specialized pollen-collecting mouthparts, most with gymnosperm pollen on their bodies (Ren et al., 2009; Peñalver et al., 2012, 2015; Peris et al., 2017; Cai et al., 2018). Of these ancient gymnosperm pollinators, some inevitably went extinct, while others gave rise to extant gymnosperm pollinators (while most modern gymnosperms are primarily wind-pollinated, some taxa – including cycads and many species in Gnetales – are insect pollinated; Rydin and Bolinder, 2015). However, some ancient gymnosperm pollinators may have transitioned to angiosperms with the advent of the angiosperm radiation in the Cretaceous; along with insects that formed novel associations with angiosperms (such as the Anthophila, bees), they became the predecessors of modern angiosperm pollinators (Peris et al., 2017). By the mid-Cretaceous, insects were the dominant pollen vectors of angiosperms (Dilcher, 2000; Gandolfo et al., 2004; Hu et al., 2008), with the earliest direct evidence of insect-angiosperm pollination relationships dating to 99 Mya (Bao et al., 2019). Phylogenetic studies suggest that insect pollination is the ancestral state of angiosperms, with anemophily and ambophily (dual wind and insect pollination) evolving as a response to pollinator limitation and other abiotic changes in the environment (Culley et al., 2002).

4.2 Floral traits and the ecology of pollination

Flower characteristics, or *floral traits*, in angiosperms are as diverse as the taxa itself – Figure 2 displays some examples in crops. Floral traits encompass the morphology of flowers (including the shape, size, and structure of the corolla; the quantity and arrangement of stamens and stigmas; and the form of the ovary), floral architecture (the arrangement of flowers on a plant), and secondary chemicals (color and fragrance). Floral traits serve chiefly as 'advertisements' that guide pollinators to floral rewards (pollen and nectar) and as facilitators of pollen transfer. The floral trait differences in monkeyflowers (*Mimulus* spp.) highlight the variation and role of several floral traits in attracting pollinators, even within a single genus. The pink flowers of the predominantly bee-pollinated *Mimulus lewisii* produce small amounts of nectar and have wide-petalled corollas that form 'landing platforms' for bees. In contrast, the red, slender, nectar-rich flowers of *Mimulus cardinalis* are more attractive to their hummingbird pollinators (Schemske and Bradshaw, 1999). Furthermore, both species produced different fragrance profiles, with the scent of *M. lewisii* evoking significantly stronger neural responses in bumblebees (Byers, 2014).

Floral rewards - primarily nectar and pollen - provide the incentives for animal pollinators to visit. Floral rewards provide nutrients that are essential for the growth, development, and reproduction of the pollinators that consume them. Nectar is primarily a carbohydrate source but can also be a source of inorganic ions and amino acids (Nicolson and Thornburg, 2007); pollen is rich in lipids, proteins, and amino acids (Roulston et al., 2000; Michener, 2007; reviewed in Brodschneider and Crailsheim, 2010; Vaudo et al., 2015). Nectar, which is derived from photosynthetic products, is relatively more economical to produce than pollen, whose production requires nitrogen and phosphorus, which are often limiting resources (Proctor, 1996b). Furthermore, pollen comes at a reproductive cost since it serves as both a floral reward and a carrier of the plant's male gametes. Despite this, some plants, such as poppies (genus *Papaver*), solely produce pollen as a floral reward, which may have evolved to recruit pollen-dependent bees (Proctor, 1996b). Finally, in place of pollen and nectar, some plants excrete oily substances or resins, which are collected by pollinators as an energy source or a nest construction material (Armbruster, 1984). Some bee species have niche uses for such rewards - honey bees display a rare instance of self-medication in invertebrates and collect plant resins to improve resistance to pathogens (Simone-Finstrom and Spivak, 2010). Male euglossine bees gather fragrant secretions of orchids, presumably to attract mates (Williams and Whitten, 1983; Ray and Vendrame, 2018). In *brood-site pollination*, the host plant serves as an ideal mating site or a place to lay eggs. For example, the decaying pods of the cacao tree (*Theobroma cacao*) are primary mating sites for the tree's primary biting midge pollinators (Winder and Silva, 1972). A highly specialized brood-site pollination relationship occurs in most fig species (*Ficus* spp.), which bear male and female flowers in hollow receptacles. The wasp pollinators of figs lay their eggs inside galls that form from the ovaries of neutered female flowers. Wingless and eyeless males emerge first, locating and fertilizing females before promptly dying. Fertilized adult females leave their galls and the receptacle, collecting pollen from male flowers positioned at the exit. They then find another receptacle to lay their eggs in, fertilizing fertile female flowers in the process (Kerdelhué et al., 1996).

Differences in floral reward quality and quantity can vary widely. For example, nectar volume in individual flowers can differ across and within species (Cruden et al., 1983; Potts et al., 2004; Farkas et al., 2012). For most angiosperms, sucrose, fructose, and glucose are the predominant sugars in nectar. But variation in the activity of the enzyme invertase, which cleaves disaccharide sucrose into monosaccharide fructose and glucose, results in interspecific differences in nectar ratios for the three sugars (Nicolson and Thornburg, 2007). Nectar concentrations – the amount of sugar in nectar – are also variable across species: for example, the nectar of pepper (*Capsicum annum*) has sugar concentrations of less than 6% weight by weight (w/w) compared to the average 76% w/w concentrations found in oregano (*Origanum vulgare*) (Pamminger et al., 2019). Nectar concentrations within species can also be highly variable: for example, sugar levels in Pima cotton, *Gossypium barbadense*, can range from 34% w/w to 62% w/w (Pamminger et al., 2019). Pollen displays similar inconsistencies in nutritional content across species, with interspecific differences in concentrations and diversity in amino acids, lipids (2–60%), and proteins (1–20%) (Roulston et al., 2000; Weiner et al., 2010; Russo et al., 2019; Vaudo et al., 2020).

Differences in the quality and quantity of floral rewards can have major ecological consequences. Pollinator species have specific nutritional requirements, and so their foraging decisions are affected by the nutritional quality of floral rewards (Behmer and Joern, 2008). The sugar ratios and concentrations of nectar, for instance, are often associated with the specific pollinators that visit the plant (Baker and Baker, 1990; van Wyk et al., 1993; Perret et al., 2001; Krömer et al., 2008; Abrahamczyk et al., 2017). Sugar concentrations may affect pollinator choice because it influences the viscosity of nectar. Nectar-eating birds and lepidopterans, which are suction feeders, prefer nectar concentrations of 30–40%, while most bees, which ingest nectar by 'dipping' their tongues, prefer more viscous nectars of 50–60% (reviewed in Kim et al., 2011). The evolution of sugar ratios is also closely tied to pollinator groups, but other factors, such as plant physiological constraints, may also play a role (Abrahamczyk et al., 2017).

Because pollen is the primary source of macronutrients for most bee species, its nutritional content can be a major force in shaping the foraging patterns and decisions of generalist bees seeking to optimize foraging bouts and balance nutritional intake (reviewed in Vaudo et al., 2020). For example, when presented with a variety of pollen sources, bumble bee workers more frequently visited plant species with higher pollen protein-to-lipid ratios (Vaudo et al., 2016). *Oligolectic* bees, which are obligate specialists that rely on pollen from closely related plants, even further demonstrate the interconnectedness between pollen nutrition and pollinator-plant relationships: the pollen of plants that host a particular oligolectic bee often have relatively uniform nutritive content, and pollen nutrition is thought to be a limiting factor in these obligate specializations (Weiner et al., 2010; Vanderplanck et al., 2017).

The properties of pollen can also act as limiting factors that impede pollen collection and digestion. The pollen grains of Malvaceae plants (such as cotton and okra), for example, are oily, exceptionally large (with grain diameters over 100 μm), and covered in spines (Roberts and Vallespir, 1978; Gaglianone, 2000). While the pollen-collecting scopa of corbiculate generalist bees generally work well with most other pollen types, the physical properties of Malvaceae are an impediment to efficient pollen packing, and corbiculate bees have difficulty collecting such pollen (Vansell, 1944; Buchmann and Shipman, 1990; Vaissière and Vinson, 1994; Lunau et al., 2015; but see Konzmann et al., 2019). Such large, sticky pollen is also common in Cucurbitaceae (which includes squash, melons, and cucumbers), Convolvulaceae (sweet potato), Onagraceae (evening primrose), and Cactaceae (prickly pear) (Roberts and Vallespir, 1978; Danforth et al., 2019a). Specialist pollinators that collect pollen from these taxa have evolved modified scopal hairs to efficiently collect unwieldy pollen. Oligolectic *Eucera* bees, for example, have widely spaced, smooth scopal hairs to collect the large, spiked, and oily pollen of *Cucurbita* plants (Roberts and Vallespir, 1978). Pollen's chemical composition can also limit pollinator visitors: the pollen of Asteraceae, Ranunculaceae, Boraginaceae, Fabaceae, and Cucurbitaceae is toxic or nutritionally limiting to the larvae of generalist bees, and Asteraceae pollen is also often spiny. Unsurprisingly, the specialist pollinators of these plants have physiological modifications to digest such pollen (Danforth et al., 2019b). Pollen properties can thus restrict the diversity of pollinators that can access a host plant, and specialist pollinators that evolve mechanisms to overcome such physical and chemical impediments gain exclusive access to food sources unavailable to many other pollinators.

Other inherent traits of floral rewards can also have ecological effects on pollinators. Secondary chemicals, for example, are ubiquitously found in nectar and pollen. Though floral secondary chemicals are primarily related to attracting pollinators (i.e. pigments and fragrances), defense-related chemicals are often expressed in pollen and nectar, sometimes in concentrations similar to non-floral tissue (Adler, 2000; Adler and Irwin, 2005; Manson et al., 2012; Cook et al., 2013; Stevenson et al., 2017; Palmer-Young et al., 2019). Consumption of these floral secondary compounds can have various beneficial effects on pollinators: for example, caffeine and nicotine in nectar improve pollinator memory of floral rewards (Wright et al., 2013; Baracchi et al., 2017), while a variety of floral chemicals ameliorate infections from pathogens and parasites (Baracchi et al., 2015; Palmer-Young et al., 2016; Richardson et al., 2016; Egan et al., 2018). However, at higher concentrations, a range of compounds can have deterrent or toxic effects (Detzel and Wink, 1993). Similarly, the nutritional content of pollen can have repercussions on pollinator health – in honey bees, higher quality pollen diets increase resistance to parasites and regulate the development of hypopharyngeal glands and vitellogenin expression in nurse bees (Keller et al., 2005; Di Pasquale et al., 2013). The consumption of sunflower (*Helianthus annus*) pollen displays a trade-off between medicinal effects and nutrition: while consumption of pollen dramatically reduces infections of parasites in bumblebees (Giacomini et al., 2018), monofloral diets of sunflower pollen are considered a poor nutritional food source for honey and bumble bees (McAulay and Forrest, 2019).

4.3 The co-evolution of plants and pollinators

Certain floral traits, including colors, structures, fragrances, and rewards, can be associated with attracting specific pollinators (reviewed in Sheehan et al., 2012; Rosas-Guerrero et al., 2014). Floral traits that predictably correspond to pollinator specificity are referred to as *pollination syndromes*, though the reliability of syndromes in floral ecology and evolution has been debated, especially when considering the roles of generalist pollinators (Waser et al., 1996; Fenster et al., 2004; Ollerton et al., 2009; but see Rosas-Guerrero et al., 2014). Regardless, it is undeniable that pollinators play a strong role in the evolution of their flowering plant partners. Pollinator preference is a strong selective force affecting floral traits, driving diversification at several stages of the speciation process, and acting synergistically with other pressures and evolutionary mechanisms – such as herbivory and genetic constraints – to promote reproductive isolation (Schemske and Bradshaw, 1999; Morgan, 2007; Kay and Sargent, 2009; van der Niet et al., 2014; Ramos and Schiestl, 2020).

Like all mutualisms, plant-pollinator relationships may be mediated by a conflict between two mutualistic partners; i.e. each partner in the interaction benefits by reducing the costs of the interaction. Such adaptations by one partner to minimize costs can drive reciprocal changes in the other, resulting in an 'arms race' and directional selection for more extreme traits. Flower *nectar spurs* reflect how this coevolutionary conflict between pollinators and plants can bilaterally shape their evolution. Flowers with nectar spurs limit access so that only pollinators with sufficiently long tongues can reach floral rewards. Selection favors longer nectar spurs because it forces pollinators to optimally contact the reproductive organs of the flower. The pollinator, in turn, is subject to selective pressure to evolve longer and longer tongues. Darwin famously referenced this 'arms race' hypothesis after the discovery of the Madagascan orchid *Angraecum sesquipidale* and its exceptional 22-cm-long nectar spur, predicting the existence of a pollinator with an equivalently long tongue (reviewed Arditti et al., 2012). (Such a pollinator was conclusively found 22 years after his death: a hawkmoth aptly named *Xanthopan morganii praedicta*.) Alternatively, in populations where nectar spurs rapidly evolve to be longer, a novel mutualism can form as the plants' original short-tongued pollinators are excluded and longer-tongued pollinators become the predominant pollinator visitors. Whittall and Hodges (2007) found evidence supporting both scenarios in the evolution of nectar spurs and speciation of columbines (*Aquilegia* spp.).

Instances of cheating in both plants and pollinators further illustrate potential conflicts of interest. Nectar larceny occurs when a pollinator takes floral rewards without providing any pollination services (reviewed in Irwin et al., 2010). For example, in the flowers of alfalfa, the stamens and pistil are held under tension and must be unfurled or 'tripped,' by a visiting bee to be successfully pollinated (Fig. 2e; Bohart, 1957). Honey bees, having learned to take nectar while avoiding this 'tripping' mechanism, act as nectar thieves (Cane, 2002). Pollinators can also act as nectar robbers by chewing holes in flowers to access nectar (Maloof and Inouye, 2000). In blueberry flowers, honey bees can act as 'secondary' robbers by stealing nectar from holes already chewed into flowers by robbing carpenter bees (Free, 1968). Robbing has various reproductive consequences for plants, who in response have evolved a suite of defenses to deter robbers; for instance, the thickening of calyxes and the allocation of secondary compounds in nectar (Irwin et al., 2010). On the other hand, orchids (family Orchidaceae) rely on a suite of tactics to deposit pollen on insects while minimally investing in floral rewards (Ray and Vendrame, 2018). Some orchid species, such as *Ophrys apifera*, are pollinated solely by male bees or wasps. These orchids bear structures that resemble the females of their respective pollinators, with many species enhancing the deception by mimicking sex pheromones (Schiestl et al., 2000; Bohman et al., 2014). As males attempt to mate with the female mimics, pollen is deposited on their bodies. Males then complete the pollination process when they are tricked by another orchid and deposit the pollen on its stigmas (Ray and Vendrame, 2018).

5 Domestication and its impact on plant-pollinator relationships

The development of agriculture was facilitated by human selection that resulted in a rapid evolutionary change in cultivated plant and animal species (*artificial selection*). During cultivation, species are further affected by natural selection, to some degree, through environmental factors such as resource limitation and disturbance. The result of such forces is *domestication* – the gradual geographical and/or genetic isolation of a founding population that eventually distinguishes it from its wild relatives (Purugganan and Fuller, 2009; Milla et al., 2015). An estimated 2500 plant species have undergone domestication to some degree (Zeven and de Wit, 1982; Dirzo and Raven, 2003). Some crops are domesticated to a light degree, divergent from their wild relatives by only a few major traits: for example, the only major genetic changes in African oil palms (*Elaeis guineensis*) relate to kernel development inside fruits (Gepts, 2004). But some crops, including the world's major cereals and crops in the *Cucurbita* family, are *heavily domesticated* and are almost unrecognizable from their wild relatives (Fuller et al., 2010; Lira et al., 2016). In this section, we focus on food crops; horticultural domestication also has interesting consequences for pollinators but is beyond our scope.

5.1 Domestication and its effect on plant traits

Artificial selection leading to domestication often favors traits that ease harvest work, increase yield, or enhance flavor and nutrition. Modern wheat (*Triticum* spp.) highlights this: it produces 4.63 tons of grain per hectare (Langemeier and Yeager, 2019) compared to the 0.5–0.8 t/ha produced by their relatives in the Fertile Crescent (Harlan, 1967). In this section, we briefly review commonly shared traits in major crops (collectively referred to as the *domestication syndrome*), though we note that crops display a wide array of domesticated traits that are not mentioned in this section (Meyer et al., 2012). Crops are usually selected to have larger harvest organs: for example, compared to their wild relatives, grain crops dedicate up to 60% of their aboveground biomass to grains (versus 20–30% typical in wild relatives) and have seeds that are 5–10 times larger (Evans, 1993). Many crops display relatively compact growth forms compared to their wild predecessors, resulting in improved productivity and higher densities (Donald, 1968). Maize (*Z. mays*) grows on a single stem, a stark departure from the highly branched growth forms of its wild teosinte (*Zea* spp.) relatives. Domestication shifts many crops toward increased self-fertilization (Rick, 1950; Sapir, 2009; Velázquez-López et al., 2018) or a switch to asexual vegetative reproduction, such as in bananas and cassava (Simmonds, 1960; Elias and McKey, 2000). This increases the ease of propagation and allows for crops to exist in areas where their native pollinators may be lacking (Gepts, 2004). Furthermore, the removal of seeds increases palatability (e.g. bananas). Some crops (primarily cereals) are typically *nonshattering*, as opposed to their wild relatives that disperse their seeds at maturity. Such loss or reduction of seed dispersal allows for effective harvest by ensuring that crops hold on to grains when they ripen (Di Vittori et al., 2019). Finally, the flavor, quality, and handling of harvested organs can be improved through reduced defensive structures or secondary chemicals that are toxic or antinutritive (Vanderborght, 1983; Benrey et al., 1998; Wilson and Dufour, 2002; Gols et al., 2008; Sujana et al., 2012; Olsen and Wendel, 2013). The domestication of *Cucurbita*, for example, resulted in the decrease of toxic and bitter substances in the fruit and the reduction of silicified hairs in the leaves, stems, and fruit (Lira et al., 2016).

5.2 Domestication and its effect and plant-animal interactions

Gepts (2004) describes humans and crop plants in a 'mutualistic' relationship – crops have transformed the trajectory of human development, and in turn, crop plants have been propagated far beyond their native ranges but at a physiological cost. In transforming crops to favor harvest and yield, domestication often reduces their resilience, and many crops are unable to survive without human upkeep (Gepts, 2004; Purugganan and Fuller, 2011). Artificial selection is propagated through multiple genetic bottlenecks, and consequently crops are much less genetically diverse than their wild relatives (van de Wouw et al., 2010; Meyer et al., 2012).

Crop fitness is also affected by changing dynamics of their natural enemies. Cultivated plants are less herbivore-resistant than their wild counterparts (Whitehead et al., 2017) and are often subject to intense pressure from herbivore pests (reviewed in Macfadyen and Bohan, 2010; Chen et al., 2015). Increased vulnerability to herbivores may be the consequence of several outcomes of domestication: lower genetic diversity; human selection against distasteful, antinutritive, or unwieldy defensive structures and compounds; and increased quality and nutrition of harvested organs that herbivores favor (Newell-McGloughlin, 2008). Domestication-induced changes in secondary chemicals can even deter natural enemies of crop herbivores: unlike their wild relatives, domesticated maize and cabbage do not produce volatile attractants of parasitoid wasps and entomopathogenic nematodes (Gouinguené et al., 2001; Köllner et al., 2008; Gols et al., 2011).

Many harvest organs of crops are flower-derived (e.g. fruits, nuts, grains, or flowers themselves), and strong selection can alter floral traits from their ancestral states. Fruit quality traits such as size, color, and antioxidant content are targeted during artificial selection (Kalt et al., 2001). Plants that are harvested for their flowers tend to have larger or different-colored flowers: many cultivars of domesticated sunflowers, for example, grow taller and have larger flower heads (Heiser, 1978). Major reproductive structures can also be changed: domesticated tomatoes (*Solanum lycopersicum*) and cowpeas (*Vigna ungiculata*) have shortened styles and stigmas, perhaps associated with higher inbreeding and more reliable fruit setting (Rick and Dempsey, 1969; Lush, 1979); while domesticated sunflowers used as cut flowers are often bred to have cytoplasmic male sterility and thus produce no pollen (Heiser, 1978). Furthermore, selection for larger fruits often relates to selection for larger flower traits – several floral morphological traits of domesticated squash (*Cucurbita* spp.) were larger than their wild relatives (Glasser, 2020). Finally, the chemicals of flowers themselves may also be affected – plant-wide differences in volatile chemical profiles are responsible for the distinct flavors of different basil cultivars (*Ocimum* spp.) (Dudai and Belanger, 2008; Muráriková et al., 2017).

Domestication can affect a plant's relationships with its herbivores, but do similar altered dynamics occur between crops and their pollinators? This would be expected, especially for specialist pollinators that heavily rely on their crop partners. For instance, squash and its relatives (*Cucurbita* spp.) are all monoecious and self-incompatible, relying heavily on pollinators to reproduce. The oligolectic squash bee *Eucera pruinosa* relies solely on the pollen of *Cucurbita* plants, and its expansion out of Mexico into temperate North America closely coincides with the domestication and subsequent wide-range cultivation of squash, pumpkins, and gourds in the Holocene (López-Uribe et al., 2016). Floral architecture is paramount in plant–pollinator dynamics, and changes in the floral structures of crops can also alter their ecological roles. For example, the taller growth forms and enlarged floral heads of domesticated sunflowers increase the likelihood of visitation from large- and medium-sized bees (Sapir, 2009). Changes in secondary chemicals are the most common modifications associated with domestication (Meyer et al., 2012) and such alterations can affect pollinator-plant relationships. Pollen and nectar of cultivated blueberry plants (*Vaccinium corymbosum*) had markedly different chemical compositions than floral rewards of wild *V. corymbosum* (Egan et al., 2018). Furthermore, these differences could have consequences on pollinator health: while nectar from both wild and domesticated blueberries contained 4-*O*-caffeoylshikimic acid, a compound effective against the bumble bee parasite *Crithidia bombi*, only wild blueberries contained the compound in sufficient concentrations to reduce infections in bees. To our knowledge, these studies are the only published work directly measuring the effects of domestication on pollinators, and more work must be done to elucidate the influence of crop domestication on pollinator dynamics.

6 How do pollinators impact agriculture?

In terms of total production volume, 60% of the world's crops (mostly grains and root crops) are not pollinator-dependent (IPBES, 2016). However, insect pollinators are crucial to the production of 70% of the world's 124 main cultivated crops (about 35% of total agricultural output), including high commodity crops such as oilseed rape, cocoa, and coffee, which are worth an estimated 100–351 billion USD per year (Delaplane and Mayer, 2000; Klein et al., 2007; Aizen et al., 2009; Lautenbach et al., 2012; Regan et al., 2015; Potts et al., 2016; IPBES, 2016). Pollinator-dependent crops are a key source of micronutrients, including vitamins A and C, calcium, fluoride, and folic acid (Smith et al., 2015).

Furthermore, the economic benefit of pollination continues to rise each year since 1995 (Klein et al., 2007; Aizen et al., 2009; Lautenbach et al., 2012).

Pollinators may increase yield through multiple mechanisms. Pollinators can increase a plant's *fruit set*, the proportion of the plant's flowers that successfully develop into mature fruits or seeds. Increased outcrossing through pollination can also increase the size, attractiveness, uniformity, and quality of crops (Klatt et al., 2014; Cusser et al., 2016), as well as reducing the time to harvest. Reduced pollinator activity can often translate to decreased yield (Ashman et al., 2004; Benjamin and Winfree, 2014; Reilly et al., 2020).

Wild pollinators in or around agricultural land can provide pollination services to the crops they visit. However, pollinators are often *managed* and purposefully introduced in agricultural land to enhance pollination. While other non-bee insects contribute significantly to the pollination of several agricultural crops (Rader et al., 2016), bees are by far the most ubiquitous and commonly managed pollinators. The western honey bee (*Apis mellifera*) – with its high densities, history of domestication, and ease of management – is the most widely managed pollinator used to improve crop yield globally (National Research Council, 2007) and is present on every continent except Antarctica (Whitfield et al., 2006). *A. mellifera* can supplement the production of 96% of the world's major crops, with yield increases as high as 90% in some crops (Southwick and Southwick, 1992; Klein et al., 2007) and may even supplement the pollination services to surrounding wild plants (Klein et al., 2007). Other honey bees, such as the eastern honey bee (*A. cerana*), are also managed due to their resistance to certain pathogens and pesticides. Bumble bees (e.g. *Bombus terrestris and B. impatiens*) are much easier to use in enclosed spaces and are employed in the greenhouse production of a variety of fruits and vegetables, such as tomatoes and cucurbits (Velthuis and van Doorn, 2006). Solitary bee species are also used for specialty crop production; for example, the leafcutter bee *Megachile rotunda* is primarily managed for alfalfa, blueberry, and clover production, and mason bees (*Osmia* spp.) are managed for a variety of orchard crops, including almonds and cherries (Delaplane and Mayer, 2000). Finally, *Nomia melanderi* is the only solitary ground-nesting bee that is intensely managed for agriculture; it is used to pollinate alfalfa in the western United States (Cane, 2002).

Crop pollination by wild pollinators is a significant ecosystem service to agriculture. Wild bees, for example, can complement pollination services provided by managed pollinators, improving production and yield (Morandin and Winston, 2005; Greenleaf and Kremen, 2006; Winfree et al., 2007; Garibaldi et al., 2013; Mallinger et al., 2015, 2019; Nicholson and Ricketts, 2019; Reilly et al., 2020). Pollination by wild pollinators can compensate for losses in pollination services associated with declines in managed pollinators (Winfree et al., 2007) and provide equivalent or greater pollination than managed pollinators (MacKenzie and Winston, 1984; Garibaldi et al., 2013; Mallinger et al., 2015, 2019; Averill et al., 2018; Reilly et al., 2020). Wild pollinators can often provide sufficient pollination without the use of managed pollinators (Winfree et al., 2007; Reilly et al., 2020), and in such cases the addition of managed pollinators can result in *pollen saturation*, with increasingly negligible or even negative effects on yield (Chacoff et al., 2008; Morris et al., 2010; Garibaldi et al., 2011). Furthermore, the reliance on a few managed bee species to produce pollination-dependent crops is risky, tying a third of global crop production to a few managed bee species (Klein et al., 2007). Wild pollinator species are also in decline, especially in agricultural lands (Kevan et al., 1990; Potts et al., 2010; Garibaldi et al., 2011). While threatened bee species are rarely seen on crops and only a few wild pollinator species provide significant pollination services, the abundance of wild species that benefit yield is improved by management strategies that promote wild pollinator diversity overall (Kleijn et al., 2015). Wild pollinator diversity can also have synergistic effects on managed pollinators, increasing the effectiveness of their pollination (Brittain et al., 2013). Thus, promoting a diverse assemblage of wild pollinators serves the dual benefit of preserving biodiversity and increasing the efficiency of overall pollinator services (Burkle et al., 2017).

The relationship between agriculture and pollinators is deeply intertwined, and a loss of pollinators compromises the many benefits of pollination services. In the United States, a substantial portion of pollinator-dependent crops (39% of pollinator-dependent crop area) already lacks sufficient pollinator populations to fill the demand for pollinator services (Koh et al., 2016). The absence of pollinators would result in a 40–90% decrease in production for 91 of the world's 124 leading crops (IPBES, 2016), an associated economic loss of about US$200 billion annually (Gallai et al., 2009), and a wildly disproportionate increase in the amount of land used for crop production; these effects will be of a greater magnitude in the developing world (Aizen et al., 2009). Hand-pollinated crops highlight the potential difficulties imposed by future pollinator losses – crops such as vanilla bean and oil palms require intense, time-consuming labor, often resulting in higher costs for consumers and poor working conditions for laborers (Wurz et al., 2021). The loss of pollination services would threaten global food security, and resulting deficiencies in macronutrient availability are estimated to increase global non-communicable and malnutrition-related deaths by 1.7 million annually (Tscharntke et al., 2012; Smith et al., 2015). Finally, pollinator decline is a threat to invaluable cultural institutions and practices that are dependent on pollinators (Potts et al., 2016; Dicks et al., 2021).

7 Modern agriculture and pollinators

The meteoric increase in human population in the mid-twentieth century was sustained by an equivalently rapid intensification of the world's food production.

In 1960, global grain production was 1 billion tons, but in a mere 40 years, that figure had doubled (Khush, 2001). This so-called green revolution was facilitated by revolutionary agricultural technologies, including the development of high-yielding crop varieties, efficient irrigation, and increased use of agrochemicals (Matson et al., 1997; Khush, 2001). While recent agricultural intensification has prevented widespread famine in an increasingly populated world, it has also had serious consequences on ecosystems in and around agricultural lands (Matson et al., 1997; Ricketts et al., 2008). One such consequence is pollinator decline – global losses in wild insect pollinator abundance and diversity (Powney et al., 2019; van Klink et al., 2020; Wagner, 2020) and increased colony mortality in managed pollinators (vanEngelsdorp et al., 2008; Williams et al., 2010). While pollinator decline is a multifaceted issue with many causal factors, agriculture and its landscape effects have been implicated as key (Kremen et al., 2002; Ollerton et al., 2014; IPBES, 2016; Dicks et al., 2021). In this section, we briefly discuss some of these factors; subsequent chapters will review individual threats to pollinators in depth. Furthermore, detailed reviews are provided in González-Varo et al., 2013; Potts et al., 2016; and Dicks et al., 2021.

7.1 Land use and management

A body of evidence has linked the global intensification of human land use and its environmental consequences as a major contributor to pollinator decline (Kennedy et al., 2013; Basu et al., 2016; Kovács-Hostyánszki et al., 2017). In 2015, approximately 19.3 million km^2 of land was used for crop production, comprising about 14.7% of Earth's terrestrial surface. Total agricultural land (cropland and pastures) covered 54.3 million km^2, or 41.2% of Earth's land (Ellis et al., 2020). Agricultural land is only expected to increase further, with expected increases of about 10% of current ranges by 2030 (Tilman et al., 2001; Haines-Young, 2009). Much of this new agricultural land will likely be created at the expense of uninhabited wildlands or through intensified use of currently inhabited landscapes (Ellis et al., 2020). The resulting habitat loss can disrupt pollination networks by limiting resources, foraging sites, and brood sites for pollinators (Potts et al., 2010; Roulston and Goodell, 2011; Baude et al., 2016). Human disturbance has generally reduced the richness and abundance of pollinator communities (Taki et al., 2007; Ricketts et al., 2008; Steffan-Dewenter and Schiele, 2008; Spiesman and Inouye, 2013; Connelly et al., 2015); although the intensity of these effects can vary among different pollinator groups (Sjödin et al., 2007; Traveset et al., 2018). Furthermore, the large-scale planting of one or two crops, or *monocultures*, simplifies the landscape around them and decreases pollinator foraging resources, especially during periods when crops are not in bloom (Tscharntke et al., 2005; Kremen and Miles, 2012). Landscape simplification can also decrease the pollination services provided by pollinators (Connelly et al., 2015) and have physiological effects: when the plant species diversity is limited, or only suboptimal resources are available, bees can experience nutritional stress that impacts their body size, population sizes, pathogen loads, immunocompetence, microbiota, and fertility (Carvell et al., 2006; Renauld et al., 2016; Parreño et al., 2021).

7.2 Pathogens and pesticides

Pathogens are attributed as another major force in pollinator decline (vanEngelsdorp et al., 2008; Goulson et al., 2015; Dicks et al., 2021); the impacts of diseases on pollinators will be further discussed elsewhere in this volume. In this section, we briefly review the impact of *pathogen spillover*. In generalist pathogens that infect multiple hosts, some host species display higher resistance or tolerance and can thus be 'reservoirs' of infection that maintain high pathogen prevalence. Such reservoir species can then spread the pathogen to other hosts; this phenomenon is called *pathogen spillover* (Daszak et al., 2000; Power and Mitchell, 2004). With the increased use of managed pollinators, evidence has shown that they can serve as pathogen reservoirs, spreading the infection to wild conspecifics (Graystock et al., 2013, 2015). Managed bumble bees used in greenhouses, for example, can serve as reservoirs for the gut parasites *C. bombi* and *Nosema bombi*, spreading the pathogens to wild bumble bee populations, who can then radiate the infection away from the source (Colla et al., 2006; Krömer et al., 2008; Otterstatter and Thomson, 2008; Graystock et al., 2013; Murray et al., 2013; Schmid-Hempel et al., 2014). *A. mellifera* is another prime candidate for a reservoir species: it is widely distributed, dwells in dense colonies, is highly mobile due to migratory beekeeping operations, and can host a range of pathogens that can infect other Apoidea species (Fürst et al., 2014; Ravoet et al., 2014; Goulson and Hughes, 2015; Glenny et al., 2017).

Concerningly, pathogen spread in pollinators is not limited to conspecifics. Bumble bees can spread the neogregarine parasite *Apicystis bombi* to honey bees, while a wide variety of pathogens, including *N. ceranae*, deformed wing virus, and black queen cell virus, can be transmitted from honey bees to bumble bees (Otterstatter and Whidden, 2004; Goka et al., 2006; Chen and Siede, 2007; Plischuk et al., 2009; Graystock et al., 2014; McMahon et al., 2015). Honey bee-specific diseases, including *Varroa destructor* mites, can spread from *A. mellifera* to other Apidae species (Plischuk and Lange, 2009; Guzman-Novoa et al., 2015; Manley et al., 2015; Purkiss and Lach, 2019; but see Gusachenko et al., 2020). Furthermore, evidence on the effects of pathogen spillover from managed bees to native bees is overwhelmingly negative (reviewed in Mallinger et al., 2017).

The impacts of pesticide use have further compounded the pressure imposed by pathogens (vanEngelsdorp et al., 2008; Sanchez-Bayo and Goka, 2014; Goulson et al., 2015) and will be discussed in detail elsewhere in this volume. In brief, herbicides can reduce floral abundance and biodiversity of flowers in agricultural landscapes (Albrecht et al., 2007), while insecticides can cause a variety of direct effects on pollinators even at sublethal doses (reviewed in Sanchez-Bayo and Goka, 2014) and reduce pollinator abundance and species richness (Brittain et al., 2010; Mallinger et al., 2015; Woodcock et al., 2017; Main et al., 2020).

7.3 Introductions of non-native bee species

Humans have inadvertently increased the range of managed pollinator species by introducing them to agricultural systems outside of their native ranges. Honey bees are established all over the world except Antarctica and are arguably the most widely distributed species after humans (Goulson and Hughes, 2015). Outside of western honey bees, *Bombus terrestris* was the earliest managed pollinator released outside of its native range, introduced in New Zealand in 1886 and 1906 to assist in the production of red clover. In 1982, New Zealand populations of *B. terrestris* were used as a source for an introduction in Chile (Arretz and Macfarlane, 1986). The species then spread to Argentina, and today its range is expanding northward and southward (Schmid-Hempel et al., 2014). This introduction resulted in the widespread and precipitous decline of the native *B. dahlbombi*. More generally, *B. terrestris* became widespread after its popularity in greenhouse production; its escape from greenhouses established them in Japan, for example (Inoue et al., 2008). In North America, escaped *B. impatiens* from managed greenhouse colonies have been established in the Pacific Northwest and are expanding their range (Looney et al., 2019).

8 Conclusion

The mutualism between flowering plants and their pollinators is deeply intertwined: outcrossing flowering plants rely on pollinators to reproduce, and pollinators are provided with resources to survive and proliferate. Humanity is similarly dependent on crop plants; human society and its ever-increasing advancement are primarily fueled by agriculture. Humans are thus indebted to pollinators, and indeed, pollination ecosystem services have aided humanity even before the inception of agriculture – when humans relied on fruit from wild pollinator-dependent plants. But in recent decades, exponential human population growth has placed greater demand on agricultural production, with the potential to strain relationships between crops and pollinators. Mitigating threats that impact pollinators and employing pollinator-friendly management strategies on agricultural lands are necessary to safeguard pollinators in the face of agricultural intensification. The remainder of this book is dedicated to recent advances in understanding and preserving relationships that link humans, crops, and pollinators. Promoting pollinators and maintaining sufficient production to sustain human population growth are not mutually exclusive goals. In fact, both interests can be fittingly mutualistic: humans can quite literally reap the rewards from the services that healthy pollinator communities provide (Kleijn et al., 2015).

9 Where to look for further information

The Natural History of Pollination is a summary of the mechanisms that drive pollination, covering the floral adaptations of plants, the diversity of animals that pollinate plants, the evolution of plant–pollinator relationships, and much more. While the book is over 20 years old, it is still an excellent resource to begin understanding the rich relationships between flowering plants and the pollinators on which they rely. (https://www.harpercollins.com/products/the-natural-history-of-pollination-collins-new-naturalist-library-book-83-michael-proctorpeter-yeoandrew-lack?variant=32207722381346).

The Intergovernmental Platform on Biodiversity and Ecosystem Services has published a detailed report that covers the importance of pollinators for global food production, pollinator decline and its drivers, and suggestions to policymakers for addressing risks and improving conditions for pollinators in agroecosystems. It is also an excellent source for papers regarding such topics: https://ipbes.net/assessment-reports/pollinators.

Finally, the US Department of Agriculture publishes an annual Strategic Pollinator Priorities and Goals Report. These reports synthesize current pollinator research and feedback from stakeholders and identify key priorities for promoting pollinators in agroecosystems. The most recent report (at the time of writing) can be accessed here: https://www.usda.gov/sites/default/files/documents/pollinator-priorities-2021R4-508-version.pdf.

10 References

Abrahamczyk, S., Kessler, M., Hanley, D., Karger, D. N., Müller, M. P. J., Knauer, A. C., Keller, F., Schwerdtfeger, M. and Humphreys, A. M. (2017). Pollinator adaptation and the evolution of floral nectar sugar composition. *Journal of Evolutionary Biology* 30(1), 112–127. https://doi.org/10.1111/jeb.12991.

Ackerman, J. D. (2000). Abiotic pollen and pollination: ecological, functional, and evolutionary perspectives. *Plant Systematics and Evolution* 222(1–4), 167–185. https://doi.org/10.1007/BF00984101.

Adler, L. S. (2000). The ecological significance of toxic nectar. *Oikos* 91(3), 409–420. https://doi.org/10.1034/j.1600-0706.2000.910301.x.

Adler, L. S. and Irwin, R. E. (2005). Ecological costs and benefits of defenses in nectar. *Ecology* 86(11), 2968–2978. https://doi.org/10.1890/05-0118.

Aizen, M. A., Garibaldi, L. A., Cunningham, S. A. and Klein, A. M. (2009). How much does agriculture depend on pollinators? Lessons from long-term trends in crop production. *Annals of Botany* 103(9), 1579–1588. https://doi.org/10.1093/aob/mcp076.

Aizen, M. A. and Harder, L. D. (2007). Expanding the limits of the pollen-limitation concept: effects of pollen quantity and quality. *Ecology* 88(2), 271–281. https://doi.org/10.1890/06-1017.

Albrecht, M., Duelli, P., Müller, C., Klein, D. and Schmid, B. (2007). The Swiss agri-environment scheme enhances pollinator diversity and plant reproductive success in nearby intensively managed farmland: agri-environment schemes and pollination. *Journal of Applied Ecology* 44(4), 813–822. https://doi.org/10.1111/j.1365-2664.2007.01306.x.

Arditti, J., Elliott, J., Kitching, I. J. and Wasserthal, L. T. (2012). 'Good Heavens what insect can suck it'- Charles Darwin, *Angraecum sesquipedale* and *Xanthopan morganii praedicta*: Darwin, *Angraecum* and *Xanthopan*. *Botanical Journal of the Linnean Society* 169(3), 403–432. https://doi.org/10.1111/j.1095-8339.2012.01250.x.

Armbruster, W. S. (1984). The role of resin in angiosperm pollination: ecological and chemical considerations. *American Journal of Botany* 71(8), 1149–1160. https://doi.org/10.1002/j.1537-2197.1984.tb11968.x.

Arretz, P. V. and Macfarlane, R. P. (1986). The introduction of *Bombus ruderatus* to Chile for red clover pollination. *Bee World* 67(1), 15–22. https://doi.org/10.1080/0005772X.1986.11098855.

Ascher, J. S. and Pickering, J. (2018). Discover Life bee species guide and world checklist (Hymenoptera: Apoidea: Anthophila). Available at: http://www.discoverlife.org/mp/20q?guide=Apoidea_species.

Ashman, T.-L., Knight, T. M., Steets, J. A., Amarasekare, P., Burd, M., Campbell, D. R., Dudash, M. R., Johnston, M. O., Mazer, S. J., Mitchell, R. J., Morgan, M. T. and Wilson, W. G. (2004). Pollen limitation of plant reproduction: ecological and evolutionary causes and consequences. *Ecology* 85(9), 2408–2421. https://doi.org/10.1890/03-8024.

Augusto, L., Davies, T. J., Delzon, S. and De Schrijver, A. (2014). The enigma of the rise of angiosperms: can we untie the knot? *Ecology Letters* 17(10), 1326–1338. https://doi.org/10.1111/ele.12323.

Averill, A. L., Sylvia, M. M., Hahn, N. and Couto, A. V. (2018). Bees (Hymenoptera: Apoidea) foraging on American cranberry in Massachusetts. *Northeastern Naturalist* 25(3), 502–512. https://doi.org/10.1656/045.025.0314.

Baker, H. G. and Baker, I. (1990). The predictive value of nectar chemistry to the recognition of pollinator types. *Israel Journal of Plant Sciences* 39(1–2), 57–166. https://doi.org/10.1080/0021213X.1990.10677140.

Bao, T., Wang, B., Li, J. and Dilcher, D. (2019). Pollination of Cretaceous flowers. *Proceedings of the National Academy of Sciences of the United States of America* 116(49), 24707–24711. https://doi.org/10.1073/pnas.1916186116.

Baracchi, D., Brown, M. J. F. and Chiktta, L. (2015). Behavioural evidence for self-medication in bumblebees? *F1000Research* 4(73), 73. https://doi.org/10.12688/f1000research.6262.2.

Baracchi, D., Marples, A., Jenkins, A. J., Leitch, A. R. and Chittka, L. (2017). Nicotine in floral nectar pharmacologically influences bumblebee learning of floral features. *Scientific Reports* 7(1), Article 1951. https:/doi.org/10.1038/s41598-017-01980-1.

Barrett, S. C. H. (2003). Mating strategies in flowering plants: the outcrossing-selfing paradigm and beyond. *Philosophical Transactions of the Royal Society of London. Series B, Biological Sciences* 358(1434), 991–1004. https://doi.org/10.1098/rstb.2003.1301.

Basu, P., Parui, A. K., Chatterjee, S., Dutta, A., Chakraborty, P., Roberts, S. and Smith, B. (2016). Scale dependent drivers of wild bee diversity in tropical heterogeneous agricultural landscapes. *Ecology and Evolution* 6(19), 6983–6992. https://doi.org/10.1002/ece3.2360.

Baude, M., Kunin, W. E., Boatman, N. D., Conyers, S., Davies, N., Gillespie, M. A. K., Morton, R. D., Smart, S. M. and Memmott, J. (2016). Historical nectar assessment reveals the fall and rise of floral resources in Britain. *Nature* 530(7588), 85–88. https://doi.org/10.1038/nature16532.

Behmer, S. T. and Joern, A. (2008). Coexisting generalist herbivores occupy unique nutritional feeding niches. *Proceedings of the National Academy of Sciences of the United States of America* 105(6), 1977–1982. https://doi.org/10.1073/pnas.0711870105.

Benjamin, F. E. and Winfree, R. (2014). Lack of pollinators limits fruit production in commercial blueberry (*Vaccinium corymbosum*). *Environmental Entomology* 43(6), 1574–1583. https://doi.org/10.1603/EN13314.

Benrey, B., Callejas, A., Ríos, L., Oyama, K. and Denno, R. F. (1998). The effects of domestication of *Brassica* and *Phaseolus* on the interaction between phytophagous insects and parasitoids. *Biological Control* 11(2), 130–140. https://doi.org/10.1006/bcon.1997.0590.

Bohart, G. E. (1957). Pollination of alfalfa and red clover. *Annual Review of Entomology* 2(1), 355–380. https://doi.org/10.1146/annurev.en.02.010157.002035.

Bohman, B., Phillips, R. D., Menz, M. H. M., Berntsson, B. W., Flematti, G. R., Barrow, R. A., Dixon, K. W. and Peakall, R. (2014). Discovery of pyrazines as pollinator sex pheromones and orchid semiochemicals: implications for the evolution of sexual deception. *New Phytologist* 203(3), 939–952. https://doi.org/10.1111/nph.12800.

Bradley, R. (1731). New Improvements in Planting and Gardening. J. and J. Knapton in St. Paul's Church-Yard, A. Bettesworth and C. Hitch in Pater-Noster-Row, J. Pemberton in Fleet-Street, and D. Browne without Temple Bar.

Brittain, C., Williams, N., Kremen, C. and Klein, A. M. (2013). Synergistic effects of non-*Apis* bees and honey bees for pollination services. *Proceedings of the Royal Society. Series B,: Biological Sciences* 280(1754), 2012-2767. https://doi.org/10.1098/rspb.2012.2767.

Brittain, C. A., Vighi, M., Bommarco, R., Settele, J. and Potts, S. G. (2010). Impacts of a pesticide on pollinator species richness at different spatial scales. *Basic and Applied Ecology* 11(2), 106–115. https://doi.org/10.1016/j.baae.2009.11.007.

Brodschneider, R. and Crailsheim, K. (2010). Ernährung und Gesundheit bei Honigbienen [Nutrition and health in honey bees]. *Apidologie* 41(3), 78–294. https://doi.org/10.1051/apido/2010012.

Buchmann, S. L. and Shipman, C. W. (1990). Pollen harvesting rates for *Apis mellifera* L. on *Gossypium* (Malvaceae) flowers. *Journal of the Kansas Entomological Society* 63(1), 92–100.

Buggs, R. J. A. (2021). The origin of Darwin's "abominable mystery". *American Journal of Botany* 108(1), 22–36. https://doi.org/10.1002/ajb2.1592.

Burkle, L. A., Delphia, C. M. and O'Neill, K. M. (2017). A dual role for farmlands: food security and pollinator conservation. *Journal of Ecology* 105(4), 890–899. https://doi.org/10.1111/1365-2745.12784.

Busch, J. W. and Delph, L. F. (2012). The relative importance of reproductive assurance and automatic selection as hypotheses for the evolution of self-fertilization. *Annals of Botany* 109(3), 553–562. https://doi.org/10.1093/aob/mcr219.

Byers, K. (2014). Floral Volatiles in Mimulus: Chemical Ecology, Insect Olfaction, Genetics, and Reproductive Isolation [Doctoral dissertation, University of Washington]. Available at: https://hdl.handle.net/1773/26218.

Cai, C., Escalona, H. E., Li, L., Yin, Z., Huang, D. and Engel, M. S. (2018). Beetle pollination of cycads in the Mesozoic. *Current Biology: CB* 28(17), 2806–2812.e1 https://doi.org/10.1016/j.cub.2018.06.036.

Cane, J. H. (2002). Pollinating bees (Hymenoptera: Apiformes) of U.S. alfalfa compared for rates of pod and seed set. *Journal of Economic Entomology* 95(1), 22–27. https://doi.org/10.1603/0022-0493-95.1.22.

Carvell, C., Roy, D. B., Smart, S. M., Pywell, R. F., Preston, C. D. and Goulson, D. (2006). Declines in forage availability for bumblebees at a national scale. *Biological Conservation* 132(4), 481–489. https://doi.org/10.1016/j.biocon.2006.05.008.

Chacoff, N. P., Aizen, M. A. and Aschero, V. (2008). Proximity to forest edge does not affect crop production despite pollen limitation. *Proceedings. Biological Sciences* 275(1637), 907–913. https://doi.org/10.1098/rspb.2007.1547.

Charlesworth, D. (2006). Evolution of plant breeding systems. *Current Biology: CB* 16(17), R726–R735. https://doi.org/10.1016/j.cub.2006.07.068.

Chen, Y. H., Gols, R. and Benrey, B. (2015). Crop domestication and its impact on naturally selected trophic interactions. *Annual Review of Entomology* 60(1), 35–58. https://doi.org/10.1146/annurev-ento-010814-020601.

Chen, Y. P. and Siede, R. (2007). Honey bee viruses. *Advances in Virus Research* 70, 33–80. https://doi.org/10.1016/S0065-3527(07)70002-7.

Christenhusz, M. J. M. and Byng, J. W. (2016). The number of known plants species in the world and its annual increase. *Phytotaxa* 261(3), 201–217. https://doi.org/10.11646/phytotaxa.261.3.1.

Coiro, M., Doyle, J. A. and Hilton, J. (2019). How deep is the conflict between molecular and fossil evidence on the age of angiosperms? *New Phytologist* 223(1), 83–99. https://doi.org/10.1111/nph.15708.

Colla, S. R., Otterstatter, M. C., Gegear, R. J. and Thomson, J. D. (2006). Plight of the bumble bee: pathogen spillover from commercial to wild populations. *Biological Conservation* 129(4), 461–467. https://doi.org/10.1016/j.biocon.2005.11.013.

Connelly, H., Poveda, K. and Loeb, G. (2015). Landscape simplification decreases wild bee pollination services to strawberry. *Agriculture, Ecosystems and Environment* 211, 51–56. https://doi.org/10.1016/j.agee.2015.05.004.

Cook, D., Manson, J. S., Gardner, D. R., Welch, K. D. and Irwin, R. E. (2013). Norditerpene alkaloid concentrations in tissues and floral rewards of larkspurs and impacts on pollinators. *Biochemical Systematics and Ecology* 48, 123–131. https://doi.org/10.1016/j.bse.2012.11.015.

Cox, P. A. (1988). Hydrophilous pollination. *Annual Review of Ecology and Systematics* 19(1), 261–279. https://doi.org/10.1146/annurev.es.19.110188.001401.

Crane, E. (1986). Rock paintings related to honey hunting. *Bee World* 67(1), 23–25. https://doi.org/10.1080/0005772X.1986.11098856.

Crane, E. (1999). *The World History of Beekeeping and Honey Hunting*. Routledge, New York, NY.

Cronk, Q. and Ojeda, I. (2008). Bird-pollinated flowers in an evolutionary and molecular context. *Journal of Experimental Botany* 59(4), 715–727. https://doi.org/10.1093/jxb/ern009.

Cruden, R., Hermann, S. M. and Peterson, S. (1983). Patterns of nectar production and plant-pollinator coevolution. In: Elias, T. and Bentley, B. (Eds), *The Biology of Nectaries* (pp. 80–125). Columbia University Press.

Culley, T. M., Weller, S. G. and Sakai, A. K. (2002). The evolution of wind pollination in angiosperms. *Trends in Ecology and Evolution* 17(8), 361–369. https://doi.org/10.1016/S0169-5347(02)02540-5.

Cusser, S., Neff, J. L. and Jha, S. (2016). Natural land cover drives pollinator abundance and richness, leading to reductions in pollen limitation in cotton agroecosystems. *Agriculture, Ecosystems and Environment* 226, 33–42. https://doi.org/10.1016/j.agee.2016.04.020.

Danforth, B. N. (2007). Bees. *Current Biology: CB* 17(5), R156–R161. https://doi.org/10.1016/j.cub.2007.01.025.

Danforth, B. N., Minckley, R. L. and Neff, J. L. (2019a). *Bee Phylogeny, Bee Diversity, and the Distinction Between Solitary and Social Bees. The Solitary Bees: Biology, Evolution, Conservation* (1st edn.). Princeton University Press, Princeton, NJ.

Danforth, B. N., Minckley, R. L. and Neff, J. L. (2019b). *Bees and Plants: Love Story, Arms Race, or Something in Between? The Solitary Bees: Biology, Evolution, Conservation* (1st edn.). Princeton University Press, Princeton, NJ.

Darwin, C. R. (1877). *The Various Contrivances by Which Orchids Are Fertilised* (2nd edn.). John Murray, London.

Darwin, C. R. (1879). Letter 12167-Darwin, C. R., to Hooker, J.D., 22 July 1879. https://www.darwinproject.ac.uk/letter/?docId=letters/DCP-LETT-12167.xml. Accessed on 7 December 2021.

Daszak, P., Cunningham, A. A. and Hyatt, A. D. (2000). Emerging infectious diseases of wildlife: threats to biodiversity and human health. *Science* 287(5452), 443–449. https://doi.org/10.1126/science.287.5452.443.

Delaplane, K. S. and Mayer, D. F. (2000). *Crop Pollination by Bees*. CABI Publishing.

Detzel, A. and Wink, M. (1993). Attraction, deterrence, or intoxication of bees (*Apis mellifera*) by plant allelochemicals. *Chemoecology* 4(1), 8–18. https://doi.org/10.1007/BF01245891.

Di Pasquale, G., Salignon, M., Le Conte, Y., Belzunces, L. P., Decourtye, A., Kretzschmar, A., Suchail, S., Brunet, J. L. and Alaux, C. (2013). Influence of pollen nutrition on honey bee health: do pollen quality and diversity matter? *PLoS ONE* 8(8), Article e72016. https://doi.org/10.1371/journal.pone.0072016.

Di Vittori, V., Gioia, T., Rodriguez, M., Bellucci, E., Bitocchi, E., Nanni, L., Attene, G., Rau, D. and Papa, R. (2019). Convergent evolution of the seed shattering trait. *Genes* 10(1), Article 68. https://doi.org/10.3390/genes10010068.

Dicks, L. V., Breeze, T. D., Ngo, H. T., Senapathi, D., An, J., Aizen, M. A., Basu, P., Buchori, D., Galetto, L., Garibaldi, L. A., Gemmill-Herren, B., Howlett, B. G., Imperatriz-Fonseca, V. L., Johnson, S. D., Kovács-Hostyánszki, A., Kwon, Y. J., Lattorff, H. M. G., Lungharwo, T., Seymour, C. L., Vanbergen, A. J. and Potts, S. G. (2021). A global-scale expert assessment of drivers and risks associated with pollinator decline. *Nature Ecology and Evolution* 5(10), 1453–1461. https://doi.org/10.1038/s41559-021-01534-9.

Dilcher, D. (2000). Toward a new synthesis: major evolutionary trends in the angiosperm fossil record. *Proceedings of the National Academy of Sciences of the United States of America* 97(13), 7030–7036. https://doi.org/10.1073/pnas.97.13.7030.

Dirzo, R. and Raven, P. H. (2003). Global state of biodiversity and loss. *Annual Review of Environment and Resources* 28(1), 137–167. https://doi.org/10.1146/annurev.energy.28.050302.105532.

Donald, C. M. (1968). The breeding of crop ideotypes. *Euphytica* 17(3), 385–403. https://doi.org/10.1007/BF00056241.

Doyle, J. A. (2012). Molecular and fossil evidence on the origin of angiosperms. *Annual Review of Earth and Planetary Sciences* 40(1), 301–326. https://doi.org/10.1146/annurev-earth-042711-105313.

Dudai, N. and Belanger, F. C. (2008). Aroma as a factor in breeding process of fresh herbs – the case of basil. In: Havkin-Frenkel, D. and Belanger, F. C. (Eds), *Biotechnology in Flavor Production* (pp. 161–179). Blackwell Publishing.

Egan, P. A., Adler, L. S., Irwin, R. E., Farrell, I. W., Palmer-Young, E. C. and Stevenson, P. C. (2018). Crop domestication alters floral reward chemistry with potential consequences for pollinator health. *Frontiers in Plant Science* 9, Article 1357. https://doi.org/10.3389/fpls.2018.01357.

Elias, M. and McKey, D. (2000). The unmanaged reproductive ecology of domesticated plants in traditional agroecosystems: an example involving cassava and a call for data. *Acta Oecologica* 21(3), 223–230. https://doi.org/10.1016/S1146-609X(00)00053-9.

Ellis, E. C., Beusen, A. H. W. and Goldewijk, K. K. (2020). Anthropogenic biomes: 10,000 BCE to 2015 CE. *Land* 9(5), Article 129. https://doi.org/10.3390/land9050129.

Evans, L. T. (1993). *Crop Evolution, Adaptation, and Yield*. Cambridge University Press.

Faegri, K. and van der Pijl, L. (1979). *The Principles of Pollination Ecology* (3rd edn.). Pergamon.

Farkas, A., Molnár, R., Morschhauser, T. and Hahn, I. (2012). Variation in nectar volume and sugar concentration of *Allium ursinum* L. ssp. *ucrainicum* in three habitats. *TheScientificWorldJournal* 2012, Article 138579. https://doi.org/10.1100/2012/138579.

Fenster, C. B., Armbruster, W. S., Wilson, P., Dudash, M. R. and Thomson, J. D. (2004). Pollination syndromes and floral specialization. *Annual Review of Ecology, Evolution, and Systematics* 35(1), 375–403. https://doi.org/10.1146/annurev.ecolsys.34.011802.132347.

Fleming, T. H. and Sosa, V. J. (1994). Effects of nectarivorous and frugivorous mammals on reproductive success of plants. *Journal of Mammalogy* 75(4), 845–851. https://doi.org/10.2307/1382466.

Free, J. B. (1968). The behaviour of bees visiting runner beans (*Phaseolus multiflorus*). *Journal of Applied Ecology* 5(3), 631–638. Available at: https://doi.org/2401637.

Free, J. B. (1970). *Insect Pollination of Crops*. Academic Press, New York.

Friedman, J. and Barrett, S. C. H. (2009). Wind of change: new insights on the ecology and evolution of pollination and mating in wind-pollinated plants. *Annals of Botany* 103(9), 1515–1527. https://doi.org/10.1093/aob/mcp035.

Friedman, W. E. (2009). The meaning of Darwin's 'abominable mystery'. *American Journal of Botany* 96(1), 5–21. https://doi.org/10.3732/ajb.0800150.

Friis, E. M., Crane, P. R., and Pedersen, K. R. (1987). *Early Flowers and Angiosperm Evolution*. Cambridge University Press, Cambridge.

Fuller, D. Q., Allaby, R. G. and Stevens, C. (2010). Domestication as an innovation: the entanglement of techniques, technologies, and chance in the domestication of cereal crops. *World Archaeology* 42(1), 13–28. https://doi.org/10.1080/00438240903429680.

Fürst, M. A., McMahon, D. P., Osborne, J. L., Paxton, R. J. and Brown, M. J. F. (2014). Disease associations between honeybees and bumblebees as a threat to wild pollinators. *Nature* 506(7488), 364–366. https://doi.org/10.1038/nature12977.

Gaglianone, M. C. (2000). Behavior on flowers, structures associated to pollen transport and nesting biology of *Perditomorpha brunerii* and *Cephalurgus anomalus* (Hymenoptera:Colletidae, Andrenidae). *Revista de Biología Tropical* 48(1), 89–99. Available at: http://www.scielo.sa.cr/scielo.php?script=sci_arttext&pid=S0034-77442000000100010&lng=en.

Gallai, N., Salles, J. M., Settele, J. and Vassière, B. E. (2009). Economic valuation of the vulnerability of world agriculture confronted with pollinator decline. *Ecological Economics* 68(3), 810–821. https://doi.org/10.1016/j.ecolecon.2008.06.014.

Gandolfo, M. A., Nixon, K. C. and Crepet, W. L. (2004). Cretaceous flowers of Nymphaeaceae and implications for complex insect entrapment pollination mechanisms in early Angiosperms. *Proceedings of the National Academy of Sciences of the United States of America* 101(21), 8056–8060. https://doi.org/10.1073/pnas.0402473101.

Garibaldi, L. A., Steffan-Dewenter, I., Kremen, C., Morales, J. M., Bommarco, R., Cunningham, S. A., Carvalheiro, L. G., Chacoff, N. P., Dudenhöffer, J. H., Greenleaf, S. S., Holzschuh, A., Isaacs, R., Krewenka, K., Mandelik, Y., Mayfield, M. M., Morandin, L. A., Potts, S. G., Ricketts, T. H., Szentgyörgyi, H., Viana, B. F., Westphal, C., Winfree, R. and Klein, A. M. (2011). Stability of pollination services decreases with isolation from natural areas despite honey bee visits. *Ecology Letters* 14(10), 1062–1072. https://doi.org/10.1111/j.1461-0248.2011.01669.x.

Garibaldi, L. A., Steffan-Dewenter, I., Winfree, R., Aizen, M. A., Bommarco, R., Cunningham, S. A., Kremen, C., Carvalheiro, L. G., Harder, L. D., Afik, O., Bartomeus, I., Benjamin, F., Boreux, V., Cariveau, D., Chacoff, N. P., Dudenhöffer, J. H., Freitas, B. M., Ghazoul, J., Greenleaf, S., Hipólito, J., Holzschuh, A., Howlett, B., Isaacs, R., Javorek, S. K., Kennedy, C. M., Krewenka, K. M., Krishnan, S., Mandelik, Y., Mayfield, M. M., Motzke, I., Munyuli, T., Nault, B. A., Otieno, M., Petersen, J., Pisanty, G., Potts, S. G., Rader, R., Ricketts, T. H., Rundlöf, M., Seymour, C. L., Schüepp, C., Szentgyörgyi, H., Taki, H., Tscharntke, T., Vergara, C. H., Viana, B. F., Wanger, T. C., Westphal, C., Williams, N. and Klein, A. M. (2013). Wild pollinators enhance fruit set of crops regardless of honey bee abundance. *Science* 339(6127), 1608–1611. https://doi.org/10.1126/science.1230200.

Gepts, P. (2004). Crop domestication as a long-term selection experiment. In: Janick, J. (Ed.), *Plant Breeding Reviews* (vol. 24, pp. 1–37). John Wiley & Sons, Inc.

Giacomini, J. J., Leslie, J., Tarpy, D. R., Palmer-Young, E. C., Irwin, R. E. and Adler, L. S. (2018). Medicinal value of sunflower pollen against bee pathogens. *Scientific Reports* 8(1), Article 14394. https://doi.org/10.1038/s41598-018-32681-y.

Glasser, S. (2020). *Influence of Plant Domestication on Plant-Pollinator Interactions: Floral Characteristics and Floral Visitor Behavior in Wild and Cultivated Cucurbita Species*. [Poster, Ecological Society of America Virtual Meeting, August 3 – August 6, 2020].

Glenny, W., Cavigli, I., Daughenbaugh, K. F., Radford, R., Kegley, S. E. and Flenniken, M. L. (2017). Honey bee (*Apis mellifera*) colony health and pathogen composition in migratory beekeeping operations involved in California almond pollination. *PLoS ONE* 12(8), Article e0182814. https://doi.org/10.1371/journal.pone.0182814.

Glover, B. J. (2014). *Understanding Flowers and Flowering: An Integrated Approach* (2nd edn.). Oxford University Press, Oxford.

Goka, K., Okabe, K. and Yoneda, M. (2006). Worldwide migration of parasitic mites as a result of bumblebee commercialization: special feature: population ecology of biological invasion. *Population Ecology* 48(4), 285–291. https://doi.org/10.1007/s10144-006-0010-8.

Goldberg, E. E., Kohn, J. R., Lande, R., Robertson, K. A., Smith, S. A. and Igić, B. (2010). Species selection maintains self-incompatibility. *Science* 330(6003), 493–495. https://doi.org/10.1126/science.1194513.

Gols, R., Bukovinszky, T., van Dam, N. M., Dicke, M., Bullock, J. M. and Harvey, J. A. (2008). Performance of generalist and specialist herbivores and their endoparasitoids differs on cultivated and wild brassica populations. *Journal of Chemical Ecology* 34(2), 132–143. https://doi.org/10.1007/s10886-008-9429-z.

Gols, R., Bullock, J. M., Dicke, M., Bukovinszky, T. and Harvey, J. A. (2011). Smelling the wood from the trees: non-linear parasitoid responses to volatile attractants produced by wild and cultivated cabbage. *Journal of Chemical Ecology* 37(8), 795–807. https://doi.org/10.1007/s10886-011-9993-5.

González-Varo, J. P., Biesmeijer, J. C., Bommarco, R., Potts, S. G., Schweiger, O., Smith, H. G., Steffan-Dewenter, I., Szentgyörgyi, H., Woyciechowski, M. and Vilà, M. (2013). Combined effects of global change pressures on animal-mediated pollination. *Trends in Ecology and Evolution* 28(9), 524–530. https://doi.org/10.1016/j.tree.2013.05.008.

Goodwillie, C., Kalisz, S. and Eckert, C. G. (2005). The evolutionary enigma of mixed mating systems in plants: occurrence, theoretical explanations, and empirical evidence. *Annual Review of Ecology, Evolution, and Systematics* 36(1), 47–79. https://doi.org/10.1146/annurev.ecolsys.36.091704.175539.

Gouinguené, S., Degen, T. and Turlings, T. C. J. (2001). Variability in herbivore-induced odour emissions among maize cultivars and their wild ancestors (teosinte). *Chemoecology* 11(1), 9–16. https://doi.org/10.1007/PL00001832.

Goulson, D., Nicholls, E., Botías, C. and Rotheray, E. L. (2015). Bee declines driven by combined stress from parasites, pesticides, and lack of flowers. *Science* 347(6229), Article 1255957. https://doi.org/10.1126/science.1255957.

Goulson, D. and Hughes, W. O. H. (2015). Mitigating the anthropogenic spread of bee parasites to protect wild pollinators. *Biological Conservation* 191, 10–19. https://doi.org/10.1016/j.biocon.2015.06.023.

Graystock, P., Goulson, D. and Hughes, W. O. H. (2014). The relationship between managed bees and the prevalence of parasites in bumblebees. *PeerJ* 2, e522. https://doi.org/10.7717/peerj.522.

Graystock, P., Goulson, D. and Hughes, W. O. H. (2015). Parasites in bloom: flowers aid dispersal and transmission of pollinator parasites within and between bee species. *Proceedings. Biological Sciences* 282(1813), Article 20151371. https://doi.org/10.1098/rspb.2015.1371.

Graystock, P., Yates, K., Evison, S. E. F., Darvill, B., Goulson, D. and Hughes, W. O. H. (2013). The Trojan hives: pollinator pathogens, imported and distributed in bumblebee colonies. *Journal of Applied Ecology* 50(5), 1207–1215. https://doi.org/10.1111/1365-2664.12134.

Greenleaf, S. S. and Kremen, C. (2006). Wild bees enhance honey bees' pollination of hybrid sunflower. *Proceedings of the National Academy of Sciences of the United States of America* 103(37), 13890–13895. https://doi.org/10.1073/pnas.0600929103.

Grimaldi, D. (1999). The co-radiations of pollinating insects and angiosperms in the Cretaceous. *Annals of the Missouri Botanical Garden* 86(2), 373–406. https://doi.org/10.2307/2666181.

Gusachenko, O. N., Woodford, L., Balbirnie-Cumming, K., Ryabov, E. V. and Evans, D. J. (2020). Evidence for and against deformed wing virus spillover from honey bees to bumble bees: a reverse genetic analysis. *Scientific Reports* 10(1), Article 16847. https://doi.org/10.1038/s41598-020-73809-3.

Guzman-Novoa, E., Hamiduzzaman, M. M., Anguiano-Baez, R., Correa-Benítez, A., Castañeda-Cervantes, E. and Arnold, N. I. (2015). First detection of honey bee viruses in stingless bees in North America. *Journal of Apicultural Research* 54(2), 93–95. https://doi.org/10.1080/00218839.2015.1100154.

Hahn, M. and Brühl, C. A. (2016). The secret pollinators: an overview of moth pollination with a focus on Europe and North America. *Arthropod-Plant Interactions* 10(1), 21–28. https://doi.org/10.1007/s11829-016-9414-3.

Haines-Young, R. (2009). Land use and biodiversity relationships. *Land Use Policy* 26, S178–S186. https://doi.org/10.1016/j.landusepol.2009.08.009.

Harlan, J. (1967). A wild wheat harvest in Turkey. *Archaeology* 20, 197–201. Available at: https://www.jstor.org/stable/41667737.

Heiser, C. B. (1978). Taxonomy of *Helianthus* and origin of domesticated sunflower. *Agronomy Monographs* 19, 31–53. https://doi.org/10.2134/agronmonogr19.c2.

Hu, S., Dilcher, D. L., Jarzen, D. M. and Tayler, D. W. (2008). Early steps of angiosperm pollinator coevolution. *Proceedings of the National Academy of Sciences of the United States of America* 105(1), 240–245. https://doi.org/10.1073/pnas.0707989105.

Husband, B. C. and Schemske, D. W. (1996). Evolution of the magnitude and timing of inbreeding depression in plants. *Evolution; International Journal of Organic Evolution* 50(1), 54–70. https://doi.org/10.1111/j.1558-5646.1996.tb04472.x.

Igić, B. and Kohn, J. R. (2006). The distribution of plant mating systems: study bias against obligately outcrossing species. *Evolution; International Journal of Organic Evolution* 60(5), 1098–1103. https://doi.org/10.1111/j.0014-3820.2006.tb01186.x.

IPBES (Intergovermental Science-Policy Platform on Biodiversity and Ecosystem Services) (2016). *Summary for Policymakers of the Assessment Report of the Intergovernmental Science-Policy Platform on Biodiversity and Ecosystem Services on Pollinators, Pollination and Food Production*. Potts, S. G., Imperatriz-Fonseca, V. L. and Ngo, H. T. (Eds). Secretariat of the IPBES.

Inoue, M. N., Yokoyama, J. and Washitani, I. (2008). Displacement of Japanese native bumblebees by the recently introduced *Bombus terrestris* (L.) (Hymenoptera: Apidae). *Journal of Insect Conservation* 12(2), 135–146. https://doi.org/10.1007/s10841-007-9071-z.

Inouye, D. W., Larson, B. M. H., Ssymank, A. and Kevan, P. G. (2015). Flies and flowers III: ecology of foraging and pollination. *Journal of Pollination Ecology* 16(16), 115–133. https://doi.org/10.26786/1920-7603(2015)15.

Irvine, A. K. and Armstrong, J. E. (1990). Beetle pollination in tropical rainforests of Australia. In: Bawa, K. S. and Hadley, M. (Eds), *Reproductive* Ecology *of* Tropical Forest Plants (pp. 135–149).

Irwin, R. E., Bronstein, J. L., Manson, J. S. and Richardson, L. (2010). Nectar robbing: ecological and evolutionary perspectives. *Annual Review of Ecology, Evolution, and Systematics* 41(1), 271–292. https://doi.org/10.1146/annurev.ecolsys.110308.120330.

Itino, T., Kato, M. and Hotta, M. (1991). Pollination ecology of the two wild bananas, *Musa acuminata* subsp. *halabanensis* and *M. salaccensis*: chiropterophily and Ornithophily. *Biotropica* 23(2), 151–158. https://doi.org/10.2307/2388300.

Kalisz, S., Vogler, D. W. and Hanley, K. M. (2004). Context-dependent autonomous self-fertilization yields reproductive assurance and mixed mating. *Nature* 430(7002), 884–887. https://doi.org/10.1038/nature02776.

Kalt, W., Ryan, D. A. J., Duy, J. C., Prior, R. L., Ehlenfeldt, M. K. and Vander Kloet, S. P. (2001). Interspecific variation in anthocyanins, phenolics, and antioxidant capacity among genotypes of highbush and lowbush blueberries (*Vaccinium cyanococcus* spp.). *Journal of Agricultural and Food Chemistry* 49(10), 4761–4767. https://doi.org/10.1021/jf010653e.

Kay, K. M. and Sargent, R. D. (2009). The role of animal pollination in plant speciation: integrating ecology, geography, and genetics. *Annual Review of Ecology, Evolution, and Systematics* 40(1), 637–656. https://doi.org/10.1146/annurev.ecolsys.110308.120310.

Keller, I., Fluri, P. and Imdorf, A. (2005). Pollen nutrition and colony development in honey bees–Part II. *Bee World* 86(2), 27–34. https://doi.org/10.1080/0005772X.2005.11099650.

Kennedy, C. M., Lonsdorf, E., Neel, M. C., Williams, N. M., Ricketts, T. H., Winfree, R., Bommarco, R., Brittain, C., Burley, A. L., Cariveau, D., Carvalheiro, L. G., Chacoff, N. P., Cunningham, S. A., Danforth, B. N., Dudenhöffer, J. H., Elle, E., Gaines, H. R., Garibaldi, L. A., Gratton, C., Holzschuh, A., Isaacs, R., Javorek, S. K., Jha, S., Klein, A. M., Krewenka, K., Mandelik, Y., Mayfield, M. M., Morandin, L., Neame, L. A., Otieno, M., Park, M., Potts, S. G., Rundlöf, M., Saez, A., Steffan-Dewenter, I., Taki, H., Viana, B. F., Westphal, C., Wilson, J. K., Greenleaf, S. S. and Kremen, C. (2013). A global quantitative synthesis of local and landscape effects on wild bee pollinators in agroecosystems. *Ecology Letters* 16(5), 584–599. https://doi.org/10.1111/ele.12082.

Kerdelhué, C., Rasplus, J.-Y. and Kerdelhue, C. (1996). The evolution of dioecy among ficus (Moraceae): an alternative hypothesis involving non-pollinating fig wasp pressure on the fig-pollinator mutualism. *Oikos* 77(1), 163–166. https://doi.org/10.2307/3545597.

Kevan, P. G., Clark, E. A. and Thomas, V. G. (1990). Insect pollinators and sustainable agriculture. *American Journal of Alternative Agriculture* 5(1), 13–22. http://doi.org/10.1017/S0889189300003179.

Khush, G. S. (2001). Green revolution: the way forward. *Nature Reviews. Genetics* 2(10), 815–822. https://doi.org/10.1038/35093585.

Kim, W., Gilet, T. and Bush, J. W. M. (2011). Optimal concentrations in nectar feeding. *Proceedings of the National Academy of Sciences of the United States of America* 108(40), 16618–16621. https://doi.org/10.1073/pnas.1108642108.

Klatt, B. K., Holzschuh, A., Westphal, C., Clough, Y., Smit, I., Pawelzik, E. and Tscharntke, T. (2014). Bee pollination improves crop quality, shelf life and commercial value.

Proceedings. Biological Sciences 281(1775), Article 20132440. https://doi.org/10.1098/rspb.2013.2440.

Kleijn, D., Winfree, R., Bartomeus, I., Carvalheiro, L. G., Henry, M., Isaacs, R., Klein, A. M., Kremen, C., M'Gonigle, L. K., Rader, R., Ricketts, T. H., Williams, N. M., Lee Adamson, N., Ascher, J. S., Báldi, A., Batáry, P., Benjamin, F., Biesmeijer, J. C., Blitzer, E. J., Bommarco, R., Brand, M. R., Bretagnolle, V., Button, L., Cariveau, D. P., Chifflet, R., Colville, J. F., Danforth, B. N., Elle, E., Garratt, M. P. D., Herzog, F., Holzschuh, A., Howlett, B. G., Jauker, F., Jha, S., Knop, E., Krewenka, K. M., Le Féon, V., Mandelik, Y., May, E. A., Park, M. G., Pisanty, G., Reemer, M., Riedinger, V., Rollin, O., Rundlöf, M., Sardiñas, H. S., Scheper, J., Sciligo, A. R., Smith, H. G., Steffan-Dewenter, I., Thorp, R., Tscharntke, T., Verhulst, J., Viana, B. F., Vaissière, B. E., Veldtman, R., Ward, K. L., Westphal, C. and Potts, S. G. (2015). Delivery of crop pollination services is an insufficient argument for wild pollinator conservation. *Nature Communications* 6(1), Article 7414. https://doi.org/10.1038/ncomms8414.

Klein, A. M., Vassière, B. E., Cane, J. H., Steffan-Dewenter, I., Cunningham, S. A., Kremen, K. and Tscharntke, T. (2007). Importance of pollinators in changing landscapes for world crops. *Proceedings of the Royal Society. Series B,: Biological Sciences* 274(1608), 303–313. https://doi.org/10.1098/rspb.2006.3721.

Knuth, P. (1906–1909). *Handbook of Flower Pollination*. Davis, J. R. A. (Trans.) (vols. 1–3). Clarendon Press, London.

Koh, I., Lonsdorf, E. V., Willliams, N. M., Brittain, C., Isaacs, R., Gibbs, J. and Ricketts, T. H. (2016). Modeling the status, trends, and impacts of wild bee abundance in the United States. *Proceedings of the National Academy of Sciences of the United States of America* 113(1), 140–145.

Köllner, T. G., Held, M., Lenk, C., Hiltpold, I., Turlings, T. C. J., Gershenzon, J. and Degenhardt, J. (2008). A maize (*E*)-β-caryophyllene synthase implicated in indirect defense responses against herbivores is not expressed in most American Maize varieties. *The Plant Cell* 20(2), 482–494. https://doi.org/10.1105/tpc.107.051672.

Konzmann, S., Koethe, S. and Lunau, K. (2019). Pollen grain morphology is not exclusively responsible for pollen collectability in bumble bees. *Scientific Reports* 9(1), Article 4705. https://doi.org/10.1038/s41598-019-41262-6.

Kovács-Hostyánszki, A., Espíndola, A., Vanbergen, A. J., Settele, J., Kremen, C. and Dicks, L. V. (2017). Ecological intensification to mitigate impacts of conventional intensive land use on pollinators and pollination. *Ecology Letters* 20(5), 673–689. https://doi.org/10.1111/ele.12762.

Kremen, C. and Miles, A. (2012). Ecosystem services in biologically diversified versus conventional farming systems: benefits, externalities, and trade-offs. *Ecology and Society* 17(4), Article 40. https://doi.org/10.5751/ES-05035-170440.

Kremen, C., Williams, N. M. and Thorp, R. W. (2002). Crop pollination from native bees at risk from agricultural intensification. *Proceedings of the National Academy of Sciences of the United States of America* 99(26), 16812–16816. https://doi.org/10.1073/pnas.262413599.

Krömer, T., Kessler, M., Lohaus, G. and Schmidt-Lebuhn, A. N. (2008). Nectar sugar composition and concentration in relation to pollination syndromes in Bromeliaceae. *Plant Biology* 10(4), 502–511. https://doi.org/10.1111/j.1438-8677.2008.00058.x.

Kunz, T. H., de Torrez, E. B., Bauer, D., Lobova, T. and Fleming, T. H. (2011). Ecosystem services provided by bats. *Annals of the New York Academy of Sciences* 1223(1), 1–38. https://doi.org/10.1111/j.1749-6632.2011.06004.x.

Lande, R. and Schemske, D. W. (1985). The evolution of self-fertilization and inbreeding depression in plants. I. GENETIC MODELS. *Evolution; International Journal of Organic Evolution* 39(1), 24–40. https://doi.org/10.1111/j.1558-5646.1985.tb04077.x.

Langemeier, M. and Yeager, E. (2019). *International Benchmarks for Wheat Production*. [Report, Purdue University Center for Commercial Agriculture].

Larson, B. M. H., Kevan, P. G. and Inouye, D. W. (2001). Flies and flowers: taxonomic diversity of anthophiles and pollinators. The Canadian Entomologist 133(4), 439–465.

Lautenbach, S., Seppelt, R., Liebscher, J. and Dormann, C. F. (2012). Spatial and temporal trends of global pollination benefit. *PLoS ONE* 7(4), Article e35954. https://doi.org/10.1371/journal.pone.0035954.

Lira, R., Eguiarte, L., Montes, S., Zizumbo-Villarreal, D., Marín, P. C. G. and Quesada, M. (2016).Homo sapiens–Cucurbita interaction in Mesoamerica: domestication, dissemination, and diversification. In: Lira, R., Casas, A. and Blancas, J. (Eds), *Ethnobotany of Mexico* (pp. 389–401). Springer.

Looney, C., Strange, J. P., Freeman, M. and Jennings, D. (2019). The expanding Pacific Northwest range of *Bombus impatiens* and its establishment in Washington State. *Biological Invasions* 21(6), 1879–1885.

López-Uribe, M. M., Cane, J. H., Minckley, R. L. and Danforth, B. N. (2016). Crop domestication facilitated rapid geographical expansion of a specialist pollinator, the squash bee *Peponapis pruinosa*. *Proceedings. Biological Sciences* 283(1833), Article 20160443. https://doi.org/10.1098/rspb.2016.0443.

Lunau, K., Piorek, V., Krohn, O. and Pacini, E. (2015). Just spines – mechanical defence of malvaceous pollen against collection by corbiculate bees. *Apidologie* 46(2), 144–149. https://doi.org/10.1007/s13592-014-0310-5.

Lush, W. M. (1979). Floral morphology of wild and cultivated cowpeas. *Economic Botany* 33(4), 442–447. https://doi.org/10.1007/BF02858340.

Macfadyen, S. and Bohan, D. A. (2010). Crop domestication and the disruption of species interactions. *Basic and Applied Ecology* 11(2), 116–125. https://doi.org/10.1016/j.baae.2009.11.008.

Macfarlane, R. P. (1995). Applied pollination on tropical oceanic islands. In: Roubik, D. W. (Ed.), *Pollination of Cultivated Plants in the Tropics*. FAO Agricultural Services Bulletin.

MacKenzie, K. E. and Winston, M. L. (1984). Diversity and abundance of native bee pollinators on berry crops and natural vegetation in the Lower Fraser Valley, British Columbia. *The Canadian Entomologist* 116(7), 965–974. https://doi.org/10.4039/Ent116965-7.

Magallón, S. and Castillo, A. (2009). Angiosperm diversification through time. *American Journal of Botany* 96(1), 349–365. https://doi.org/10.3732/ajb.0800060.

Main, A. R., Webb, E. B., Goyne, K. W. and Mengel, D. (2020). Reduced species richness of native bees in field margins associated with neonicotinoid concentrations in non-target soils. *Agriculture, Ecosystems and Environment* 287, Article 106693. https://doi.org/10.1016/j.agee.2019.106693.

Mallinger, R. E., Bradshaw, J., Varenhorst, A. J. and Prasifka, J. R. (2019). Native solitary bees provide economically significant pollination services to confection sunflowers (*Helianthus annuus* L.) (Asterales: Asteraceae) grown across the northern Great Plains. *Journal of Economic Entomology* 112(1), 40–48. https://doi.org/10.1093/jee/toy322.

Mallinger, R. E., Gaines-Day, H. R. and Gratton, C. (2017). Do managed bees have negative effects on wild bees?: a systematic review of the literature. *PLoS ONE* 12(12), Article e0189268. https://doi.org/10.1371/journal.pone.0189268.

Mallinger, R. E., Werts, P. and Gratton, C. (2015). Pesticide use within a pollinator-dependent crop has negative effects on the abundance and species richness of sweat bees, *Lasioglossum* spp., and on bumble bee colony growth. *Journal of Insect Conservation* 19(5), 999–1010. https://doi.org/10.1007/s10841-015-9816-z.

Maloof, J. E. and Inouye, D. W. (2000). Are nectar robbers cheaters or mutualists? *Ecology* 81(10), 2651–2661. https://doi.org/10.1890/0012-9658(2000)081[2651:ANRCOM]2.0.CO;2.

Manley, R., Boots, M. and Wilfert, L. (2015). Emerging viral disease risk to pollinating insects: ecological, evolutionary and anthropogenic factors. *Journal of Applied Ecology* 52(2), 331–340. https://doi.org/10.1111/1365-2664.12385.

Manson, J. S., Rasmann, S., Halitschke, R., Thomson, J. D. and Agrawal, A. A. (2012). Cardenolides in nectar may be more than a consequence of allocation to other plant parts: a phylogenetic study of *Asclepias*. *Functional Ecology* 26(5), 1100–1110. https://doi.org/10.1111/j.1365-2435.2012.02039.x.

Matson, P. A., Parton, W. J., Power, A. G. and Swift, M. J. (1997). Agricultural intensification and ecosystem properties. *Science* 277(5325), 504–509. https://doi.org/10.1126/science.277.5325.504.

McAulay, M. K. and Forrest, J. R. K. (2019). How do sunflower pollen mixtures affect survival of queenless microcolonies of bumblebees (Bombus impatiens)? *Arthropod-Plant Interactions* 13(3), 517–529. https://doi.org/10.1007/s11829-018-9664-3.

McMahon, D. P., Fürst, M. A., Caspar, J., Theodorou, P., Brown, M. J. F. and Paxton, R. J. (2015). A sting in the spit: widespread cross-infection of multiple RNA viruses across wild and managed bees. *Journal of Animal Ecology* 84(3), 615–624. https://doi.org/10.1111/1365-2656.12345.

Meyer, R. S., DuVal, A. E. and Jensen, H. R. (2012). Patterns and processes in crop domestication: an historical review and quantitative analysis of 203 global food crops. *New Phytologist* 196(1), 29–48. https://doi.org/10.1111/j.1469-8137.2012.04253.x.

Michener, C. D. (2007). *The Bees of the World* (2nd edn.). JHU Press.

Milla, R., Osborne, C. P., Turcotte, M. M. and Violle, C. (2015). Plant domestication through an ecological lens. *Trends in Ecology and Evolution* 30(8), 463–469. https://doi.org/10.1016/j.tree.2015.06.006.

Miller, P. (1724). *The Gardener's and Florist's Dictionary*, Charles Rivington, London.

Morandin, L. A. and Winston, M. L. (2005). Wild bee abundance and seed production in conventional, organic, and genetically modified canola. *Ecological Applications* 15(3), 871–881. https://doi.org/10.1890/03-5271.

Morgan, M. T. (2007). Selection on reproductive characters: conceptual foundations and their extension to pollinator interactions. In: Harder, L. D. and Barrett, S. C. H. (Eds), *Ecology and Evolution of Flowers* (pp. 25–40). Oxford University Press, Oxford.

Morris, W. F., Vázquez, D. P. and Chacoff, N. P. (2010). Benefit and cost curves for typical pollination mutualisms. *Ecology* 91(5), 1276–1285. https://doi.org/10.1890/08-2278.1.

Muráriková, A., Ťažký, A., Neugebauerová, J., Planková, A., Jampílek, J., Mučaji, P. and Mikuš, P. (2017). Characterization of essential oil composition in different basil species and pot cultures by a GC-MS Method. *Molecules* 22(7), Article 1221. https://doi.org/10.3390/molecules22071221.

Murray, T. E., Coffey, M. F., Kehoe, E. and Horgan, F. G. (2013). Pathogen prevalence in commercially reared bumble bees and evidence of spillover in conspecific populations. *Biological Conservation* 159, 269-276. https://doi.org/10.1016/j.biocon.2012.10.021.

National Research Council. (2007). Status of pollinators in North America. National Academies Press, Washington, D.C.

Neff, J. L. and Simpson, B. B. (1996). *Bees, Pollination Systems and Plant Diversity*. CAB International.

Newell-McGloughlin, M. (2008). Nutritionally improved agricultural crops. *Plant Physiology* 147(3), 939–953. https://doi.org/10.1104/pp.108.121947.

Nicholson, C. C. and Ricketts, T. H. (2019). Wild pollinators improve production, uniformity, and timing of blueberry crops. *Agriculture, Ecosystems and Environment* 272, 29–37. https://doi.org/10.1016/j.agee.2018.10.018.

Nicolson, S. W. and Thornburg, R. W. (2007). Nectar chemistry. In: Nicolson, S. W., Nepi, M., and Pacini, E. (Eds), *Nectaries and Nectar* (pp. 215-264). Springer Netherlands. https://doi.org/10.1007/978-1-4020-5937-7_5.

Ollerton, J. (2017). Pollinator diversity: distribution, ecological function, and conservation. *Annual Review of Ecology, Evolution, and Systematics* 48(1), 353–376. https://doi.org/10.1146/annurev-ecolsys-110316-022919.

Ollerton, J., Alarcón, R., Waser, N. M., Price, M. V., Watts, S., Cranmer, L., Hingston, A., Peter, C. I. and Rotenberry, J. (2009). A global test of the pollination syndrome hypothesis. *Annals of Botany* 103(9), 1471–1480. https://doi.org/10.1093/aob/mcp031.

Ollerton, J., Erenler, H., Edwards, M. and Crockett, R. (2014). Extinctions of aculeate pollinators in Britain and the role of large-scale agricultural changes. Science 346(6215), 1360–1362.

Ollerton, J., Winfree, R. and Tarrant, S. (2011). How many flowering plants are pollinated by animals? *Oikos* 120(3), 321-326. https://doi.org/10.1111/j.1600-0706.2010.18644.x.

Olsen, K. M. and Wendel, J. F. (2013). A bountiful harvest: genomic insights into crop domestication phenotypes. *Annual Review of Plant Biology* 64(1), 47–70. https://doi.org/10.1146/annurev-arplant-050312-120048.

Otterstatter, M. C. and Thomson, J. D. (2008). Does pathogen spillover from commercially reared bumble bees threaten wild pollinators? *PLoS ONE* 3(7), Article e2771. https://doi.org/10.1371/journal.pone.0002771.

Otterstatter, M. C. and Whidden, T. L. (2004). Patterns of parasitism by tracheal mites (*Locustacarus buchneri*) in natural bumble bee populations. *Apidologie* 35(4), 351-357. https://doi.org/10.1051/apido:2004024.

Palmer-Young, E. C., Farrell, I. W., Adler, L. S., Milano, N. J., Egan, P. A., Irwin, R. E. and Stevenson, P. C. (2019). Secondary metabolites from nectar and pollen: a resource for ecological and evolutionary studies. *Ecology* 100(4), Article e02621. https://doi.org/10.1002/ecy.2621.

Palmer-Young, E. C., Sadd, B. M., Stevenson, P. C., Irwin, R. E. and Adler, L. S. (2016). Bumble bee parasite strains vary in resistance to phytochemicals. *Scientific Reports* 6(1), Article 37087. https://doi.org/10.1038/srep37087.

Pamminger, T., Becker, R., Himmelreich, S., Schneider, C. W. and Bergtold, M. (2019). The nectar report: quantitative review of nectar sugar concentrations offered by bee visited flowers in agricultural and non-agricultural landscapes. *PeerJ* 7, Article e6329. https://doi.org/10.7717/peerj.6329.

Parreño, M. A., Alaux, C., Brunet, J. L., Buydens, L., Filipiak, M., Henry, M., Keller, A., Klein, A. M., Khulmann, M., Leroy, C., Meeus, I., Palmer-Young, E., Oiot, N., Requier, F., Ruedenauer, F., Smagghe, G., Stevenson, P. C. and Leonhardt, S. D. (2021). Critical links between biodiversity and health in wild bee conservation. *Trends in Ecology and Evolution* 37(4), Article 2945. https://doi.org/10.1016/j.tree.2021.11.013.

Peñalver, E., Arillo, A., Pérez-de la Fuente, R., Riccio, M. L., Delclòs, X., Barrón, E. and Grimaldi, D. A. (2015). Long-proboscid flies as pollinators of Cretaceous gymnosperms. *Current Biology: CB* 25(14), 1917–1923. https://doi.org/10.1016/j.cub.2015.05.062.

Peñalver, E., Labandeira, C. C., Barrón, E., Delclòs, X., Nel, P., Nel, A., Tafforeau, P. and Soriano, C. (2012). Thrips pollination of Mesozoic gymnosperms. *Proceedings of the National Academy of Sciences of the United States of America* 109(22), 8623–8628. https://doi.org/10.1073/pnas.1120499109.

Peris, D., Pérez-de la Fuente, R., Peñalver, E., Delclòs, X., Barrón, E. and Labandeira, C. C. (2017). False blister beetles and the expansion of gymnosperm-insect pollination modes before angiosperm dominance. *Current Biology: CB* 27(6), 897–904. https://doi.org/10.1016/j.cub.2017.02.009.

Perret, M., Chautems, A., Spichiger, R., Peixoto, M. and Savolainen, V. (2001). Nectar sugar composition in relation to pollination syndromes in Sinningieae (Gesneriaceae). *Annals of Botany* 87(2), 267–273. https://doi.org/10.1006/anbo.2000.1331.

Piotrowska, K. (2008). Pollen production in selected species of anemophilous plants. *Acta Agrobotanica* 61(1), 41–52. https://doi.org/10.5586/aa.2008.005.

Plischuk, S. and Lange, C. E. (2009). Invasive *Bombus terrestris* (Hymenoptera: Apidae) parasitized by a flagellate (Euglenozoa: Kinetoplastea) and a neogregarine (Apicomplexa: Neogregarinorida). *Journal of Invertebrate Pathology* 102(3), 263-265. https://doi.org/10.1016/j.jip.2009.08.005.

Plischuk, S., Martín-Hernández, R., Prieto, L., Lucía, M., Botías, C., Meana, A., Abrahamovich, A. H., Lange, C. and Higes, M. (2009). South American native bumblebees (Hymenoptera: Apidae) infected by *Nosema ceranae* (*Microsporidia*), an emerging pathogen of honeybees (*Apis mellifera*). *Environmental Microbiology Reports* 1(2), 131–135. https://doi.org/10.1111/j.1758-2229.2009.00018.x.

Potts, S. G., Biesmeijer, J. C., Kremen, C., Neumann, P., Schweiger, O. and Kunin, W. E. (2010). Global pollinator declines: trends, impacts and drivers. *Trends in Ecology and Evolution* 25(6), 345–353. https://doi.org/10.1016/j.tree.2010.01.007.

Potts, S. G., Imperatriz-Fonseca, V., Ngo, H. T., Aizen, M. A., Biesmeijer, J. C., Breeze, T. D., Dicks, L. V., Garibaldi, L. A., Hill, R., Settele, J. and Vanbergen, A. J. (2016). Safeguarding pollinators and their values to human well-being. *Nature* 540(7632), 220–229. https://doi.org/10.1038/nature20588.

Potts, S. G., Vulliamy, B., Roberts, S., O'Toole, C., Dafni, A., Ne'eman, G. and Willmer, P. G. (2004). Nectar resource diversity organises flower-visitor community structure. *Entomologia Experimentalis et Applicata* 113(2), 103–107. https://doi.org/10.1111/j.0013-8703.2004.00212.x.

Power, A. G. and Mitchell, C. E. (2004). Pathogen spillover in disease epidemics. *The American Naturalist* 164 (Suppl. 5), S79–S89. https://doi.org/10.1086/424610.

Powney, G. D., Carvell, C., Edwards, M., Morris, R. K. A., Roy, H. E., Woodcock, B. A. and Isaac, N. J. B. (2019). Widespread losses of pollinating insects in Britain. *Nature Communications* 10(1), Article 1018. https://doi.org/10.1038/s41467-019-08974-9.

Proctor, M. C. (1996a). The study of pollination: a short history. In: Corbet, S. A., Walters, S. M., West, R., Streeter, D. and Ratcliffe, D. A. (Eds), *The Natural History of Pollination* (pp.15–43). Collins, London, UK.

Proctor, M. C. (1996b). Pollination and fertilisation. In: Corbet, S. A., Walters, S. M., West, R., Streeter, D. and Ratcliffe, D. A. (Eds), *The Natural History of Pollination* (pp. 44–108). Collins, London, UK.

Proctor, M. C. (1996c). Pollination by wind and water. In: Corbet, S. A., Walters, S. M., West, R., Streeter, D. and Ratcliffe, D. A. (Eds), *The Natural History of Pollination* (pp. 689–767). Collins, London, UK.

Purkiss, T. and Lach, L. (2019). Pathogen spillover from *Apis mellifera* to a stingless bee. *Proceedings. Biological Sciences* 286(1908), Article 20191071. https://doi.org/10.1098/rspb.2019.1071.

Purugganan, M. D. and Fuller, D. Q. (2009). The nature of selection during plant domestication. *Nature* 457(7231), 843–848. https://doi.org/10.1038/nature07895.

Purugganan, M. D. and Fuller, D. Q. (2011). Archaeological data reveal slow rates of evolution during plant domestication. *Evolution; International Journal of Organic Evolution* 65(1), 171–183. https://doi.org/10.1111/j.1558-5646.2010.01093.x.

Rader, R., Bartomeus, I., Garibaldi, L. A., Garratt, M. P. D., Howlett, B. G., Winfree, R., Cunningham, S. A., Mayfield, M. M., Arthur, A. D., Andersson, G. K. S., Bommarco, R., Brittain, C., Cavalheiro, L. G., Chacoff, N. P., Entling, M. H., Foully, B., Freitas, B. M., Gemmill-Herren, B., Ghazoul, J., Griffin, S. R., Gross, C. L., Herbertsson, L., Herzog, F., Hipólito, J., Jaggar, S., Jauker, F., Klein, A. M., Kleijn, D., Krishnan, S., Lemos, C. Q., Lindström, S. A., Mandelik, Y., Monteiro, V. M., Nelson, W., Nilsson, L., Pattemore, D. E., Pereira, Nde O., Pisanty, G., Potts, S. G., Reemer, M., Rundlöf, M., Sheffield, C. S., Scheper, J., Schüepp, C., Smith, H. G., Stanley, D. A., Stout, J. C., Szentgyörgyi, H., Taki, H., Vergara, C. H., Viana, B. F. and Woyciechowski, M. (2016). Non-bee insects are important contributors to global crop pollination. *Proceedings of the National Academy of Sciences of the United States of America* 113(1), 146–151. https://doi.org/10.1073/pnas.1517092112.

Rader, R., Howlett, B. G., Cunningham, S. A., Westcott, D. A. and Edwards, W. (2012). Spatial and temporal variation in pollinator effectiveness: do unmanaged insects provide consistent pollination services to mass flowering crops? *Journal of Applied Ecology* 49(1), 126–134. https://doi.org/10.1111/j.1365-2664.2011.02066.x.

Ramos, S. E. and Schiestl, F. P. (2020). Evolution of floral fragrance is compromised by herbivory. *Frontiers in Ecology and Evolution* 8, Article 30. https://doi.org/10.3389/fevo.2020.00030.

Ravoet, J., De Smet, L., Meeus, I., Smagghe, G., Wenseleers, T. and de Graaf, D. C. (2014). Widespread occurrence of honey bee pathogens in solitary bees. *Journal of Invertebrate Pathology* 122, 55–58. https://doi.org/10.1016/j.jip.2014.08.007.

Ray, H. A. and Vendrame, W. A. (2018). *Orchid Pollination Biology*. [Report. University of Florida/IFAS Extension].

Regan, E. C., Santini, L., Ingwall-King, L., Hoffmann, M., Rondinini, C., Symes, A., Taylor, J. and Butchart, S. H. M. (2015). Global trends in the status of bird and mammal pollinators: bird and mammal pollinator trends. *Conservation Letters* 8(6), 397–403. https://doi.org/10.1111/conl.12162.

Reilly, J. R., Artz, D. R., Biddinger, D., Bobiwash, K., Boyle, N. K., Brittain, C., Brokaw, J., Campbell, J. W., Daniels, J., Elle, E., Ellis, J. D., Fleischer, S. J., Gibbs, J., Gillespie, R. L., Gundersen, K. B., Gut, L., Hoffman, G., Joshi, N., Lundin, O., Mason, K., McGrady,

C. M., Peterson, S. S., Pitts-Singer, T. L., Rao, S., Rothwell, N., Rowe, L., Ward, K. L., Williams, N. M., Wilson, J. K., Isaacs, R. and Winfree, R. (2020). Crop production in the USA is frequently limited by a lack of pollinators. *Proceedings of the Royal Society. Series B,: Biological Sciences* 287, Article 20200922. https://doi.org/10.1098/rspb.2020.0922.

Ren, D., Labandeira, C. C., Santiago-Blay, J. A., Rasnitsyn, A., Shih, C., Bashkuev, A., Logan, M. A. V., Hotton, C. L. and Dilcher, D. (2009). A probable pollination mode before angiosperms: Eurasian, long-proboscid scorpionflies. *Science* 326(5954), 840–847. https://doi.org/10.1126/science.1178338.

Renauld, M., Hutchinson, A., Loeb, G., Poveda, K. and Conelly, H. (2016). Landscape simplification constrains adult size in a native ground-nesting bee. *PLoS ONE* 11(3), Article e0150946. https://doi.org/10.1371/journal.pone.0150946.

Richardson, L. L., Bowers, M. D. and Irwin, R. E. (2016). Nectar chemistry mediates the behavior of parasitized bees: consequences for plant fitness. *Ecology* 97(2), 325-337. https://doi.org/10.1890/15-0263.1.

Rick, C. M. (1950). Pollination relations of *Lycopersicon esculentum* in native and foreign regions. *Evolution* 4(2), 110–122. https://doi.org/10.1111/j.1558-5646.1950.tb00046.x.

Rick, C. M. and Dempsey, W. H. (1969). Position of the stigma in relation to fruit setting of the tomato. *Botanical Gazette. Botanical Gazette* 130(3), 180–186.

Ricketts, T. H., Regetz, J., Steffan-Dewenter, I., Cunningham, S. A., Kremen, C., Bogdanski, A., Gemmill-Herren, B., Greenleaf, S. S., Klein, A. M., Mayfield, M. M., Morandin, L. A., Ochieng', A., Potts, S. G. and Viana, B. F. (2008). Ochieng. *Ecology Letters* 11(5), 499–515. https://doi.org/10.1111/j.1461-0248.2008.01157.x.

Roberts, R. B. and Vallespir, S. R. (1978). Specialization of hairs bearing pollen and oil on the legs of bees (Apoidea: Hymenoptera). *Annals of the Entomological Society of America* 71(4), 619–627. https://doi.org/10.1093/aesa/71.4.619.

Rosas-Guerrero, V., Aguilar, R., Martén-Rodríguez, S., Ashworth, L., Lopezaraiza-Mikel, M., Bastida, J. M. and Quesada, M. (2014). A quantitative review of pollination syndromes: do floral traits predict effective pollinators? *Ecology Letters* 17(3), 388-400. https://doi.org/10.1111/ele.12224.

Roulston, T. H., Cane, J. H. and Buchmann, S. L. (2000). What governs protein content of pollen: pollinator preferences, pollen–pistil interactions, or phylogeny? *Ecological Monographs* 70(4), 617–643. https://doi.org/10.1890/0012-9615(2000)070[0617:WGPCOP]2.0.CO;2.

Roulston, T. H. and Goodell, K. (2011). The role of resources and risks in regulating wild bee populations. *Annual Review of Entomology* 56(1), 293–312. https://doi.org/10.1146/annurev-ento-120709-144802.

Russo, L., Vaudo, A. D., Fisher, C. J., Grozinger, C. M. and Shea, K. (2019). Bee community preference for an invasive thistle associated with higher pollen protein content. *Oecologia* 190(4), 901–912. https://doi.org/10.1007/s00442-019-04462-5.

Rydin, C. and Bolinder, K. (2015). Moonlight pollination in the gymnosperm *Ephedra* (Gnetales). *Biology Letters* 11(4), Article 20140993. https://doi.org/10.1098/rsbl.2014.0993.

Sanchez-Bayo, F. and Goka, K. (2014). Pesticide residues and bees – A risk assessment. *PLoS ONE* 9(4), Article e94482. https://doi.org/10.1371/journal.pone.0094482.

Sapir, Y. (2009). Effects of floral traits and plant genetic composition on pollinator behavior. *Arthropod-Plant Interactions* 3(2), 115–129. https://doi.org/10.1007/s11829-009-9062-y.

Saunders, M. E. (2018). Insect pollinators collect pollen from wind-pollinated plants: implications for pollination ecology and sustainable agriculture. Insect conservation and diversity 11(1), 13–31.

Sayers, T. D. J., Steinbauer, M. J. and Miller, R. E. (2019). Visitor or vector? The extent rove beetle (Coleopteara: Staphylinidae) pollination and floral interactions. *Arthropod-Plant Interactions* 13(5), 685–701. https://doi.org/10.1007/s11829-019-09698-9.

Schemske, D. W. and Bradshaw, H. D. (1999). Pollinator preference and the evolution of floral traits in monkeyflowers (Mimulus). *Proceedings of the National Academy of Sciences of the United States of America* 96(21), 11910–11915. https://doi.org/10.1073/pnas.96.21.11910.

Schiestl, F. P., Ayasse, M., Paulus, H. F., Löfstedt, C., Hansson, B. S., Ibarra, F. and Francke, W. (2000). Sex pheromone mimicry in the early spider orchid (*Ophrys sphegodes*): patterns of hydrocarbons as the key mechanism for pollination by sexual deception. *Journal of Comparative Physiology. A, Sensory, Neural, and Behavioral Physiology* 186(6), 567–574. https://doi.org/10.1007/s003590000112.

Schmid-Hempel, R., Eckhardt, M., Goulson, D., Heinzmann, D., Lange, C., Plischuk, S., Escudero, L. R., Salathé, R., Scriven, J. J. and Schmid-Hempel, P. (2014). The invasion of southern South America by imported bumblebees and associated parasites. *Journal of Animal Ecology* 83(4), 823–837. https://doi.org/10.1111/1365-2656.12185.

Sheehan, H., Hermann, K. and Kuhlemeier, C. (2012). Color and scent: how single genes influence pollinator attraction. *Cold Spring Harbor Symposia on Quantitative Biology* 77, 117–133. https://doi.org/10.1101/sqb.2013.77.014712.

Simmonds, N. W. (1960). *Bananas* (1st edn.). Longmans.

Simone-Finstrom, M. and Spivak, M. (2010). Propolis and bee health: the natural history and significance of resin use by honey bees. *Apidologie* 41(3), 295–311. https://doi.org/10.1051/apido/2010016.

Sjödin, N. E., Bengtsson, J. and Ekbom, B. (2007). The influence of grazing intensity and landscape composition on the diversity and abundance of flower-visiting insects: flower-visitor diversity and grazing intensity. *Journal of Applied Ecology* 45(3), 763–772. https://doi.org/10.1111/j.1365-2664.2007.01443.x.

Smith, M. R., Singh, G. M., Mozaffarian, D. and Myers, S. S. (2015). Effects of decreases of animal pollinators on human nutrition and global health: a modelling analysis. *The Lancet* 386(10007), 1964–1972. https://doi.org/10.1016/S0140-6736(15)61085-6.

Southwick, E. E. and Southwick, L., Jr. (1992). Estimating the economic value of honey bees (Hymenoptera: Apidae) as agricultural pollinators in the United States. *Journal of Economic Entomology* 85(3), 621–633. https://doi.org/10.1093/jee/85.3.621.

Spiesman, B. J. and Inouye, B. D. (2013). Habitat loss alters the architecture of plant-pollinator interaction networks. *Ecology* 94(12), 2688–2696. https://doi.org/10.1890/13-0977.1.

Steffan-Dewenter, I. and Schiele, S. (2008). Do resources or natural enemies drive bee population dynamics in fragmented habitats? *Ecology* 89(5), 1375–1387. https://doi.org/10.1890/06-1323.1.

Stevenson, P. C., Nicolson, S. W. and Wright, G. A. (2017). Plant secondary metabolites in nectar: impacts on pollinators and ecological functions. *Functional Ecology* 31(1), 65–75. https://doi.org/10.1111/1365-2435.12761.

Sujana, G., Sharma, H. C. and Manohar Rao, D. (2012). Pod surface exudates of wild relatives of pigeonpea influence the feeding preference of the pod borer,

Helicoverpa armigera. *Arthropod-Plant Interactions* 6(2), 231–239. https://doi.org/10.1007/s11829-011-9179-7.

Taki, H., Kevan, P. G. and Ascher, J. S. (2007). Landscape effects of forest loss in a pollination system. *Landscape Ecology* 22(10), 1575–1587. https://doi.org/10.1007/s10980-007-9153-z.

Tilman, D., Fargione, J., Wolff, B., D'Antonio, C., Dobson, A., Howarth, R., Schindler, D., Schlesinger, W. H., Simberloff, D. and Swackhamer, D. (2001). Forecasting agriculturally driven global environmental change. *Science* 292(5515), 281–284. https://doi.org/10.1126/science.1057544.

Traveset, A., Castro-Urgal, R., Rotllàn-Puig, X. and Lázaro, A. (2018). Effects of habitat loss on the plant-flower visitor network structure of a dune community. *Oikos* 127(1), 45–55. https://doi.org/10.1111/oik.04154.

Tscharntke, T., Clough, Y., Wanger, T. C., Jackson, L., Motzke, I., Perfecto, I., Vandermeer, J. and Whitbread, A. (2012). Global food security, biodiversity conservation and the future of agricultural intensification. *Biological Conservation* 151(1), 53–59. https://doi.org/10.1016/j.biocon.2012.01.068.

Tscharntke, T., Klein, A. M., Kruess, A., Steffan-Dewenter, I. and Thies, C. (2005). Landscape perspectives on agricultural intensification and biodiversity – ecosystem service management. *Ecology Letters* 8(8), 857–874. https://doi.org/10.1111/j.1461-0248.2005.00782.x.

Vaillant, S. (1718). *Discours sur la structure des fleurs, leurs differences et l'usage de leurs parties [On the Structure of Flowers, Their Differences, and the Use of Their Parts]*. Pierre Vander, Leiden.

Vaissière, B. E. and Vinson, S. B. (1994). Pollen morphology and its effect on pollen collection by honey bees. *Grana* 33(3), 128–138. https://doi.org/10.1080/00173139409428989.

van de Wouw, M., Kik, C., van Hintum, T., van Treuren, R. and Visser, B. (2010). Genetic erosion in crops: concept, research results and challenges. *Plant Genetic Resources* 8(1), 1–15. https://doi.org/10.1017/S1479262109990062.

van der Niet, T., Peakall, R. and Johnson, S. D. (2014). Pollinator-driven ecological speciation in plants: new evidence and future perspectives. *Annals of Botany* 113(2), 199–211. https://doi.org/10.1093/aob/mct290.

van Klink, R., Bowler, D. E., Gongalsky, K. B., Swengel, A. B., Gentile, A. and Chase, J. M. (2020). Meta-analysis reveals declines in terrestrial but increases in freshwater insect abundances. *Science* 368(6489), 417–420. https://doi.org/10.1126/science.aax9931.

van Wyk, B.-E., Whitehead, C. S., Glen, H. F., Hardy, D. S., Van Jaarsveld, E. J. and Smith, G. F. (1993). Nectar sugar composition in the subfamily Alloideaec (Aphodelaceae). *Biochemical Systematics and Ecology* 21(2), 249–253. https://doi.org/10.1016/0305-1978(93)90042-P.

Vanderborght, T. (1983). Evaluation of *Phaseolus vulgaris* wild types and weedy forms. *Plant Genetics Research Newsletter* 54, 18–25.

Vanderplanck, M., Vereecken, N. J., Grumiau, L., Esposito, F., Lognay, G., Wattiez, R. and Michez, D. (2017). The importance of pollen chemistry in evolutionary host shifts of bees. *Scientific Reports* 7, Article 43508. https://doi.org/10.1038/srep43058.

vanEngelsdorp, D., Hayes, J., Jr., Underwood, R. M. and Pettis, J. (2008). A survey of honey bee colony losses in the U.S., fall 2007 to spring 2008. *PLoS ONE* 3(12), Article e4071. https://doi.org/10.1371/journal.pone.0004071.

Vansell, G. H. (1944). Cotton nectar in relation to bee activity and honey production. *Journal of Economic Entomology* 37(4), 528–530.

Vaudo, A. D., Patch, H. M., Mortensen, D. A., Tooker, J. F. and Grozinger, C. M. (2016). Macronutrient ratios in pollen shape bumble bee (*Bombus impatiens*) foraging strategies and floral preferences. *Proceedings of the National Academy of Sciences of the United States of America* 113(28), E4035–E4042. https://doi.org/10.1073/pnas.1606101113.

Vaudo, A. D., Tooker, J. F., Grozinger, C. M. and Patch, H. M. (2015). Bee nutrition and floral resource restoration. *Current Opinion in Insect Science* 10, 133–141. https://doi.org/10.1016/j.cois.2015.05.008.

Vaudo, A. D., Tooker, J. F., Patch, H. M., Biddinger, D. J., Coccia, M., Crone, M. K., Fiely, M., Francis, J. S., Hines, H. M., Hodges, M., Jackson, S. W., Michez, D., Mu, J., Russo, L., Safari, M., Treanore, E. D., Vanderplanck, M., Yip, E., Leonard, A. S. and Grozinger, C. M. (2020). Pollen protein: lipid macronutrient ratios may guide broad patterns of bee species floral preferences. *Insects* 11(2), Article 132. https://doi.org/10.3390/insects11020132.

Velázquez-López, R., Wegier, A., Alavez, V., Pérez-López, J., Vázquez-Barrios, V., Arroyo-Lambaer, D., Ponce-Mendoza, A. and Kunin, W. E. (2018). The mating system of the wild-to-domesticated complex of *Gossypium hirsutum* L. is mixed. *Frontiers in Plant Science* 9, Article 574. https://doi.org/10.3389/fpls.2018.00574.

Velthuis, H. H. W. and van Doorn, A. (2006). A century of advances in bumblebee domestication and the economic and environmental aspects of its commercialization for pollination. *Apidologie* 37(4), 421–451. https://doi.org/10.1051/apido:2006019.

Vogel, S. and Westerkamp, C. (1991). Pollination: an integrating factor of biocenoses. In: Seitz, A. and Loeschcke, V. (Eds), *Species Conservation: A Population-Biological Approach* (pp. 159–170). Birkhäuser, Basel.

Wagner, D. L. (2020). Insect declines in the Anthropocene. *Annual Review of Entomology* 65(1), 457–480. https://doi.org/10.1146/annurev-ento-011019-025151.

Wardhaugh, C. W. (2015). How many species of arthropods visit flowers? *Arthropod-Plant Interactions* 9(6), 547–565. https://doi.org/10.1007/s11829-015-9398-4.

Waser, N. M. (2006). Specialization and generalization in plant-pollinator interactions: a historical perspective. In: Waser, N. M. and Ollerton, J. (Eds), *Plant-Pollinator Interactions: From Specialization to Generalization* (pp. 3–18). The University of Chicago Press, Chicago, IL.

Waser, N. M., Chittka, L., Price, M. V., Williams, N. M. and Ollerton, J. (1996). Generalization in pollination systems, and why it matters. *Ecology* 77(4), 1043–1060. https://doi.org/10.2307/2265575.

Weiner, C. N., Hilpert, A., Werner, M., Linsenmair, K. E. and Blüthgen, N. (2010). Pollen amino acids and flower specialisation in solitary bees. *Apidologie* 41(4), 476–487. https://doi.org/10.1051/apido/2009083.

Westerkamp, C. and Gottsberger, G. (2000). Diversity pays in crop pollination. *Crop Science* 40(5), 1209–1222. https://doi.org/10.2135/cropsci2000.4051209x.

Whitehead, S. R., Turcotte, M. M. and Poveda, K. (2017). Domestication impacts on plant-herbivore interactions: a meta-analysis. *Philosophical Transactions of the Royal Society of London. Series B, Biological Sciences* 372(1712), Article 20160034. https://doi.org/10.1098/rstb.2016.0034.

Whitfield, C. W., Behura, S. K., Berlocher, S. H., Clark, A. G., Johnston, J. S., Sheppard, W. S., Smith, D. R., Suarez, A. V., Weaver, D. and Tsutsui, N. D. (2006). Thrice out of Africa:

ancient and recent expansions of the honey bee, Apis mellifera. *Science* 314(5799), 642–645. https://doi.org/10.1126/science.1132772.
Whittall, J. B. and Hodges, S. A. (2007). Pollinator shifts drive increasingly long nectar spurs in columbine flowers. *Nature* 447(7145), 706–709. https://doi.org/10.1038/nature05857.
Willmer, P. (2011). *Pollination and Floral Ecology*. Princeton University Press, Princeton, NJ.
Williams, G. R., Tarpy, D. R., van Engelsdorp, D., Chauzat, M. P., Cox-Foster, D. L., Delaplane, K. S., Neumann, P., Pettis, J. S., Rogers, R. E. L. and Shutler, D. (2010). Colony Collapse Disorder in context. *BioEssays: News and Reviews in Molecular, Cellular and Developmental Biology* 32(10), 845–846. https://doi.org/10.1002/bies.201000075.
Williams, N. H. and Whitten, W. M. (1983). Orchid floral fragrances and male euglossine bees: methods and advances in the last sesquidecade. *The Biological Bulletin* 164(3), 355–395. https://doi.org/10.2307/1541248.
Wilson, W. M. and Dufour, D. L. (2002). Why "bitter" cassava? Productivity of "bitter" and "sweet" cassava in a Tukanoan Indian settlement in the northwest Amazon. *Economic Botany* 56(1), 49–57. https://doi.org/10.1663/0013-0001(2002)056[0049:WBCPOB]2.0.CO;2.
Winder, J. A. and Silva, P. (1972). Cacao pollination: microdiptera of cacao plantations and some of their their breeding places. *Bulletin of Entomological Research* 61(4), 651–655. https://doi.org/10.1017/S0007485300047465.
Winfree, R., Bartomeus, I. and Cariveau, D. P. (2011). Native pollinators in anthropogenic habitats. *Annual Review of Ecology, Evolution, and Systematics* 42(1), 1–22. https://doi.org/10.1146/annurev-ecolsys-102710-145042.
Winfree, R., Williams, N. M., Dushoff, J. and Kremen, C. (2007). Native bees provide insurance against ongoing honey bee losses. *Ecology Letters* 10(11), 1105–1113. https://doi.org/10.1111/j.1461-0248.2007.01110.x.
Woodcock, B. A., Bullock, J. M., Shore, R. F., Heard, M. S., Pereira, M. G., Redhead, J., Ridding, L., Dean, H., Sleep, D., Henrys, P., Peyton, J., Hulmes, S., Hulmes, L., Sárospataki, M., Saure, C., Edwards, M., Genersch, E., Knäbe, S. and Pywell, R. F. (2017). Country-specific effects of neonicotinoid pesticides on honey bees and wild bees. *Science* 356(6345), 1393–1395. https://doi.org/10.1126/science.aaa1190.
Wright, G. A., Baker, D. D., Palmer, M. J., Stabler, D., Mustard, J. A., Power, E. F., Borland, A. M. and Stevenson, P. C. (2013). Caffeine in floral nectar enhances a pollinator's memory of reward. *Science* 339(6124), 1202–1204. https://doi.org/10.1126/science.1228806.
Wurz, A., Grass, I. and Tscharntke, T. (2021). Hand pollination of global crops – A systematic review. *Basic and Applied Ecology* 56, 299–321. https://doi.org/10.1016/j.baae.2021.08.008.
Yeo, P. F. (1996). The insect visitors I: beetles, flies, and some others. In: Corbet, S. A., Walters, S. M., West, R., Streeter, D. and Ratcliffe, D. A. (Eds), *The Natural History of Pollination* (pp. 109–194). Collins, London.
Zeven, A. C. and de Wit, J. M. J. (1982). Dictionary for cultivated plants and their regions of diversity excluding most ornamentals, forest trees and lower plants. *Center for Agricultural Publishing and Documentary*.
Zhou, X. and Meier, I. (2014). Efficient plant male fertility depends on vegetative nuclear movement mediated by two families of plant outer nuclear membrane proteins. *Proceedings of the National Academy of Sciences of the United States of America* 111(32), 11900–11905. https://doi.org/10.1073/pnas.1323104111.

Chapter 2

The role and application of olfaction in crop plant-pollinator interactions

Sarah E. J. Arnold, Natural Resources Institute, University of Greenwich, UK and Nelson Mandela African Institution of Science and Technology, Tanzania; Alison S. Scott-Brown, University of Cambridge and Royal Botanic Gardens, Kew, UK; and Philip C. Stevenson, Royal Botanic Gardens, Kew and Natural Resources Institute, University of Greenwich, UK

1 Introduction

Pollination by animals, especially insects, is important to the production of around three in four crops (Klein et al., 2007). Critical to this ecosystem service are the individual-level interactions between plants and their pollen vectors. The primary benefit of this interaction for the pollinator is obtaining a food reward, although in some plant families (and often through chemical deception) flower visitors also anticipate an opportunity to reproduce (Urru et al., 2011). Plants have developed mechanisms to distinguish themselves from others and attract an effective pollinator. In turn, pollinators must efficiently detect and decipher plant cues to compete for potentially limited resources. In this chapter we discuss some of the mechanisms that are important in underpinning these interactions in pollination systems and how understanding and manipulating these processes could contribute to more sustainable crop production.

http://dx.doi.org/10.19103/AS.2022.0111.06

1.1 Key issues

Other chapters have discussed the importance of pollination in agriculture and some of the constraints challenging pollen vectors. Recognition of a flower by a pollinator depends upon multimodal cues including odour. Understanding the role played by odour in crop pollination, how it can be manipulated and what might limit it could help establish sustainable pollination services in crop production. Some important considerations include:

- How attractive are the volatiles of crop flowers, and is pollination efficacy based on attractiveness *per se*?
- Has crop domestication influenced crop floral odours, and what are the implications of this?
- Which crops are most reliant on pollinators that orient or select flowers primarily via scent?
- Can the use of odour cues by pollinators be manipulated to increase visit frequency or efficacy, with the aim of improving fruit set?
- Can other non-floral odours enhance pollination?

1.2 How and why plants produce floral odours

Flowers of crops and wild plants produce odours primarily to communicate with pollinators, encouraging more visits and thus more pollen transfer, essential for fruit and seed set. Floral odour production can also confer wider benefits, such as selecting for specific pollinators or deterring herbivores, and in ornamentals, may be selected to please human consumers. The odour - usually a blend of compounds - is detected by a potential pollinator (normally with the antennae in the case of insects) and the pollinator uses this, along with other cues, to locate the flower and approach it. The effectiveness of floral odours at attracting pollinators is under the influence of environmental factors such as wind speed and direction and temperature; the result is that the odour does not disperse uniformly in all directions away from the source but moves according to air currents and can meander (Brady et al., 1990). Therefore, the detectability of a crop's floral odour to potential pollinators is influenced by the wider context, not just the plant's physiology and genetics and the identity of the pollinator. The effectiveness of agricultural technologies that use artificial or synthetic odour blends to manipulate pollinator attraction to crop flowers are similarly also susceptible to these environmental variations.

While some pollinators are innately attracted to specific volatile compounds (e.g. midges orient towards the odour blend containing extract from the headspace of cacao flowers (Arnold et al., 2019)), for many pollinators the exact nature of the odour is less important than the learned experience and associated or expected benefits. For example, generalist bumblebees such as *Bombus terrestris* will learn to visit flowers or other targets providing a food reward bearing a range of odours, from pleasant (e.g. peppermint or linalool) to less pleasant odours, as long as the cue is associated with an energetic or other reward (Sommerlandt et al., 2014). Naïve bees of various species do have innate preferences for some odours more than others (Haber et al., 2021) but almost any compound can be learned in association with food. Alternatively, some insects including bumblebees and wasps may be attracted to deceptive flowers emitting odours that mimic sex pheromones. In that case, the plants are presenting a strongly innate cue that may be difficult to avoid or adapt against because the pollinators depend upon responding to it in order to secure copulation.

Consequently, it is not possible to suggest a single, optimal odour composition for a crop flower to emit to maximise pollination. Which odours can assist in promoting flower visitation depends on the species of insect (or other organism) that pollinate it, the wider environmental context, including which other plant species are in flower and what the pollinators' prior experience may have included.

However, a distinctive scent may benefit a crop plant, in a similar way to the role of floral colour and shape which are under selective pressure to make flowers distinguishable from each other within a community, allowing pollinators to discriminate among them. In a crop, mass flowering may produce a more profound cue that can attract potential pollinators from further away, because of the scale of odour emission. Alternatively, mass of flowering crop plants may require much lower levels of floral odour emission per flower than flowers in complex communities owing to the abundance of distinguishing compounds.

Floral volatiles are usually produced as secondary metabolic by-products from pathways deriving from primary metabolism, mainly the Calvin cycle. Benzenoid and phenylpropanoid compounds are produced via the shikimate pathway; terpenoids come from the MEP pathway derived from pyruvate (Muhlemann et al., 2014). They are usually emitted directly from epidermal cells in the flower. However, different parts of the flower can emit different odour compounds - for example, the nectar of courgette (*Cucurbita pepo* var. Tosca, Chapin, Consul) emits high levels of 1,2,4-trimethoxybenzene and 1,4-dimethoxybenzene, but the petals, stigma and anthers do not; similarly, the anthers emit far more linalool and eucalyptol per mass unit compared to the stigmas (Mena Granero et al., 2005).

Some volatile organic compounds (VOCs) occur widely in floral odours (Table 1) whereas some crop flowers have distinctive and unusual flower bouquets, such as cacao (Arnold et al., 2019), which emits various straight-chain unsaturated hydrocarbons. In cacao these compounds may distinguish

Table 1 Some common floral odour compounds and selected examples of crops that produce them

Compound	Group	Example of a crop with this as part of the odour bouquet
Benzaldehyde	Benzenoid	Apricot, cherry, plum (El-Sayed et al., 2018), strawberry (Ceuppens et al., 2015)
Benzyl alcohol	Alcohol/ phenol	Courgette (Mena Granero et al., 2005), peach (El-Sayed et al., 2018), strawberry (Klatt et al., 2013)
Camphor	Terpenoid	Apricot (El-Sayed et al., 2018)
Decanal	Aldehyde/fatty acid derivative	Chinese cabbage (Mas et al., 2020)
1,4-dimethoxybenzene	Phenolic	Courgette (Mena Granero et al., 2005)
Eucalyptol	Terpenoid	Courgette (Mena Granero et al., 2005)
Eugenol	Benzenoid	Blueberry (Qian et al., 2021), courgette (Mena Granero et al., 2005)
(*E,E*)-α-farnesene	Terpenoid	Cherry (El-Sayed et al., 2018), oilseed rape (Girling et al., 2013), pear (Ma et al., 2021), strawberry (Ceuppens et al., 2015)
Geraniol	Terpenoid	Blueberry (Qian et al., 2021)
(+)-Limonene	Terpenoid	Courgette (Mena Granero et al., 2005), pear (Ma et al., 2021), strawberry (Klatt et al., 2013; Ceuppens et al., 2015)
Linalool	Terpenoid	Avocado (Mas et al., 2020), cherry (El-Sayed et al., 2018), courgette (Mena Granero et al., 2005)
3-Methyl-1-butanol	Alcohol	Pear (Ma et al., 2021)
Nonanal	Aldehyde/fatty acid derivative	Avocado (Mas et al., 2020), carrot, Chinese cabbage (Mas et al., 2018), strawberry (Klatt et al., 2013)
(*E*)-β-ocimene	Terpenoid	Courgette (Mena Granero et al., 2005), strawberry (Klatt et al., 2013)
α-pinene	Terpenoid	Courgette (Mena Granero et al., 2005), pear (Ma et al., 2021)
1,2,4-trimethoxybenzene	Phenolic	Courgette (Mena Granero et al., 2005)
Vanillin	Aldehyde/fatty acid derivative	Blueberry (Qian et al., 2021)

the flowers for pollinating Ceratopogonid midges. In orchids, alkenes are proposed to be a preadaptation in the evolution of deceptive odours (Schiestl and Cozzolino, 2008). *Serapia* orchids produce a range of long-chain alkanes that differ qualitatively among closely related species and may have a more direct influence on pollinator attraction (Pellegrino et al., 2012).

Selective breeding alters the odour composition of crops, such that modern cultivars may differ significantly from wild relatives. This could include gain-of-volatile, but unless the selective breeding specifically selects for odour or pollinator attractiveness, loss-of-volatile is a risk. There is considerable cultivar-specific variation among crop floral volatiles, which can impact attractiveness to pollinators (Klatt et al., 2013; Ceuppens et al., 2015). Enhancing pollination via odour lures should therefore consider whether the lure is simply fulfilling a function that in other cultivars would be done more effectively by the flower.

1.3 Intraspecific variation in floral volatiles in crops and crop relatives

The scent of a flower, including a crop flower, can vary within a species, over time and space, between subspecies and cultivars, the physiological condition of the plant, and between male and female flowers on the same plant.

1.3.1 Dioecious flower volatiles

Some species of plant, including some crops (cucurbits) and crop relatives (*Ficus* spp., *Fragaria virginiana*), possess separate male and female (or male/hermaphrodite or female/hermaphrodite) flowers either on the same or separate plants. Where these plants are insect pollinated, they may employ mimicry to ensure visits of pollinators to both sexes of flower, raising the question of whether visually similar male and female flowers also possess the same volatile composition. While overlaps in the major compounds present often occur (Ervik et al., 1999; Grison-Pigé et al., 2001), it is common for flowers of different sexes to have measurable differences in volatile composition or abundance (Ervik et al., 1999; Grison-Pigé et al., 2001; Ashman et al., 2005). This can affect their attractiveness to insect pollinators (Ashman et al., 2005).

1.3.2 Spatio-temporal variability

The Araceae, which includes some important crop species such as taro, cocoyam and Mexican breadfruit, includes many species with sapromyophilous flowers that attract carrion insects and flowers that attract coprophilous species. Such flowers commonly produce compounds that smell like the usual food site, typically rotting meat or dung, of the pollinating insects, most commonly Diptera. For example, oligo dimethyl sulphides are produced in Titan Arum (*Amorphophallus titanum*) (Kite and Hetterschieid, 1997). *Arum maculatum* is pollinated by female *Psychoda phalaenoides* (in Western/Central Europe) and both sexes of both *P. phalaenoides* and *Psychoda grisescens* flies in other regions (Gfrerer et al., 2021). It produces a hyperdiverse assemblage of compounds including indole and *p*-cresol which are also found in cow dung, the normal breeding microhabitat of the pollinator (Kite, 1995). *A. maculatum* floral odour varies across time and space and shows considerable variation within- and among-populations which have been proposed to be adaptations to distinct species of *Psychoda* (Gfrerer et al., 2021; Szenteczki et al., 2021). Both *P. phalaenoides* and *P. grisescens* have overlapping brood-site preferences (cow dung), but *P. phalaenoides* also uses sewage whereas *P. grisescens* does not (Satchell, 1947); so the *A. maculatum* odour variation may conceivably reflect differences in the substrate uses of the local species.

Floral odour variation in crops that are pollinator dependent can be influenced by the conflicting need to protect the plant and flower against damaging herbivory. For example, strawberry flower odours contain a range of compounds including green leaf volatiles that are associated with defence and those cultivars producing higher amounts of those compounds are visited less by bumble bees than cultivars producing lower amounts (Ceuppens et al., 2015). In elder (*Sambucus nigra*), visitation by *Thrips major* is mediated by changing concentrations of floral volatiles including linalool that attract the pollinating thrips with peaks in abundance coinciding with pollen dehiscence and then declining after pollination to be replaced by deterrent cyanogenic glycosides, prunasin and sambunigrin (Scott-Brown et al., 2019). This example is discussed in further detail as a case study.

Spatio-temporal variation in floral odours within plants is common but association with pollinator-mediated selection is not always clear and other ecological explanations such as the role of natural enemies, environmental constraints, phylogenetic inertia or biochemical constraints could be drivers of variation (Delle-Vedove et al., 2017). That said, there are examples of plant-pollinator interactions, especially with specialists, where floral odours attract pollinators (Milet-Pinheiro et al., 2013). However, these may be examples of pollinators adapting to floral cues associated with good reward rather than flowers producing odours for which pollinators have innate attraction. Floral odours enhance short-term pollinator specificity which benefits both pollinator and plant and influences the evolution of plant systems of reproduction by influencing floral visitation by pollinators (Wright and Schiestl, 2009).

1.4 Pollinator detection of floral odours

Most pollinators are insects and thus detect odours with their antennae. Within the insect brain, the mushroom bodies and antennal lobes play important roles in olfactory learning (Hu et al., 2010; Carcaud et al., 2016). The antennae have sensillae, chemical-sensitive cells, that are specific to a single compound or a set of closely related structures. In the case of vertebrate pollinators, the role of odour can vary from minor in birds to intense in rodents and bats (Bestmann et al., 1997; von Helversen et al., 2000; Fleming et al., 2009; Dellinger et al., 2019). Relatively few crop flowers are pollinated by vertebrates (the main examples are fruits such as dragon fruit and durian, and also *Parkia* spp. beans (Piechowski et al., 2010; Ratto et al., 2018)), but in cases where the pollinator is nocturnal or has poor eyesight, odour can play a key role in ensuring pollination (Ratto et al., 2018).

1.4.1 Assessing the ability of an insect to detect crop floral odours and respond to them

Electro-antennography (EAG), especially when coupled with mass spectroscopy (GC-EAG, also known as GC-EAD) enables researchers to evaluate an insect's ability to detect the individual components of an odour blend taken from a headspace extraction from a crop flower. While EAG tests the response of the antenna overall, single-sensillum recording will indicate the presence and activity of individual chemical-sensitive cells on an insect's antenna. Similar electrophysiology can take place on other organs involved in chemoreception, such as tarsae and mouthparts. However, although electrophysiological studies can provide information about whether an odour is detectable to an insect, they reveal nothing about the ecological significance of specific volatile organic compounds (VOCs). A detectable odour could induce attraction, deterrence, gustation, oviposition, aggression or can elicit different behaviours depending on context, concentration and the other volatiles present.

Once it has been established that an insect can detect specific floral volatiles, behavioural assay methods can permit evaluation of the insect's responses to it. Various apparatus can be employed, including Y-tube and four-arm olfactometers, wind tunnels, feeding and flight arenas, and Y-mazes (Ceuppens et al., 2015; Solís-Montero et al., 2018). These allow the insect to make choices about where they choose to orient and land, providing insight into what they consider most attractive both with and without learning experience, and the data collected provide a starting point for determining attraction.

1.5 How insects learn about floral odours and the relevance to crops

Most insects pollinating crop flowers are not naïve to flowers. They are experienced foragers who have visited and sampled a range of flowers. Their decision to visit the crop in some cases may be based on instinct or innate attraction as is assumed to be the case for cocoa midges (Arnold et al., 2019) or *Elaeidobius kamerunicus* (Adaigbe et al., 2011), the pollinator of oil palm. In the case of insects showing sophisticated learning abilities such as generalist bees and hawkmoths, the decision to visit the crop is fine-tuned based on previous rewarding experiences with those flowers (Riffell et al., 2008; Wright and Schiestl, 2009). Where an insect receives nectar or pollen from the flower, it forms a memory association between that reward and the cues associated with the flower, including its specific odour, and will seek out that odour more often in future. This benefits the plant, particularly for pollinators with relatively high floral constancy, because they will visit many conspecific flowers and thus transfer more conspecific pollen. Because distinctive and memorable odours are beneficial in this interaction (Wright and Schiestl, 2009), in addition to those that are innately attractive, more complex and recognisable floral blends may be favoured. Indeed, crop cultivars with more simplified odour composition can be associated with reduced visitation rates even from pollinators that learn well (Ceuppens et al., 2015). Similarly, reliable odours that do not vary too much within a species will favour more rapid learning and constancy (Wright and Schiestl, 2009).

Social insects also learn about floral odours in the nest. As nectar often bears the scent of the flower, nectar brought back to the nest will be consumed by other workers in the colony apart from the forager who collected it, thereby forming an odour-memory association without a direct flower visit. This can influence foraging preference outside the nest in stingless bees (Reichle et al., 2010), honeybees (Farina et al., 2007) and bumblebees (Molet et al., 2009).

1.5.1 Context-specific use of odour information

Where a pollinator also interacts with a plant at other points in the insect's life cycle, it may respond to odour volatiles present in the floral headspace, but in a different way according to the different context. Mated and unmated insects, insects under nutritional stress versus those that are satiated, and newly eclosed versus senescent individuals, respond differently to odour cues. Aphidophagous hoverflies may use plant wound volatiles and aphid alarm pheromones as oviposition cues (James, 2003; Yu et al., 2008). Some of those odours conceivably may also attract individuals towards the plant for flower-visiting purposes, but this cannot be assumed without testing. Therefore, efforts to develop volatile attractants for hoverflies need to be mindful that a lure blend to stimulate attraction to the crop for oviposition may not function effectively to promote flower-visiting. Equally, a lure blend derived from oviposition behaviours may be wholly unattractive to male insects and have no influence on their tendency to visit the crop of interest.

2 Key challenges

2.1 Anthropogenic disturbance of pollination

Anthropogenic drivers of environmental change, especially large-scale habitat loss and global climate change, are implicated in pollinator declines (Powney et al., 2019) with severe implications for plants which rely wholly on interactions with these organisms for reproduction in natural systems (Biesmeijer et al., 2006) or cropping systems. An insidious effect of such drivers is their potential to directly disrupt the fine balance of mutualisms between plants and pollinators at a local scale. Such effects are harder to detect than declining pollinator abundance or distribution, but may have equivalent impacts on plant reproductive success by reducing effective visitation rates.

Although anthropogenic constraints are reviewed in detail elsewhere in this book, the reliance of pollination systems on such complex interspecific plant-insect signalling mechanisms means that they are highly vulnerable to influence by their unique abiotic and biotic environments. Therefore, exploring the extent to which anthropogenic activity can influence the odour-based interactions between crops and their pollinators is worthy of consideration here.

2.1.1 Air: environmental pollution in the form anthropogenic volatile pollutants

Anthropogenic volatile pollutants (AVPs) derived from traffic and industrial processing activities can disrupt scent-based communication between flowers and pollinators (Macgregor and Scott-Brown, 2020). Increased levels of AVPs in the environment can disguise floral signals and increase degradation rates for floral volatiles (Girling et al., 2013; Fuentes et al., 2016; Cook et al., 2020), including ubiquitous scent components such as linalool (McFrederick et al., 2008), limiting signalling range and ability to decipher. Other pollutants and contaminants, particularly pesticidal drift (Williamson and Wright, 2013) and heavy metals (Meindl and Ashman, 2013), relevant to agricultural production systems can interfere with the pollinators' ability to detect and learn odours.

2.1.2 Soil: the physiological state of a crop can influence its production of floral volatiles

This can relate to factors such as nitrogen enrichment of habitats via atmospheric pollution and soil improvement processes exceeded through agricultural intensification (Galloway and Cowling, 2002; Fowler et al., 2013), which constitute a threat to biodiversity at all trophic levels (Pöyry et al., 2017). Soil nitrogen availability influences production and emission of a range of plant volatiles in some species, including those involved in plant-insect signalling (Kivimäenpää et al., 2016), though evidence of effects on VOCs in the floral scent of crop species is still limited (Majetic et al., 2017). Soil inorganic components may also influence the abundance and diversity of floral volatiles (Ormeño et al., 2008). However, studies which specifically examine how these may negatively or positively affect interactions between pollen vectors and cultivated plants are few.

2.1.3 Climate and extreme weather events

Numerous studies have focused on the negative effects of climate change and extreme weather events on major cash crops and the stability of food production systems (Cogato et al., 2019). Although the general effects of elevated global temperatures on pollinators and cropping systems are reviewed in more detail in a previous chapter, it is important to note here the effect of extreme temperatures on the stability of olfactory cues and synchronisation of detection of crop plant flowers by their pollinators. To be effective at attracting pollinators, first, signals must not be altered or become less attractive under altered environments: drought, high precipitation, extreme temperatures (and shifting crop phenologies). Given that natural scent production emissions operate by increasing respiration rates at higher ambient temperatures, climate warming may lead to physiological limits to respiration rates, preventing the continuing maintenance of volatile production and altering the efficiency of olfactory sensitivity detection as a flower-visitation mechanism. Second, pollinator preference may itself change, with flower scents becoming less detectable or attractive when conditions change.

2.2 Competition

Few crops have a single specialist pollinating taxon or species (there are exceptions such as oil palm, fig, vanilla and arguably cocoa). As a result, it is rare that the crop itself is integral to the survival of the pollinator, even when pollinator visitors are critical for the crop's fruit set (though a notable exception is the squash bees (Hymenoptera: *Eucerini*, *Peponapis* and *Xenoglossa*, which are oligolectic on *Cucurbita* pollen (Hurd et al., 1971)). In the case of most crops, the signals (visual, odour) displayed by the crop flowers to appeal to pollinators are competing with those of other plants in the surroundings. While there is benefit to the crop from mass flowering in presenting a large target area and more intense cues, the crops are still in competition with wildflowers (Holzschuh et al., 2011). Where selective breeding has unintentionally resulted in dilution or simplification of a crop's odour bouquet, effective pollination may be more of a challenge. Conversely, the observation that cocoa is pollinated by locally native Ceratopogonid midges wherever the crop has been introduced, even far outside its native range, demonstrates that the plasticity of pollinator behaviour can be of great benefit to the crop.

3 Case studies

3.1 Study 1: critical timing for optimisation of pollination in a tree crop

Coinciding the timing of the release of floral scent with the foraging activity of efficient pollen vectors is crucial for pollination precision, and studies have revealed that this energy-efficient strategy exists in many pollinator-reliant plant taxa (Hoballah et al., 2005; Prieto-Benítez et al., 2016; Chapurlat et al., 2018; Powers et al., 2020). In natural systems multiple evolutionary pressures on plants and pollinators may determine the optimal timing for the attraction of a pollinator via release of floral scent and fertilisation to occur. For example, flowers that open at night may avoid pollinator-mediated competition, thus reducing the danger of sharing actively foraging pollinators with co-flowering species and potentially reducing pollination success (Macgregor and Scott-Brown, 2020). Competitors for floral rewards may include less efficient pollen-vectoring organisms and heterospecific pollen collectors (Morales and Traveset, 2008), herbivores (e.g. Theis and Adler, 2012; Schiestl, 2015; Nunes et al., 2016) and pollinator predators.

Thus, through processes of crop selection and breeding programmes and introduction of domesticated cultivars into new geographic ranges, the desirable components of floral scent may be downregulated, or timing of emissions may mismatch the activity of native pollinator guilds, resulting in less successful pollination in the short window of opportunity when plants are fertile. Furthermore, in such situations, crop plants may be detected and monopolised by ineffective pollinators and potentially destructive florivores.

In cases where crops are sustainably harvested from wild sources or landraces are retained there is evidence that crop plants may co-evolve partnerships where benefits for production arise. In this context, thrips (Thysanoptera) are of interest, with growing evidence that they are brood-site pollinators (Sakai, 2002), effective in transferring pollen within species that possess adaptations associated with wind pollination, such as many flowers, abundant powdery pollen and large stigmas (Tal, 2009). To demonstrate this, a recent study addressed the broad question, 'Do native flower thrips have a role as pollinators in UK landscapes?' Researchers used a combination of field and laboratory experiments to examine the behaviour and life cycle of native flower thrips, mapping these with changes in floral chemistry throughout the flower development stages in European elder, *Sambucus nigra* (Scott-Brown et al., 2019). Elderflower scent emissions containing β-(±)-linalool and (*E*)-β-ocimene peaked as the flowers were opening (pre-anthesis stage), while levels of these VOCs were reduced or absent at bud and senescence stages, with the highest abundance of compounds corresponding with increased numbers of thrips present in inflorescences. Further, recording the circadian emissions from young elderflowers over 24 h demonstrated that the abundance of volatile compounds increased between midnight to early morning, attracting the thrips to pollen-receptive flowers at night. Thrips, which are characteristically brood-site pollinators (Sakai, 2002), produced offspring which remained in the elderflowers feeding on pollen until concentrations of specific nitrogen-storage/defence secondary metabolites (cyanogenic glycosides) increased in floral tissues to levels shown to deter the thrips from senescing flowers, thereby protecting newly forming fruit. The study additionally used pollinator-exclusion experiments to confirm thrips' visitation at the critical point in floral development increased the chances of fruit set in this species. In return for pollination services, the plant provides thrips with sites for oviposition and a food source for developing larvae.

The study provided new insight into the previously ambiguously defined pollination strategies of *Sambucus nigra*, a species in high global demand for elderflower syrup, herbal medicines and natural food colouring (Jones, 2019) and draws attention to gaps in knowledge about the contribution of diverse pollinators to primary and secondary agricultural production systems. Elucidating olfactory mechanisms which underpin pollinator interactions in wild populations may steer future development of new breeding technologies, leading to the enhancement of pollination in high-value cropping systems, through focused studies on floral volatile organic compound emissions during floral development.

3.2 Study 2: caffeine to enhance learning of crop floral odours in commercial pollinators

The odours that social bees (both honeybees and bumblebees) encounter in their nest can influence their floral targets of choice outside the nest (Molet et al., 2009). This implies a mechanism in which the odours of worker-collected floral nectar or pollen brought back to the nest are detected by other workers, including those not yet actively foraging, and learned in association with food.

Several patents have been produced that seek to harness this process, by providing crop-specific floral odour blends to honeybees in the nest, with the intention that honeybees would be exposed to the crop floral odours early in their adult lives and then depart the nest already with a memory of these odours and a positive association. This would, then, be expected to result in enhanced forage focus on the crop of interest. Blends exist for apple (Farina et al., 2011a,b), pear (Farina and Díaz, 2020), almond (Farina et al., 2020b) and sunflower (Farina et al., 2012). There are also products that work using some mixture of floral odours and unspecified 'bee attractants' (Sasaki et al., 2008) that are applied both to the nest and the crop, for soybean (Apicultural Institute of Jilin Province, 2011) and linseed (Xishuangbanna Tropical Botanical Garden of CAS, 2009).

This can be taken still further. Research over the last decade, building on earlier work, has demonstrated that both honeybees and bumblebees have behavioural responses to caffeine (Couvillon et al., 2015; Thomson et al., 2015). Tethered honeybees formed longer-lasting memory associations with odour when provide with caffeine during training (Wright et al., 2013); caffeinated artificial flowers received more visits by bumblebees (Thomson et al., 2015). Thus, not only could caffeine enhance learning and memory of floral odours, but it could furthermore still have this effect when ingested by commercial pollinators in the nest, away from the target crop. It should be possible to enhance nest-based odour learning with caffeine, priming these pollinators with the scent of a crop's flowers and resulting in commercial bumblebees or possibly honeybees having an initial stronger interest in visiting crop flowers than other flowers in the environment.

Indeed, this was demonstrated in Arnold et al. (2021) in a laboratory experiment, in which bumblebees primed with a combination of synthetic strawberry crop floral odour, sugar solution and 0.1 mM caffeine proceeded to choose robotic flowers bearing that same odour more often, relative to distractor flowers with a different odour, than control bumblebees that had received uncaffeinated sugar solution alone, or the floral odour. While the effect was short-lived, this was to be expected in such a simplified setup where the targets were robotic, automatically refilling flowers, presented against an uncomplicated background, and both the strawberry floral scented and distractor targets provided an equal food reward in exchange for equal effort from the bees.

In a field situation, these types of commercial bumblebee boxes would be deployed in the crop, partway along polytunnels. A current challenge for growers of soft fruit and some salad products (e.g. tomatoes, bell peppers) in Europe is that commercial pollinators are insufficiently motivated to visit the crop; some bees will not forage at all, while other bees leave the tunnels and visit hedgerows, wildflowers and other non-crop plants around the site. As a result, on small farms, fewer pollination services are provided to the grower and wild and commercial bees compete for the same wild plant nectar and pollen resources, potentially impacting wild pollinators.

This may be less severe a problem in the United States, where field and farm sizes tend to be larger, and therefore bumblebees' ability to leave the crop and visit alternative forage is limited. Caffeine technology has the potential to ameliorate some of these problems. It is likely that in such a field deployment, the short effect duration noted in the lab will be less pronounced because the field has a more complex environment, where the memory of odour may be more critical in locating rewarding flowers. Unlike in a laboratory, where the target and distractor flowers are close, interspersed, easy to find and switch between and energetically require the same input to visit, the crop flowers in the field are closer to the nest and energetically less costly to visit, although they may be less rewarding depending on what other flowers are present in the broader vicinity.

A similar approach has been deployed in honeybees (Farina et al., 2020a; Marchi et al., 2021). This has the advantage of training many more workers to visit flowers and could be particularly valuable for this species that has such a large foraging range, and thus could choose to stray further from the crop. The disadvantage with honeybees is that if caffeine is applied to hives to enhance learning of crop floral odours, the honey in the hive could potentially become contaminated with caffeine and may be unsaleable, a consideration of special importance in regions where honey sales are a significant part of apiculturists' income. This contrasts with monofloral honey from plants such as coffee and citrus, which contain naturally occurring caffeine.

3.3 Study 3: creation of odour lures

The concept of deploying odour lures to attract pollinators to a crop is one that must be considered carefully. Given the issues with spatial scales over which odour information is used, and the fact that the lure is a point source that cannot reflect the precise locations of the target flowers, there is a chance that it could result in pollinating insects visiting the lure more than the surrounding flowers.

The relevance of odour lures also depends on the foraging ranges of the insects. If odour lures draw insects away from other competing food sources (e.g. wildflowers) and on to the crop, their use only makes sense if non-crop foraging resources are present within the range of the targeted pollinating insect.

Lures and attractants that have been trialled broadly fall into two classes: those based on a pheromone relevant to the insect (e.g. a sex, foraging or queen pheromone), and those based on host plant odours. The main species that has been targeted is *Apis mellifera*, on pollinator-dependent fruit crops such as orchard fruit and melons. Examples of products and lures include Fruit-Boost™ (based on the honeybee queen mandibular pheromone (Ellis and Delaplane, 2009)) and Bee-Scent (based on Nasonov pheromone - a citral-geranial blend (Adams, 2014) - used by honeybee workers to communicate about foraging, alongside sugars). A few also use sugar mixtures (e.g. BeeLure™), but spraying sugar on to crops risks proliferation of fungal antagonists in a similar that honeydew deposits left by sucking pests do, and sugar sprays can promote bees foraging from leaves rather than from the flowers themselves (Free, 1965).

Where lures and attractants (e.g. Nasonov pheromone-based dispensers or Fruit-Boost™) have been deployed over several trials in the late 1980s to the mid-1990s, there has often been no significant benefit to pollination (usually by honeybees, *A. mellifera*) and fruit set (Table 2). As such, this approach is seldom adopted in current agriculture and seems to yield unreliable benefits. Some more positive findings have come from Indian studies in *Apis cerana* and *Apis dorsata*, and stingless bees (*Trigona* spp.) have also been tested for responses (Patil et al., 2010).

More recently, numerous patents suggesting similar approaches (including with non-*Apis* bees) have been lodged (Guangde County Fotang Strawberry Planting Professional Cooperatives, 2015; Weifang Yourong Industry Co Ltd., 2015a,b,c), reporting to improving pollination on crops such as raspberry. Some products also suggest providing the attractant to honeybees in the hive before deployment, using an odour-priming method similar to that previously discussed (Apicultural Institute of Jilin Province, 2011). These appear to be mostly based on honey-plant extract mixtures. While some of these new products may provide a reliable bee attractant with consistent benefits to the crop, results from previous research trials (Table 2) imply the performance of such products can be inconsistent. There are several potential reasons, in addition to the issues with spraying sugar solution indiscriminately on to plants, including:

- Bees' formidable learning ability – if the odour itself does not reliably signal presence of a preferred food reward, bees will learn to stop responding to it.
- Concentration issues – volatiles can be interpreted differently at different concentrations so pheromones, may not induce behavioural responses at higher concentrations (close to the crop) or lower concentrations (edge of the field) compared to the ecologically relevant concentrations.
- Spatial distribution – encouraging foraging on the crop's leaves, stems, etc. may not translate to visitation to the flowers.

However, there may be potential applications in more unusual pollinators that have not yet been exploited. Cacao suffers from low fruit set and visitation by a relatively small group of taxa that can effectively pollinate. The odour of the flowers is not strong, the insects most often do not disperse far, and the pollinating midges are often active early in the morning when the light levels around flowers may be low. Potentially, deploying odour lures on trees at the appropriate height may increase activity in the vicinity of flowers, or may result in the above-mentioned issues. While attempts have been made to synthesise the odour of cacao flowers (Arnold et al., 2019), the response of midges was unpromising; so creating an effective volatile blend would require more work.

Table 2 Summary of some of the studies carried out on honeybee attractants that are sprayed on to the crop

Crop	Country	Attractant tested	Type	Result	Reference
Apple	USA	BeeLure	Sugars + fruit odour	NS	Tew and Ferree (1984)
Apple	USA	SPLAT Bloom Polynate	Pheromone – Nasonov	NS	Adams (2014)
Blueberry	USA	SPLAT Bloom Polynate	Pheromone – Nasonov	NS	Adams (2014)
Cherry	USA	SPLAT Bloom Polynate	Pheromone – Nasonov	NS	Adams (2014)
Cherry	USA	Fruit-Boost	Pheromone – queen mandibular	NS	Naumann et al. (1994)
Cucumber	USA	Bee-Scent	Pheromone	NS	Schultheis et al. (1994)
Kiwi	Greece	Bee-here	Pheromone	NS	Tsirakoglou et al. (1995)
Onion	India	BeeQ	Sugars	Positive – increased visitation by *A. dorsata*, *Trigona* spp.	Naik et al. (2019)
Pear	USA	Fruit-Boost	Pheromone – queen mandibular	Positive – larger fruit	Naumann et al. (1994)
Watermelon	USA	Bee-Scent Beeline	Bee-Scent = pheromone Beeline = sugars and other food	NS	Schultheis et al. (1994)
Watermelon	USA	Fruit-Boost	Pheromone – queen mandibular	NS	Ellis and Delaplane (2009)

4 Summary

Pollination efficiency in crops needing pollen transfer via insect vectors requires first an understanding of the biochemical (olfactory) mechanisms of interactions between plant and pollinating organism. Unravelling all outcomes of interacting biotic/abiotic stresses on these mechanisms and translating this research to enable development of remedial biotechnologies (digital and gene-based) to alleviate pressure on pollination services is a crucial step to enable maintaining and supporting these valuable services for food production. Further, devising management options and harnessing nature-based solutions alongside the implementation of remedial biotechnologies will serve to ensure these important olfactory signalling mechanisms between pollinating taxa and crop plants are not compromised and food production levels are sufficiently and reliably maintained.

Improvements in the mechanistic understanding of a wider diversity of pollination contributors is needed. Visual and olfactory systems, and their relationship to flower-visiting activity, are well understood in some taxa (e.g. honeybee and bumblebee species) but not in others, so a broader mechanistic understanding would be valuable. In particular, the relative importance of different mechanisms in low-light conditions at night is not known and is likely to differ from mechanisms at play in diurnal pollination systems.

Importantly, all specialised olfactory detection and signalling mechanisms are liable to be directly or indirectly disrupted by anthropogenic environmental change (including air pollution, changes in soil chemistry, artificial light at night and climate change), with potential detrimental effects on pollination services, which have not been fully quantified. Increased understanding of the functioning of nocturnal pollination systems is critical for future efforts to conserve their ecological service.

5 Future trends

Future developments in the field of harnessing olfactory stimuli in crop pollination include methods of modifying the pollinators' behaviour, including building on principles explained in the case studies presented earlier and employing precision techniques such as CRISPR to alter the crops themselves.

Future pollination attractants applied to the crop will need to overcome previously observed issues with point sources, pollinator learning and target misdirection; this will require precise regulation of concentration and understanding how that changes over space and time. To date, attractants focused on non-*Apis* bees have been rare, which may be a missed opportunity since honeybees are often not the optimal pollinator of many crops. However, new products are continually being developed and trialled and some more recent attractants have indeed focused on non-*Apis* bees.

Deployment of floral odours in contexts (primarily in the nest) where the aim is to prime or induce an odour bias, rather than attract directly, is a strategy showing considerable promise. This has now been demonstrated in principle in both honeybees (Marchi et al., 2021) and bumblebees (Arnold et al., 2021). So far primarily caffeine has been tested, but some work on honeybees has demonstrated a synergistic effect of caffeine and arginine on bee performance (Marchi et al., 2021), indicating that future olfactory-priming devices could incorporate this into the technology as well.

The limitations of floral attractants and priming devices will depend on context. In a completely open environment, the environmental factors discussed earlier (wind speed and weather affecting physiological state of crops) will be particularly important in influencing odour presence and dispersal. In a polytunnel, as used for many soft- and stone-fruits, odours are retained somewhat better but there are also adverse high temperatures influencing the pollinators. Conversely, in enclosed glasshouses, certain challenges such as the pollinators' tendency to leave the crop and forage on surrounding vegetation become less significant so priming treatments are used more to increase motivation and activity on the crop.

Other technologies have the potential to grow in importance in future. Gene editing technologies offer great potential for future crop improvement, and this can extend to improving crop–insect olfactory interactions. The potential to restore functions of genes involved in VOC metabolism that may have been lost via selective breeding provides exciting opportunities to achieve better pollination activity and yield. While to date this has primarily been explored in ornamental crops (cut flowers) where floral fragrance is selected to please human consumers (Giovannini et al., 2021), the same principles could be applied to crop flowers to improve pollination.

6 Conclusion

Odour is one of a range of cue modalities that a plant can deploy in order to increase visitation and pollen transfer by floral visitors. However, the anthropogenic changes that have threatened populations of insect pollinators directly have wider implications as the very mechanisms by which crop plants and their pollinators communicate are vulnerable to disruption. Human activity has distorted pollinator–plant olfactory interactions, both by altering the plant's odour production and the environment in which it operates. Understanding the nature of such disruptions is therefore important to develop effective mitigation strategies, and ultimately conserve olfactory-driven pollination services for crop productivity. Solutions may lie in 'precision pollination', via behavioural manipulation of pollinators, or opportunities presented by biotechnology.

7 Where to look for further information

The biology of floral scent (Dudareva and Pichersky, 2006) provides a comprehensive description of the chemistry of flowers and their extraordinary diversity alongside, metabolic pathways for their biosynthesis and the genomics underpinning these biochemical processes. This book also provides a comprehensive description of the interactions floral chemicals drive especially with respect to pollinators. *Insect olfaction* (Hansson, 2010) deals with both olfaction as a model for a sensory system and the use of insect olfaction in mediating insect behaviour.

The 36th annual meeting of the International Society of Chemical Ecology (ISCE) (5–10 September 2021) provides a state of art snapshot of the science of floral volatiles and plant-pollinator interactions to give an indication of current key topics in this research area. The theme of the conference, 'Chemical Ecology and Sustainable Development', and abstracts were published online including two sessions on pollinator-flower interactions at all levels of ecological organisation in diverse taxa, and their evolutionary origins (https://isce2021.carlamani.com/abstracts/Plant%20Pollinator%20Interactions%20combined%20abstracts.pdf).

More information on the ecology of floral chemistry can be found in the *Journal of Chemical Ecology* (https://www.springer.com/journal/10886) and through the International Society of Chemical Ecology (https://www.chemecol.org/), while the Pherobase provides a comprehensive list of bioactive volatile compounds classified by functional group (http://www.pherobase.com/database/compound/compounds-index.php).

8 References

Adaigbe, V. C., Odebiyi, J. A., Omoloye, A. A., Aisagbonhi, C. I. and Iyare, O. (2011). Host location and ovipositional preference of *Elaeidobius kamerunicus* on four host palm species. *Journal of Horticulture and Forestry* 3, 163–166.

Adams, J. A. (2014). The effect of nasonov based dispensers on honey bee behavior and on pollination in blueberry, apple, and cherry. *Ann Arbor*: Michigan State University.

Apicultural Institute of Jilin Province. (2011). Bee attractant and method for field seed production and bee pollination of soybean hybrids. China: Apicultural Institute of Jilin Province.

Arnold, S. E. J., Dudenhöffer, J.-H., Fountain, M. T., James, K. L., Hall, D. R., Farman, D. I., Wäckers, F. L. and Stevenson, P. C. (2021). Bumble bees show an induced preference for flowers when primed with caffeinated nectar and a target floral odor. *Current Biology* 31, 4127-4131.

Arnold, S. E. J., Forbes, S. J., Hall, D. R., Farman, D. I., Bridgemohan, P., Spinelli, G. R., Bray, D. P., Perry, G. B., Grey, L., Belmain, S. R. and Stevenson, P. C. (2019). Floral odors and the interaction between pollinating ceratopogonid midges and cacao. *Journal of Chemical Ecology* 45(10), 869–878.

Ashman, T.-L., Bradburn, M., Cole, D. H., Blaney, B. H. and Raguso, R. A. (2005). The scent of a male: The role of floral volatiles in pollination of a gender dimorphic plant. *Ecology* 86(8), 2099–2105.

Bestmann, H. J., Winkler, L. and von Helversen, O. (1997). Headspace analysis of volatile flower scent constituents of bat-pollinated plants. *Phytochemistry* 29, 1359–1387.

Biesmeijer, J. C., Roberts, S. P. M., Reemer, M., Ohlemüller, R., Edwards, M., Peeters, T., Schaffers, A. P., Potts, S. G., Kleukers, R., Thomas, C. D., Settele, J. and Kunin, W. E. (2006). Parallel declines in pollinators and insect-pollinated plants in Britain and the Netherlands. *Science* 313(5785), 351–354.

Brady, J., Packer, M. J. and Gibson, G. (1990). Odour plume shape and host finding by tsetse. *Insect Science and Its Application* 11, 377–384.

Carcaud, J., Giurfa, M. and Sandoz, J. C. (2016). Parallel olfactory processing in the honey bee brain: Odor learning and generalization under selective lesion of a projection neuron tract. *Frontiers in Integrative Neuroscience* 9, 75.

Ceuppens, B., Ameye, M., Van Langenhove, H., Roldan-Ruiz, I. and Smagghe, G. (2015). Characterization of volatiles in strawberry varieties 'Elsanta' and 'Sonata' and their effect on bumblebee flower visiting. *Arthropod-Plant Interactions* 9(3), 281–287.

Chapurlat, E., Anderson, J., Ågren, J., Friberg, M. and Sletvold, N. (2018). Diel pattern of floral scent emission matches the relative importance of diurnal and nocturnal pollinators in populations of *Gymnadenia conopsea*. *Annals of Botany* 121(4), 711–721.

Cogato, A., Meggio, F., De Antoni Migliorati, M. and Marinello, F. (2019). Extreme weather events in agriculture: A systematic review. *Sustainability* 11(9), 2547.

Cook, B., Haverkamp, A., Hansson, B. S., Roulston, T. , Lerdau, M. and Knaden, M. (2020). Pollination in the Anthropocene: A moth can learn ozone-altered floral blends. *Journal of Chemical Ecology* 46(10), 987–996.

Couvillon, M. J., Al Toufailia, H., Butterfield, T. M., Schrell, F., Ratnieks, F. L. W. and Schürch, R. (2015). Caffeinated forage tricks honeybees into increasing foraging and recruitment behaviors. *Current Biology* 25(21), 2815–2818.

Delle-Vedove, R., Schatz, B. and Dufay, M. (2017). Understanding intraspecific variation of floral scent in light of evolutionary ecology. *Annals of Botany* 120(1), 1–20.

Dellinger, A. S., Scheer, L. M., Artuso, S., Fernández-Fernández, D., Sornoza, F., Penneys, D. S., Tenhaken, R., Dötterl, S. and Schönenberger, J. (2019). Bimodal pollination systems in Andean Melastomataceae involving birds, bats, and rodents. *The American Naturalist* 194(1), 104–116.

Dudareva, N. and Pichersky, E. (Eds.). (2006). *Biology of Floral Scent* (1st edn.). Boca Raton, FL: CRC Press.

Ellis, A. and Delaplane, K. S. (2009). An evaluation of Fruit-Boost™ as an aid for honey bee pollination under conditions of competing bloom. *Journal of Apicultural Research* 48(1), 15–18.

El-Sayed, A. M., Sporle, A., Colhoun, K., Furlong, J., White, R. and Suckling, D. M. (2018). Scents in orchards: floral volatiles of four stone fruit crops and their attractiveness to pollinators. *Chemoecology* 28, 39–49.

Ervik, F., Tollsten, L. and Knudsen, J. T. (1999). Floral scent chemistry and pollination ecology in phytelephantoid palms (Arecaceae). *Plant Systematics and Evolution* 217(3–4), 279–297.

Farina, W. M., Arenas, A. and Diaz, P. C. (2011a). A formulation that promotes direct pollination of honey bees to apple cultures. *Google Patents*.

Farina, W. M., Arenas, A. and Diaz, P. C. (2011b). Formulation that promotes targeted pollination of melliferous bees to apple crops. Argentina. *Google Patents*.

Farina, W. M., Arenas, A. and Diaz, P. C. (2012). Formulation that promotes targeted pollination of melliferous bees to sunflower crops. *Google Patents*.

Farina, W. M., Arenas, A., Díaz, P. C., Susic Martin, C. and Estravis Barcala, M. C. (2020a). Learning of a mimic odor within beehives improves pollination service efficiency in a commercial crop. *Current Biology* 30(21), 4284–4290.e5.

Farina, W. M., Barcala, C. E. and Palottini, F. (2020b). Formulation for promoting targeted pollination of almond tree crops in honey bees. *Google Patents*.

Farina, W. M. and Díaz, P. C. (2020). Formulation for promoting targeted pollination of pear tree crops in honey bees. *Google Patents*.

Farina, W. M., Grüter, C., Acosta, L. and Mc Cabe, S. (2007). Honeybees learn floral odors while receiving nectar from foragers within the hive. *Naturwissenschaften* 94(1), 55–60.

Fleming, T. H., Geiselman, C. and Kress, W. J. (2009). The evolution of bat pollination: A phylogenetic perspective. *Annals of Botany* 104(6), 1017–1043.

Fowler, D., Coyle, M., Skiba, U., Sutton, M. A., Cape, J. N., Reis, S., Sheppard, L. J., Jenkins, A., Grizzetti, B., Galloway, J. N., Vitosek, P., Leach, A., Bouwman, A. F., Butterbach-Bahl, K., Dentener, F., Stevenson, D., Amann, M. and Voss, M. (2013). The global nitrogen cycle in the twenty-first century. *Philosophical Transactions of the Royal Society of London. Series B, Biological Sciences* 368(1621), 20130164.

Free, J. B. (1965). Attempts to increase pollination by spraying crops with sugar syrup. *Journal of Apicultural Research* 4(1), 61–64.

Fuentes, J. D., Chamecki, M., Roulston, T., Chen, B. and Pratt, K. R. (2016). Air pollutants degrade floral scents and increase insect foraging times. *Atmospheric Environment* 141, 361–374.

Galloway, J. N. and Cowling, E. B. (2002). Reactive nitrogen and the world: 200 years of change. *Ambio* 31(2), 64–71.

Gfrerer, E., Laina, D., Gibernau, M., Fuchs, R., Happ, M., Tolasch, T., Trutschnig, W., Hörger, A. C., Comes, H. P. and Dötterl, S. (2021). Floral scents of a deceptive plant are hyperdiverse and under population-specific phenotypic selection. *Frontiers in Plant Science* 12, 1910.

Giovannini, A., Laura, M., Nesi, B., Savona, M. and Cardi, T. (2021). Genes and genome editing tools for breeding desirable phenotypes in ornamentals. *Plant Cell Reports* 40(3), 461–478.

Girling, R. D., Lusebrink, I., Farthing, E., Newman, T. A. and Poppy, G. M. (2013). Diesel exhaust rapidly degrades floral odours used by honeybees. *Scientific Reports* 3, 2779.

Grison-Pigé, L., Bessière, J. -M., Turlings, T. C. J., Kjellberg, F., Roy, J. and Hossaert-McKey, M. M. (2001). Limited intersex mimicry of floral odour in *Ficus carica*. *Functional Ecology* 15(4), 551–558.

Guangde County Fotang Strawberry Planting Professional Cooperatives. (2015). Honeybee Attractant and Preparation Method of Same. *Google Patents*.

Haber, A. I., Sims, J. W., Mescher, M. C., De Moraes, C. M. and Carr, D. E. (2021). A sensory bias overrides learned preferences of bumblebees for honest signals in *Mimulus guttatus*. *Proceedings of the Royal Society B. Biological Sciences* 288(1948), 20210161.

Hansson, B. S. (2010). *Insect Olfaction*. Springer, Berlin, Heidelburg.

Hoballah, M. E., Stuurman, J., Turlings, T. C. J., Guerin, P. M., Connétable, S. and Kuhlemeier, C. (2005). The composition and timing of flower odour emission by wild *Petunia axillaris* coincide with the antennal perception and nocturnal activity of the pollinator *Manduca sexta*. *Planta* 222(1), 141–150.

Holzschuh, A., Dormann, C. F., Tscharntke, T. and Steffan-Dewenter, I. (2011). Expansion of mass-flowering crops leads to transient pollinator dilution and reduced wild plant pollination. *Proceedings of the Royal Society B. Biological Sciences* 278(1723), 3444–3451.

Hu, A., Zhang, W. and Wang, Z. (2010). Functional feedback from mushroom bodies to antennal lobes in the Drosophila olfactory pathway. *Proceedings of the National Academy of Sciences of the United States of America* 107(22), 10262–10267.

Hurd, P. D., Linsley, E. G. and Whitaker, T. W. (1971). Squash and gourd bees (*Peponapis, Xenoglossa*) and the origin of the cultivated *Cucurbita*. *Evolution; International Journal of Organic Evolution* 25(1), 218–234.

James, D. G. (2003). Synthetic herbivore-induced plant volatiles as field attractants for beneficial insects. *Environmental Entomology* 32(5), 977–982.

Jones, A. (2019). Cultivating elders for the UK processing industries. Nuffield Farming Scholarship Trust Report. Available at: https://www.nuffieldscholar.org/sites/default/files/reports/2019_UK_Alice-Jones_Cultivating-Elders-For-The-Uk-Processing-Industries.pdf (accessed 21 September 2021).

Kite, G. C. (1995). The floral odour of *Arum maculatum*. *Biochemical Systematics and Ecology* 23(4), 343–354.

Kite, G. C. and Hetterschieid, W. L. A. (1997). Inflorescence odours of Amorphophallus and Pseudodracontium (Araceae). *Phytochemistry* 46(1), 71–75.

Kivimäenpää, M., Ghimire, R. P., Sutinen, S., Häikiö, E., Kasurinen, A., Holopainen, T. and Holplainen, J. K. (2016). Increases in volatile organic compound emissions of Scots pine in response to elevated ozone and warming are modified by herbivory and soil nitrogen availability. *European Journal of Forest Research* 135(2), 343–360.

Klatt, B., Burmeister, C., Westphal, C., Tscharntke, T. and von Fragstein, M. (2013). Flower volatiles, crop varieties and bee responses. *PLoS ONE* 8, e72724.

Klein, A. M., Vaissiere, B. E., Cane, J. H., Steffan-Dewenter, I., Cunningham, S. A., Kremen, C. and Tscharntke, T. (2007). Importance of pollinators in changing landscapes for world crops. *Proceedings. Biological Sciences* 274(1608), 303–313.

Ma, W., Long, D., Wang, Y., Li, X., Huang, J., Shen, J., Su, W., Jiang, Y. and Li, J. (2021). Electrophysiological and behavioral responses of Asian and European honeybees to pear flower volatiles. Journal of Asia-Pacific Entomology 24, 221–228.

Macgregor, C. J. and Scott-Brown, A. S. (2020). Nocturnal pollination: An overlooked ecosystems service vulnerable to environmental change. *Emerging Topics in Life Sciences* 4(1), 19–32.

Majetic, C. J., Fetters, A. M., Beck, O. M., Stachnik, E. F. and Beam, K. M. (2017). Petunia floral trait plasticity in response to soil nitrogen content and subsequent impacts on insect visitation. *Flora* 232, 183–193.

Marchi, I. L., Palottini, F. and Farina, W. M. (2021). Combined secondary compounds naturally found in nectars enhance honeybee cognition and survival. *Journal of Experimental Biology* 224(6), jeb239616.

Mas, F., Harper, A., Horner, R., Welsh, T., Jaksons, P. and Suckling, D. M. (2018). The importance of key floral bioactive compounds to honey bees for the detection and attraction of hybrid vegetable crops and increased seed yield. *Journal of the Science of Food and Agriculture* 98, 4445–4453.

Mas, F., Horner, R. M., Brierley, S., Butler, R. C. and Suckling, D. M. (2020). Selection of key floral scent compounds from fruit and vegetable crops by honey bees depends on sensory capacity and experience. *Journal of Insect Physiology* 121, 104002.

McFrederick, Q. S., Kathilankal, J. C. and Fuentes, J. D. (2008). Air pollution modifies floral scent trails. *Atmospheric Environment* 42(10), 2336–2348.

Meindl, G. A. and Ashman, T. L. (2013). The effects of aluminum and nickel in nectar on the foraging behavior of bumblebees. *Environmental Pollution* 177, 78–81.

Mena Granero, A., Egea Gonzalez, F. J., Guerra Sanz, J. M. and Martínez Vidal, J. L. (2005). Analysis of biogenic volatile organic compounds in zucchini flowers: Identification of scent sources. *Journal of Chemical Ecology* 31, 2309–2322.

Milet-Pinheiro, P., Ayasse, M., Dobson, H. E. M., Schlindwein, C., Francke, W. and Dötterl, S. (2013). The chemical basis of host-plant recognition in a specialized bee pollinator. *Journal of Chemical Ecology* 39(11–12), 1347–1360.

Molet, M., Chittka, L. and Raine, N. E. (2009). How floral odours are learned inside the bumblebee (*Bombus terrestris*) nest. *Naturwissenschaften* 96(2), 213–219.

Morales, C. L. and Traveset, A. (2008). Interspecific pollen transfer: Magnitude, prevalence and consequences for plant fitness. *Critical Reviews in Plant Sciences* 27(4), 221–238.

Muhlemann, J. K., Klempien, A. and Dudareva, N. (2014). Floral volatiles: From biosynthesis to function. *Plant, Cell and Environment* 37(8), 1936–1949.

Naik, J. M., Gopali, J. B., Ganiger, V. M., Patil, S. R., Mm, V. K. and Patil, B. (2019). Influence of honey bee attractants in enhancing the seed productivity and quality of onion seed crop. *International Journal of Chemical Studies* 7, 2958–2962.

Naumann, K., Winston, M. L., Slessor, K. N. and Smirle, M. J. (1994). Synthetic honey bee (Hymenoptera: Apidae) queen mandibular gland pheromone applications affect pear and sweet cherry pollination. *Journal of Economic Entomology* 87(6), 1595–1599.

Nunes, C. E. P., Peñaflor, M. F. G. V., Bento, J. M. S., Salvador, M. J. and Sazima, M. (2016). The dilemma of being a fragrant flower: The major floral volatile attracts pollinators and florivores in the euglossine-pollinated orchid *Dichaea pendula*. *Oecologia* 182(4), 933–946.

Ormeño, E., Baldy, V., Ballini, C. and Fernandez, C. (2008). Production and diversity of volatile terpenes from plants on calcareous and siliceous soils: Effect of soil nutrients. *Journal of Chemical Ecology* 34(9), 1219–1229.

Patil, J. S., Mokat, R. B., Kamate, G. S. and Mupade, R. V. (2010). Studies on the influence of bee attractants on bee visition of *Apis dorsata* and *Trigona* sp. on onion (*Allium cepa* L.). *International Journal of Plant Protection* 3, 174–178.

Pellegrino, G., Luca, A., Bellusci, F. and Musacchio, A. (2012). Comparative analysis of floral scents in four sympatric species of *Serapias* L. (Orchidaceae): Clues on their pollination strategies. *Plant Systematics and Evolution* 298(10), 1837–1843.

Piechowski, D., Dötterl, S. and Gottsberger, G. (2010). Pollination biology and floral scent chemistry of the Neotropical chiropterophilous *Parkia pendula*. *Plant Biology* 12(1), 172–182.

Powers, J. M., Secom, R., Faiola, C. L., Sakai, A. K., Weller, S. G., Campbell, D. R. and Guenther, A. (2020). Floral scent composition and fine-scale timing in two moth-pollinated Hawaiian *Schiedea* (Caryophyllaceae). *Frontiers in Plant Science* 11, 1116.

Powney, G. D., Carvell, C., Edwards, M., Morris, R. K. A., Roy, H. E., Woodcock, B. A. and Isaac, N. J. B. (2019). Widespread losses of pollinating insects in Britain. *Nature Communications* 10(1), 1018.

Pöyry, J., Carvalheiro, L. G., Heikkinen, R. K., Kühn, I., Kuussaari, M., Schweiger, O., Valtonen, A., van Bodegom, P. M. and Franzén, M. (2017). The effects of soil eutrophication propagate to higher trophic levels. *Global Ecology and Biogeography* 26(1), 18–30.

Prieto-Benítez, S., Dötterl, S. and Giménez-Benavides, L. (2016). Circadian rhythm of a *Silene* species favours nocturnal pollination and constrains diurnal visitation. *Annals of Botany* 118(5), 907–918.

Qian, Y. L., Zhang, D., An, Y., Zhou, Q. and Qian, M. C. (2021). Characterization of aroma-active compounds in northern highbush blueberries "Bluecrop" (*Vaccinium corymbosum* "Bluecrop") and "Elliott" (*Vaccinium corymbosum* "Elliott") by gas chromatography-olfactometry dilution analysis and odor activity value. *Journal of Agricultural and Food Chemistry* 69(2), 5691–5701.

Ratto, F., Simmons, B. I., Spake, R., Zamora-Gutierrez, V., MacDonald, M. A., Merriman, J. C., Tremlett, C. J., Poppy, G. M., Peh, K. S. H. and Dicks, L. V. (2018). Global importance of vertebrate pollinators for plant reproductive success: A meta-analysis. *Frontiers in Ecology and the Environment* 16(2), 82–90.

Reichle, C., Jarau, S., Aguilar, I. and Ayasse, M. (2010). Recruits of the stingless bee *Scaptotrigona pectoralis* learn food odors from the nest atmosphere. *Naturwissenschaften* 97(5), 519–524.

Riffell, J. A., Alarcón, R., Abrell, L., Davidowitz, G., Bronstein, J. L. and Hildebrand, J. G. (2008). Behavioral consequences of innate preferences and olfactory learning in hawkmoth-flower interactions. *Proceedings of the National Academy of Sciences of the United States of America* 105(9), 3404–3409.

Sakai, S. (2002). A review of brood-site pollination mutualism: Plants providing breeding sites for their pollinators. *Journal of Plant Research* 115(3), 161–168.

Sasaki, M., Ono, M. and Miyamoto, M. (2008). Method for promoting pollination of plant including induction of flower bee to floral organ of specific plant by taking advantage of floral fragrance component of flower organ of this plant. Japan. *Google Patents*.

Satchell, G. H. (1947). The ecology of the British species of *Psychoda* (Diptera: Psychodidae). *Annals of Applied Biology* 34(4), 611–621.

Schiestl, F. P. (2015). Ecology and evolution of floral volatile-mediated information transfer in plants. *New Phytologist* 206(2), 571–577.

Schiestl, F. P. and Cozzolino, S. (2008). Evolution of sexual mimicry in the orchid subtribe orchidinae: The role of preadaptations in the attraction of male bees as pollinators. *BMC Evolutionary Biology* 8, 27.

Schultheis, J. R., Ambrose, J. T., Bambara, S. B. and Mangum, W. A. (1994). Selective bee attractants did not improve cucumber and watermelon yield. *Hortscience* 29(3), 155–158.

Scott-Brown, A. S., Arnold, S. E. J., Kite, G. C., Farrell, I. W., Farman, D. I., Collins, D. W. and Stevenson, P. C. (2019). Mechanisms in mutualisms: A chemically mediated thrips pollination strategy in common elder. *Planta* 250(1), 367–379.

Solís-Montero, L., Cáceres-García, S., Alavez-Rosas, D., García-Crisóstomo, J. F., Vega-Polanco, M., Grajales-Conesa, J. and Cruz-López, L. (2018). Pollinator preferences for floral volatiles emitted by dimorphic anthers of a buzz-pollinated herb. *Journal of Chemical Ecology* 44(11), 1058-1067.

Sommerlandt, F. M. J., Rössler, W. and Spaethe, J. (2014). Elemental and non-elemental olfactory learning using PER conditioning in the bumblebee, *Bombus terrestris*. *Apidologie* 45(1), 106-115.

Szenteczki, M. A., Godschalx, A. L., Galmán, A., Espíndola, A., Gibernau, M., Alvarez, N. and Rasmann, S. (2021). Spatial and temporal heterogeneity in pollinator communities maintains within-species floral odour variation. *Oikos* 130(9), 1487-1499.

Tal, O. (2009). *Acer pseudoplatanus* (Sapindaceae): Heterodichogamy and thrips pollination. *Plant Systematics and Evolution* 278(3-4), 211-221.

Tew, J. E. and Ferree, D. C. (1984). The effects of bee lure on honey bee (*Apis mellifera*) pollination of apples. *OARDC Research Circular* 283, 31-32.

Theis, N. and Adler, L. S. (2012). Advertising to the enemy: Enhanced floral fragrance increases beetle attraction and reduces plant reproduction. *Ecology* 93(2), 430-435.

Thomson, J. D., Draguleasa, M. A. and Tan, M. G. (2015). Flowers with caffeinated nectar receive more pollination. *Arthropod-Plant Interactions* 9(1), 1-7.

Tsirakoglou, V., Thrasyvoulou, A. and Hatjina, F. (1995). Techniques to increase the attractiveness of kiwi flowers to honey bees. *Acta Horticulturae* 444, 439-452. International Symposium on Kiwifruit, Thessaloniki, Greece.

Urru, I., Stensmyr, M. C. and Hansson, B. S. (2011). Pollination by brood-site deception. *Phytochemistry* 72(13), 1655-1666.

von Helversen, O., Winkler, L. and Bestmann, H. J. (2000). Sulphur-containing 'perfumes' attract flower-visiting bats. *Journal of Comparative Physiology A, Sensory, Neural, and Behavioral Physiology* 186(2), 143-153.

Williamson, S. M. and Wright, G. A. (2013). Exposure to multiple cholinergic pesticides impairs olfactory learning and memory in honeybees. *The Journal of Experimental Biology* 216(10), 1799-1807.

Wright, G. A., Baker, D. D., Palmer, M. J., Stabler, D., Mustard, J. A., Power, E. F., Borland, A. M. and Stevenson, P. C. (2013). Caffeine in floral nectar enhances a pollinator's memory of reward. *Science* 339(6124), 1202-1204.

Wright, G. A. and Schiestl, F. P. (2009). The evolution of floral scent: The influence of olfactory learning by insect pollinators on the honest signalling of floral rewards. *Functional Ecology* 23(5), 841-851.

Xishuangbanna Tropical Botanical Garden of CAS. (2009). *Method for Inducing Bees to Pollinate Among Parents of Male Sterile Line Seed.*

Yourong, W. and Industry Co Ltd. (2015a). Biological pollination method capable of achieving high pollination rate by adopting bee attractant. *Google Patents.*

Yourong, W. and Industry Co Ltd. (2015b). Biological pollination method implemented by aid of leaf-cutter bees and application of biological pollination method to planting *Cornus officinalis* in saline-alkali soil. *Google Patents.*

Yourong, W. and Industry Co Ltd. (2015c). High pollination rate biological pollination method adopting megachilid attractant. *Google Patents.*

Yu, H., Zhang, Y., Wu, K., Gao, X. W. and Guo, Y. Y. (2008). Field-testing of synthetic herbivore-induced plant volatiles as attractants for beneficial insects. *Environmental Entomology* 37(6), 1410-1415.

Chapter 3

The role of wind pollination in crop plants

Stephen Murphy, University of Waterloo, Canada

1 Introduction

Approximately 12% of flowering plants across the globe are wind pollinated (anemophilous) - with some species having both wind and animal pollinators (ambophilous) (Ackerman, 2000; Culley et al., 2002). Anemophily is important to about 60% of the world's most important crop species - especially grasses (Poaceae) (Aizen et al., 2009; FAO, 2021). These include grasses, sedges, rushes, pecans, pistachios, hazelnuts, walnuts, coconut, mustards (e.g. canola), grapes, palms, coffee trees, and occasionally sunflowers and bamboo (see Cresswell et al., 2004; Ruiz-Sanchez et al., 2017; Saunders, 2018). Contemporary breeding programs have reduced the importance of open pollination in agricultural cultivars of some crops (e.g. wheat) (Zohary et al., 2012).

2 Fundamental mechanisms of wind pollination

Wind pollination depends on three main factors: pollen production, release/dispersal, and capture (Niklas, 1985; Ackerman and Kevan, 2005; Wilmer, 2011; Timerman and Barrett, 2018). Floral and inflorescence architecture, together with factors such as plant height and movement (collector motion), are crucial to the efficiency of pollination and mating given the aerodynamics of wind pollination (Friedman and Harder, 2004; McCombe and Ackerman, 2018; Timerman and Barrett, 2021).

http://dx.doi.org/10.19103/AS.2022.0111.07

Wind pollination is facilitated by evaporation through the stomata in response to macro- and micro-meteorological conditions (Timerman and Barrett, 2018). Evaporation will cause anthers to dehisce and release pollen. While there are other mechanisms and other influences on anther dehiscence and pollen release, evaporation probably dominates because it reflects the macro-meteorological conditions better suited to pollen release (low relative humidity, high solar radiation, high temperatures, higher wind speeds, and turbulence).

3 Wind pollination and transgenic risks in grass crop species

Because grass crop species represent a large percentage of global food supplies, this section will focus on the mechanisms and transgenic risks.

3.1 Maize/corn

In the case of maize/corn (*Zea mays*), commercial production of seed corn relies on enforced cross pollination to produce single cross-hybrids (crossing two different inbred lines). This is done by artificially removing the male inflorescence (tassel) on the pollen recipient to ensure that two different genotypes cross to produce more vigorous hybrids with desired traits such as improved yield. Pollination in the field still depends on wind pollination because of the physical distance of the tassel at the tip of the maize plants versus the ears that are produced about halfway up the plant.

Research has focussed on understanding the mechanisms and timing of pollination to help farmers optimise success (Bechoux et al., 2000; Anderson et al., 2004). To increase proximate successful pollination between adjacent 'male' and 'female' rows, cultivation practices can include deliberate planting patterns and manipulations of inflorescences. As an aside, if the goal is to harvest maize for silage/fodder, there is usually less focus on cultivation and enforced open pollination since the bulk of silage will come from the entire plant, not just the kernels.

The development of transgenic varieties of corn has renewed interest in processes of wind pollination, particularly the distances transgenic pollen might travel. This is because of the need to avoid the risk of 'contaminating' non-genetically modified corn with genetically modified pollen, given the concern of some consumers about consuming foods containing genetically modified ingredients, the risk to organic farmers of infringing certification rules for organic produce, and environmental risks such as the development of herbicide-resistant weeds due to pollination via pollen from transgenic herbicide-resistant corn varieties (Hernández-López, 2020).

These concerns have led to much work on detailed models of corn pollen emission, dispersal, and capture (e.g. Aylor et al., 2003; Bannert and Stamp, 2007). While the rule of thumb that most corn pollen dispersal is leptokurtic and lands within 20 m or less seems to hold, the wealth of information on wind pollination has been immense. Studies like Klein et al (2003) have quantified, more precisely, how much and how far *Zea* pollen disperses. Comparisons of dispersal models like those of Klein et al. (2003) show that vegetation is - surprisingly - not a major barrier to pollen dispersal and the size, density, and aerodynamics of corn pollen mean that the probability functions of density and dispersal start becoming small even 5 m away, much smaller 20 m away, and smaller still after that. However, some studies have shown dispersal as far as 4 km, suggesting that recommended buffer zones of 20–30 m between transgenic and non-transgenic fields may not always be sufficient (Hofman et al., 2014). This suggests it is not possible to achieve zero risk unless one hermetically seals a field or has enormous guard rows.

The risk of weeds acquiring herbicide resistance from *Zea* may be lower because it has fewer wild weedy relatives - with the proviso that this risk cannot be ignored in its countries of origin. Its large (90 μm) pollen grain is less likely to be dispersed to any weedy relatives, and the weedy species that are in the margins near the end rows of corn often don't flower concurrently. Further, their inflorescences and pistils will not allow corn pollen to alight, and the pollen usually will not germinate even if alight, and the species are not closely related so the chances of an intergeneric hybrid are very small.

3.2 Rice

It is not clear if rice (*Oryza*) presents a high risk for transgenic use because much of the pollination is more gravimetric - pollen falls from the male topmost flowers to the female ones below. The risk depends on how much lateral plant-to-plant pollination there actually is. There are wild congenerics that can be considered weedy, increasing the risk of gene transfer that could create even more pernicious and invasive genotypes.

3.3 Wheat and other mainly self-pollinated grass crops

Triticum (wheat) may be less of a risk because it does not rely on outcrossing via wind pollination - the commercial varieties are mainly self-pollinated. However, there has been discussion of the advantages of new developing lines with open, receptive female flowers that can be cross-pollinated to produce a more genetically diverse wheat crop via wind pollination (Dixon et al., 2018). That may mean wind pollination may soon be a more important factor in wheat production, but it means more risk for transgenic contamination. This is probably the same for other mainly self-pollinating crops like barley and oats.

3.4 Wind-pollinated hay/forage crops

Rather than focus on pollination or seed set, the trait targeted in breeding programs is usually producing palatable and nutritious leaves for livestock. Breeding focuses on reallocation of resources away from sexual reproduction, and this means allocation of resources away from flowers to improving leaves. There is a long history of plant breeding strategies and techniques, from the humbler origins of early cultivars of forage grasses to modern high-technology approaches (Barth, 2012).

Still, open wind pollination is an important consideration, and there is substantial use of crops like timothy (mainly *Phleum pratense*), fescues (*Festuca*), orchardgrasses (*Dactylis*), bromes (Bromus), Bermudagrass (*Cynodon*), bahiagrasses (*Paspalum*), blue/speargrasses (*Poa*), canarygrasses (*Phalaris*), redtops (*Agrostis*), sorghum (*Sorghum*), gamagrass (*Tripsacum*), millets/panicgrasses (*Brachiaria*, *Echinochloa*, *Urochloa*, *Panicum*), and rye (*Lolium*).

While some species and cultivars are bred for self-pollination, many are open-pollinated, making for easier management from a mechanical and regulatory perspective since the ubiquity of pollen has meant little progress on transgenic applications in forage/hay grasses (Wang and Bremmer, 2012). Studies on pollination in forage grasses are sparse but, like many cash crops, there has been a focus on technologies from crossing inbred lines to transgenic approaches.

There are typical concerns about transgenic escapes. The main difference between forage grasses and a cash crop like maize is that most of the forage grasses have smaller pollen (20–40-μm diameter vs 90 μm). This means there is a greater risk (from the perspective of transgenes in pollen) of unwanted escapes and possible cross-fertilization if the donor and recipient are non-sterile genotypes and not cleistogamous.

3.5 Wind-pollinated fruits and nuts

Studies on wind pollination in commercial production of fruits and nuts are found in the peer-reviewed literature (e.g. Polito et al., 2015), though there is an opportunity for a monograph on that subject. Chestnuts (*Castanea*), hazelnuts/filberts (*Corylus*), pecans (*Carya*), walnuts (*Juglans*), pistachios (*Pistacia*), dates (*Phoenix*), olives (*Olea*), grapes (*Vitis*) are wholly or partially wind pollinated, and many nut trees (e.g. *Macadamia*) have anecdotal evidence of some importance of wind pollination. Much of the press around pollination and transgenics has focused on rescuing the chestnuts (e.g. Steiner et al., 2017) with the usual concerns over risk less prominent in this species because of its near-extirpation by blight. As transgenic research on chestnut proceeds apace, one can expect more interest in application to other wind-pollinated nut and fruit species, but the transgenic debate will be more heated as most of those species are not imperilled.

3.6 Assessing overall transgenic risk in wind-pollinated crops

As Rong et al. (2012) noted in their study of risks of transgenic pollination and escape in rice (*Oryza sativa*), one of the challenges in experimental design and evaluation is trying to scale practical small-scale field scale studies up to the likely overall risk for crops planted over millions of hectares. If it is 1:1, then the smaller scale estimates are robust but, as Rong et al note, these can be off by a factor of 10 in those few studies that even scale up to 1–2 hectares. Rong et al scaled to ~500 m^2 and used a distance simulation model to show that there was low risk in their study on *Oryza*. This is consistent with the later study – that included size of pollen source – by Hu et al (2020).

Rong et al. (2012) note that one cannot assume a lower risk in outcrossing wind-pollinated crops like *Z. mays* or for *Secale cereale* (Rye) – which is both a crop and a weed (in *Triticum*). *Zea* is less likely to be vulnerable to transgenic escapes but *S. cereale* is a strongly outcrossing, wind-pollinated grass that has been transgenically modified, so the risk is likely greater with what smaller (40–50 μm) and lighter pollen, though Kutlu (1982) noted that about 97% of the pollen is dispersed less than 20 m. The same issues arise with many other crop species (e.g. *Hordeum*) with the same general patterns of relatively low risk.

3.7 Assessing overall transgenic risk with wind-pollinated agricultural weed species

Wind-pollinated weed species include *Ambrosia, Bromus, Setaria, Elymus, Amaranthus, Festuca,* and *Chenopodium*. These genera represent many of the problematic cash crop weeds in North America. Studies like Friedman and Harder (2004) demonstrate that the evolutionary history of weed inflorescence architecture means they are prolific at emitting and capturing conspecific pollen, despite expected stochasticity. While dispersal can be affected by the weedy individual's phenology relative to the growth of crops or other weeds and their height in taking advantage of wind, the wind-pollinated weeds seem to face few challenges in successfully setting seed in competition with crops. Since they do not depend much or at all on pollinators (though some may be ambophilous), even catastrophic acute or chronic losses of animal habitat and pollinators will not diminish their persistence.

If heterospecific pollen transfer occurs and the hybrids are formed between crops and weeds of different species or genera, the risk of transgenic escape and conference of pesticide resistance would be problematic. The null hypothesis is that this is less likely given barriers to forming intergeneric hybrids.

There is a higher risk that overuse of pesticides in resistant crops selects for pesticide-resistant weeds rather than a direct transfer of the genetic resistance from crop to seed. Since it does not appear that planting patterns of crops might sufficiently reduce pollen transfer between conspecific weeds and some of those weeds are clonal, wind pollination presents an intractable challenge in that it is not likely that farmers can thwart weed reproduction except by more interventionist management.

4 Promoting wind pollination in open-pollinated crop species

There are still many farms where open pollination to produce seeds for locally adapted genotypes is important. Traditional agriculture in Meso-America relies heavily on local land races of the three sisters – maize (anemophilous), beans (entomophilous), and squash (entomophilous) – grown in the milpa system. As an example, 30% of the farms in the author's region of southwestern Ontario in Canada still have open-pollinated forage fields, partly due to a higher proportion of farms owned by the Mennonite religious community that deliberately eschews modern in favour of traditional farming techniques.

One way to maximize open-pollinated seed set in forage grass fields is to maintain (or restore) forest tracts/corridors or tall meadow tracts/field margins around each field. These areas (which are often unproductive and unprofitable for crop production) can then harbour a diverse ecosystem, helping to meet conservation goals and deliver ecosystem services other than pollination such as promoting natural enemies of crop pests, acting as a windbreak to protect crops and preventing erosion, etc. A particular benefit is the conservation of wind-pollinated native species. Research, for example, by the author and colleagues covering multiple kilometres of margins of farm fields in southern Ontario in Canada found substantive (>500 ramets) populations of over 150 species of wind-pollinated grasses, sedges, and rushes, many of which were thought to have been locally extinct.

Marginal areas can provide a rich wind-pollination environment for native forbs and a reservoir of rare species (NSERC-CANPOLIN, 2015). From the pollination perspective, there is some risk of heterospecific pollen transfer potentially interfering with the forage crop, depending on whether the diurnal and seasonal phenologies of the forage crop overlap with other grasses. There is also a need to manage field margins so that they do not become a refuge for pest species – from crop pests like rusts that may have their alternative hosts on native plants (e.g. *Thalictrum*) to deer and crop-eating birds to weeds that may thrive in margins and then disperse into crop fields, especially no-till fields that resemble the environment of field margins.

The big 'win' is the ability of tracts – especially forest tracts – to act effectively as an aerodynamic bell jar, sealing in the pollen so it increases the chances of successful pollen transfer within a defined area rather than becoming lost to the atmosphere or adjoining fields. This is more effective for night-pollinating species with a number of forage grass species that pollinate mainly at night such as the forage grass *P. pratense* (Murphy, 2000).

In fields enclosed by forest tracts, there usually exists a surface-based radiation inversion at night, which is caused by convection-mediated high temperatures in the field during the day (Murphy, 2000). Essentially, this prevents material, such as pollen, from leaving or entering a field until the inversion burns off after sunrise. Because most individuals of forage grasses like *P. pratense* flower before sunrise, most of the pollen would be deposited within the field in which it was originally dispersed prior to the burn-off.

Pollen is generally dispersed in a leptokurtic manner: most of the pollen is dispersed within 100 m of the parent plant. Should any pollen enter the atmosphere before the inversion has stabilised or any pollen enter after it burns off, it is most likely to be carried long distances (> 20 km) as the atmosphere above the fields tends to lift particles upwards at these times. Although there is some dispute over some aspects of theories on atmospheric stability and transport of pollen, other sources support these interpretations (Ogden et al., 1969; Okubo and Levin, 1989; Di-Giovanni and Kevan, 1991; Giostra et al., 1991; Jacobs and van Boxel, 1991; Tiriabassi et al., 1991).

5 Conclusion and future trends

The general milieu of pollination and specific mechanisms of wind pollination cannot be isolated from the larger forces that affect them and they, in turn, affect. The above 'good news' vignette on wind-pollinated native species in field margins is an exemplar of that embedded and recursive relationship. Symbolically, it should remind us that while humans have bred crops and made many grass-based crops less reliant on unmanaged open wind pollination, humans owe our existence in part to wind pollination. Wind pollination is one thread in the development of human civilisation.

The biggest threat to farming - wind-pollination issues included - is anthropogenic climate change (e.g. Kjøhl et al., 2011). Given the relative dearth of papers focused on wind pollination in crops, it is not surprising that there is even less on impacts of climate change. Research has focused on climate change impacts on wind-pollination species in vulnerable e.g. alpine habitats (Hoiss et al., 2013) or on broader topics, like impacts on forests and/or wind dispersal of seeds rather than pollen (e.g. King and Ackerly, 2021).

The response of wind-pollinated crops will be as varied as the pace and intensity of climate change impact on local habitats and on the factors affecting pollination (e.g. humidity and wind). An increase in the frequency and persistence of higher local temperatures might facilitate wind pollination. However, that could be offset by increased precipitation that would wash the pollen away or if the relative humidity also increased drought or even simple rises in mean temperatures also can cause re-allocations or net decreases in biomass, hence less ability to produce even the smaller and less 'expensive' inflorescences and flowers in wind-pollinated species. It is possible that climate change might have a more immediate impact on animal pollination, potentially favouring wind-pollinated crops, but that is in the context of the overwhelmingly negative impacts if climate change continues on its present trajectory.

6 References

Ackerman, J. D. 2000. Abiotic pollen and pollination: Ecological, functional and evolutionary perspectives. *Plant Systematics and Evolution* 222(1-4):167-185.

Ackerman, J. D. and Kevan, P. G. 2005. *Abiotic Pollination*. In: Dafni, A. and Kevan, P. G. (Eds.), *Practical Pollination Biology*. Cambridge, UK: Enviroquest Ltd.

Aizen, M. A., Garibaldi, L. A., Cunningham, S. A. and Klein, A. M. 2009. How much does agriculture depend on pollinators? Lessons from long-term trends in crop production. *Annals of Botany* 103(9):1579-1588. Available at: https://academic.oup.com/aob/article/103/9/1579/146727.

Anderson, S. R., Lauer, M. J., Schoper, J. B. and Shibles, R. M. 2004. Pollination timing effects on kernel set and silk receptivity in four maize hybrids. *Crop Science* 44(2):464-473.

Aylor, D. E., Scultes, N. P. and Shields, E. J. 2003. An aerobiological framework for assessing cross-pollination in maize. *Agricultural and Forest Meteorology* 119(3-4):111-129.

Bannert, M. B. and Stamp, P. 2007. Cross-pollination of maize at long distance. *European Journal of Agronomy* 27(1):44-51. https://doi.org/10.1016/j.eja.2007.01.002.

Barth, S. 2012. Breeding strategies for forage and grass improvement. *Annals of Botany* 110(6):1261-1262.

Bechoux, N., Bernier, G. and Lejuene, P. 2000. Environmental effects on the early stages of tassel morphogenesis in maize (*Zea mays* L.). *Plant, Cell and Environment* 23(1):91-98.

Cresswell, J. E., Davies, T. W., Patrick, M. A., Russell, F., Pennel, C., Vicot, M. and Lahoubi, M. 2004. Aerodynamics of wind pollination in a zoophilous flower, *Brassica napus*. *Functional Ecology* 18(6):861-866.

Culley, T. M., Weller, S. G. and Sakai, A. K. 2002. The evolution of wind pollination in angiosperms. *Trends in Ecology and Evolution* 17(8):361-369.

Di-Giovanni, F. and Kevan, P. G. 1991. Factors affecting pollen dynamics and its importance to pollen contamination: A review. *Canadian Journal of Forest Research* 21(8):1155-1170.

Dixon, L. E., Bencivenga, S. and Boden, S. A. 2018. A new opening for wheat seed production. *Journal of Experimental Botany* 69(3):341-343.

FAO. 2021. Available at: http://www.fao.org/faostat/en/?#data.

Friedman, J. and Harder, L. D. 2004. Inflorescence architecture and wind pollination in six grass species. *Functional Ecology* 18(6):851-860.

Giostra, U., Mandrioli, P., Tampieri, F. and Trombetti, R. 1991. Model for pollen immission and transport in the evolving boundary layer. *Grana* 30(1):210–214.

Hernández-López, E. A. 2020. GMO corn in México: Precaution as law's decolonial option. *Law, Technology and Humans* 2(2):97–113.

Hofman, F., Mathias, O. and Wosniak, W. 2014. Maize pollen deposition in relation to distance from the nearest pollen source under conventional cultivation – Results of 10 years of monitoring (2001–2010). *Environmental Sciences Europe* 26:1–14.

Hoiss, B., Gaviria, J., Leingaärtner, A., Krauss, J. and Steffan-Dewenter, I. 2013. Combined effects of climate and management on plant diversity and pollination type in alpine grasslands. *Diversity and Distributions* 19:386–395. https://doi.org/10.1111/j.1472-4642.2012.00941.x.

Hu, N., Jiang, X., Yuan, Q., Liu, W., Yao, K., Long, Y. and Pei, X. 2020. Increased pollen source area does not always enhance the risk of pollen dispersal and gene flow in *Oryza sativa* L. *Scientific Reports* 10(1):6143. https://doi.org/10.1038/s41598-020-63119-z.

Jacobs, A. F. G. and van Boxel, J. Hv. 1991. Horizontal and vertical distribution of wind speed in a vegetation canopy. *Netherlands Journal of Agricultural Science* 39(3):165–178.

Kjøhl, M., Nielsen, A. and Stenseth, N. C. 2011. *Potential Effects of Climate Change on Crop Pollination*. Rome: Food and Agriculture Organization of the United Nations. Available at: http://www.fao.org/3/i2242e/i2242e.pdf.

King, M. M. and Ackerly, D. D. 2021. Global wind patterns shape genetic differentiation, asymmetric gene flow, and genetic diversity in trees. *Proceedings of the National Academy of Sciences of the United States of America* 118(17). https://doi.org/10.1073/pnas.2017317118.

Klein, E. K., Lavigne, C., Foueillassar, X., Gouyon, P. and Larédo, C. 2003. Corn pollen dispersal: Quasi-mechanistic models and field experiments. *Ecological Monographs* 73(1):131–150.

Kutlu, Y. Z. 1982. Pollen dispersal in rye (*Secale cereale* L.): Research activities on isolation problems of cross-pollinated species of GeneBank material. In: Porceddu, E. and Jenkins, G. (Eds.), *Seed Regeneration in Cross-Pollination Species. Proceedings of the C.E.C./EUCARPIA Seminar Nyborg*, Denmark, 15–17 July 1981, pp. 275–279.

McCombe, D. and Ackerman, J. D. 2018. Collector motion affects particle capture in physical models and wind pollination. *The American Naturalist* 192(1):81–93.

Murphy, S. D. 2000. Field testing for pollen allelopathy: A review. *Journal of Chemical Ecology* 26(9):2155–2172.

Niklas, K. J. 1985. The aerodynamics of wind pollination. *Botanical Review* 51(3):328–386.

NSERC. *CANPOLIN 2015. Pollination Nation: Research Highlights From the Canadian Pollination Initiative 2009–2014*. Ontario, Canada: University of Guelph.

Ogden, E. C., Hayes, J. V. and Raynor, G. S. 1969. Diurnal patterns of pollen emission in *Ambrosia, Phleum, Zea* and *Ricinus*. *American Journal of Botany* 56(1):16–21.

Okubo, A. and Levin, S. A. 1989. A theoretical framework for data analysis of wind dispersal of seeds and pollen. *Ecology* 70(2):329–338.

Polito, V. S., Pinney, K., Weinbaum, S., Aradhya, M. K., Dangl, J., Grant, Y. J. A. and Grant, J. A. 2015. Walnut pollination dynamics: Pollen flow in walnut orchards. *Acta Horticulturae* 705(705):465–472.

Rong, J., Wang, F., Song, Z., Su, J., Chen, R. and Lu, B.-R. 2012. Scale effect on rice pollen-mediated gene flow: Implications in assessing transgene flow from genetically engineered plants. *Annals of Applied Biology* 161(1):3–11.

Ruiz-Sanchez, E., Peredo, L. C., Santacruz, J. B. and Ayala-Barajas, R. 2017. Bamboo flowers visited by insects: Do insects play a role in the pollination of bamboo flowers? *Plant Systematics and Evolution* 303(1):51–59.

Saunders, M. E. 2018. Insect pollinators collect pollen from wind-pollinated plants: Implications for pollination ecology and sustainable agriculture. *Insect Conservation and Diversity* 11(1):13–31.

Timerman, D. and Barrett, S. C. H. 2018. Divergent selection on the biomechanical properties of stamens under wind and insect pollination. *Proceedings of the Royal Society, Series B* 285:1–8.

Steiner, K. C., Westbrook, J. W., Hebard, F. V., Georgi, L. L., Powell, W. A. and Fitzsimmons, S. F. 2017. Rescue of American chestnut with extraspecific genes following its destruction by a naturalized pathogen. *New Forests* 48(2):317–336.

Timerman, D. and Barrett, S. C. H. 2021.The biomechanics of pollen release: New perspectives on the evolution of wind pollination in angiosperms. *Biological Reviews of the Cambridge Philosophical Society* 96(5):2146–2163.

Tiriabassi, T., Mandrioli, P. and Manco, D. 1991. Statistical distribution of airborne pollen grains. *Grana* 30(1):255–259.

Wang, Z. Y. and Bremmer, E. C. 2012. Is genetic engineering ever going to take off in forage, turf and bioenergy crop breeding? *Annals of Botany* 110(6):1317–1325.

Wilmer, P. 2011. *Pollination and Floral Ecology*. Princeton, NJ: Princeton University Press.

Zohary, D., Hopf, M. and Weiss, E. 2012. Domestication of plants in the old world: The origin and spread of domesticated plants in southwest Asia, Europe, and the Mediterranean Basin. Available at: https://oxford.universitypressscholarship.com/view/10.1093/acprof:osobl/9780199549061.001.0001/acprof-9780199549061.

Part 2

Threats to pollinators

Chapter 4

Assessing climate change impacts on pollinators

Kit Prendergast, Curtin University, Australia

1 Introduction

Flower visiting insects are major components in ecosystems, including agricultural ecosystems, where their foraging on angiosperms results in pollination services. In most systems, the main pollinators are bees (James et al., 2008; Batra, 1995), with much research revealing a diversity of flower-visiting taxa enhances pollination services (Dainese et al., 2019; Garibaldi et al., 2013, 2016; Pérez-Méndez et al., 2020, Potts et al., 2016). As such, preserving this component of biodiversity, and its ecosystem services, is fundamental for sustainable agriculture and meeting nutritional demands of a growing human population (Kevan et al., 1990). Whilst pollinators face numerous threats in the Anthropocene, one of the major threats they face, which is a threat shared with biodiversity and human livelihoods (Reid and Huq, 2005; Battarbee et al., 2008), is that of anthropogenic climate change. We can predict that climate change impacts will be particularly acute on pollinators, with multiple impacts of climate change impinging upon their survival. The challenges that climate change may pose to pollinators will be covered in the next section, along with what evidence is available to date to support or question these theoretical predictions on the impacts of climate change on pollinators. I will then highlight the gaps in research to date and recommend future research directions. Finally, I will cover some measures that may be adopted to enable pollinators to face ongoing climate change challenges.

http://dx.doi.org/10.19103/AS.2022.0111.12

2 Challenges anthropogenic climate change poses to pollinators

Flower-visiting insects may be particularly susceptible to climate change due to their small body size which limits migration, their reliance on flowers for nutrition, and how they face many other threats such as loss of wildflowers, pesticides and fragmentation.

2.1 The impact of rising temperatures on pollinators

In most regions of the planet, climate change, by enhancing the greenhouse effect, is resulting in global warming (National Oceanic and Atmospheric Administration, 2020). The rate of warming is unprecedented, and already as of 2020, combined land and ocean temperature have increased at an average rate of 0.08°C per decade since 1880, with average rates of increase since 1981 (0.18°C) being more than twice that rate (National Oceanic and Atmospheric Administration, 2020). Pollinators are known to strongly structure communities (e.g. Mayr et al., 2020), and thus they are sensitive to increasing temperatures under climate change. It can be expected that increased temperatures caused by climate change will have numerous impacts on pollinators, and plant-pollinator assemblages (Fig. 1).

Not all regions however experience equal degrees of warming, and within a given region, different microclimates will occur due to differences in tree cover, land-surface cover and topography (Lenoir et al., 2017). At a direct level, because many pollinators are invertebrates, they are ectotherms, and their activity levels are influenced by the temperature of their ambient environment. This relationship is also influenced by body size, shape, colour and insulation (Fig. 1), whereby larger, rounder-bodied, darker-coloured, 'hairy' animals absorb and retain heat to a greater extent than smaller, slender, light-coloured, hairless animals. Consequently, big, dark-coloured, fuzzy bees like many *Bombus* would be more sensitive to hotter temperatures than tiny, light-coloured, hairless *Brachyhesma* (Fig. 1).

Research on 24 Meliponini species revealed the importance of body size, colour and behaviour of bee thermal characteristics (Pereboom and Biesmeijer, 2003). In accordance with biophysical principles, where diffusion, including of heat, occurs more rapidly for smaller objects with a higher surface area relative to volume, small bees reached lower temperature excesses and warmed up as well as lost heat more rapidly than larger bees. Both between species, and within species with colour morphs, bees that were darker (more black/brown) experienced a 28% higher temperature excess and has a 26% greater cooling constant, K, than similar-sized light-coloured (more yellow) species (Pereboom and Biesmeijer, 2003).

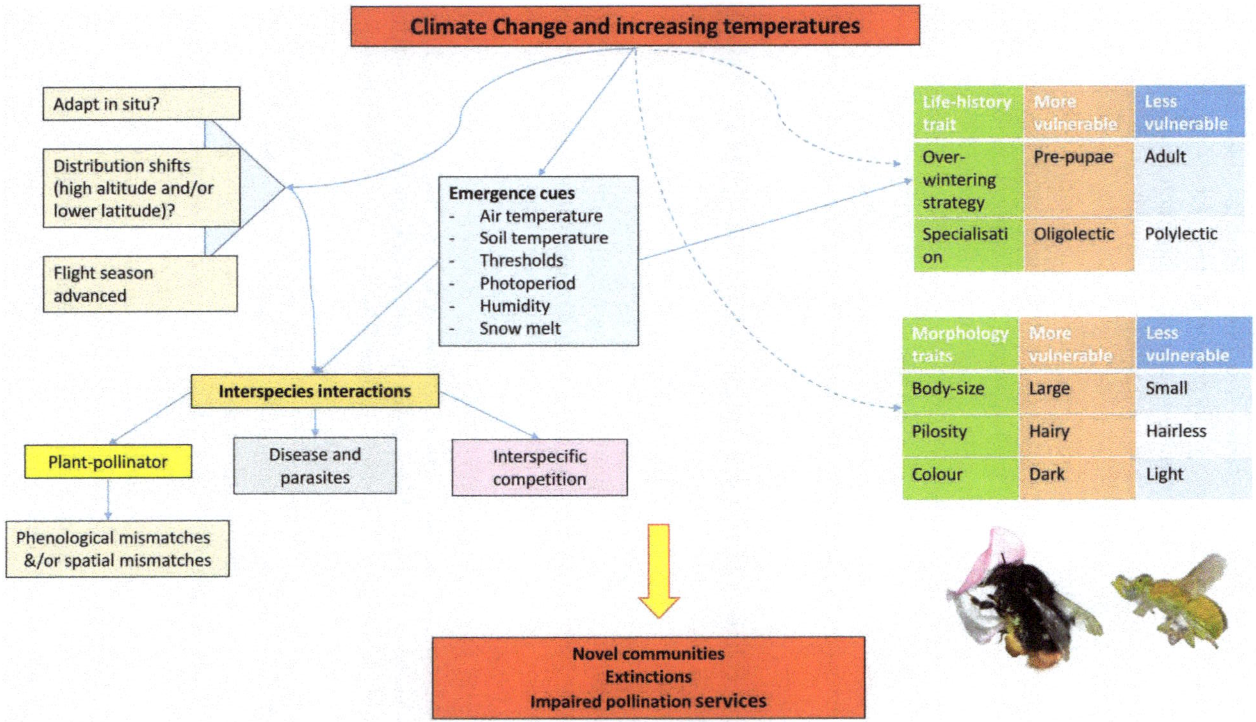

Figure 1 Potential impacts of increased temperatures as a result of anthropogenic warming on pollinators. *Brachyhesma wyndhami* photograph by Kit Prendergast, *Bombus lapidaries* photography by Claus Rasmussen.

However, these experiments were on dead bees - when looking at foraging bees, differences between colour morphs were minimal (Pereboom and Biesmeijer, 2003). This is because live bees can alter their behaviour adaptively whilst foraging e.g. forage in cooler/warmer microenvironments. Differences in convective or evaporative heat loss, or production of metabolic heat during flight, can also override the influence of body colour on heat gain/loss (Heinrich, 2013).

Behavioural adaptations also come into play and may represent a way for species to adapt to warmer temperatures: for example, large-bodied bees are known to commence foraging earlier, before the day heats up (Biesmeijer, 1997). However, there may also be risks here such as being limited by low light (Theobald et al., 2007, Dorey et al., 2020). Dark-coloured bees may also avoid overheating by foraging in shady patches; however, if smaller bees also need to forage in shadier locations, competition may occur, and shadier patches also contain flowers with less concentrated nectar (Biesmeijer, 1997, Biesmeijer et al., 1999).

The current distribution of Meliponini in Central America suggests that body size and colour explain biogeographic distribution patterns through the bees' thermal properties (Pereboom and Biesmeijer, 2003). Under climate change we may see shifts in the biogeography of species, and/or evolution towards lighter, smaller body sizes if standing genetic variation exists.

Experimental studies have demonstrated the negative impacts of warmer temperatures on bee fitness directly. Using eight rearing treatments, the effect of climate change on eight cavity-nesting solitary bees was investigated by manipulating fall and spring temperature, fall duration and timing of spring onset (Slominski and Burkle, 2019). Responses differed according to the diapausing strategy of the bees, with *Osmia*, which overwinter as adults, being more vulnerable than *Megachile*, which overwinter a pre-pupae. *Osmia* responded primarily to spring onset, with earlier spring onset causing increased pre-emergence mortality and weight-loss, reduced post-emergence lifespan and also caused unequal shifts in emergence timing for males vs. females. *Megachile* responded primarily to warmer spring temperatures, which results in decreased weight loss and increased lifespan. Variable responses to increased temperatures according to over-wintering strategy appear to be common. For example, storing cocoons of nine species of solitary, cavity-nesting bees (*Osmia bicornis, O. cornuta, Megachile ligniseca, M. versicolor, M. centuncularis, Chelostoma florisomne, Heriades truncorum,* and the cleptoparasitic *Coelioxys mandibularis* (Megachilidae) and *Hylaeus communis* (Colletidae)) at different overwintering temperatures (1.5–9.5°C), resulted in significant differences between species in weight loss, weight at emergence, and emergence date (Fründ et al., 2013). Those overwintering as adults generally exhibited decreased weight and earlier emergence with increasing temperature due to increased energy expenditure, whilst species overwintering as pre-imaginal stages exhibited weaker, or opposite responses.

These differential responses indicate that some species are particularly vulnerable to the thermal effects of climate change, whilst others may even benefit, leading to shifts in the taxonomic composition of pollinator communities.

In addition to lab-based experimental manipulations, negative fitness impacts of increased temperatures have also been demonstrated through in situ field experiments. By altering the thermal properties of nesting boxes by painting them different colours to simulate cooler temperatures (painted white, approx. 1950 temperatures), control, and future warmer temperatures (painted black, approx. 2040–2099 temperatures), it was found that the solitary bee *Osmia ribifloris* exhibited delayed emergence phenology, and increased phenological variance in the warming treatment, as well as reductions in body mass and fat content (Caradonna et al., 2018). In addition, mortality drastically increased: bees in the warming treatment suffered 30–75% mortality, whereas those in the control and cooler treatment had negligible mortality. Taken together, these direct mortality costs, as well as indirect fitness costs through reduced body condition, and phenological asynchrony, means this species may risk extinction in warmer parts of its range under climate change.

Global warming has already had major detrimental impacts on wild bees. Using a long-term database of ~550 000 georeferenced occurrence records of 66 *Bombus* species across North America and Europe, increasing frequency of hotter temperatures predicted species local extinction risk, chances of colonizing a new area, and changing species richness (Soroye et al., 2020). The impact on population extinction-colonisation dynamics of increasing frequencies of temperatures that exceed historically observed tolerances was independent of land-use change. This demonstrates that climate change represents a major driver of *Bombus* declines.

2.1.1 The impact of changing rainfall patterns

Increasing temperatures and changes in rainfall patterns are also causing increased water stress and limitations for flowers, with flow-on effects on pollinators. The impact of climate change causing reduced rainfall has been extremely evident in Australia, where there has been an increasing frequency and intensity of drought which has caused dieback of key honey sources and reduced nectar flows (Hoffmann et al., 2019), leading to yield losses for beekeepers (Somerville, 2020). Drought impacts on Australian native bees have not been published to date.

2.2 Loss of ground and moving around

Global climate change is predicted to affect habitat suitability and geographic distribution of pollinators. Under warmer temperatures, organisms can keep track of their current thermal niche by migrating to cooler regions so that they stay within their climatic envelope. This may mean shifting their distribution pole-wards, or upwards to higher altitudes. However, some species will be unable to respond adaptively, and instead climate change will lead to reductions in their fundamental niche, resulting in range reductions. A number of studies have modelled how climate change scenarios will alter the ranges of pollinator species. For example, ecological niche modelling (ENM) was used to estimate distribution changes in 18 Mesoamerican *Bombus* species using current and future climate emissions scenarios (representative concentration pathway 4.5, 6.0 and 8.5) and models (CCSM4, HadGEM2-AO, and MIROC-ESM-CHEM) (Martínez-López et al., 2021). The sobering result was that all species were predicted to undergo a reduction in their potential distribution and habitat suitability due to projected climate change; even under a low emissions scenario, distribution losses of 7–76% were predicted by 2050. Species are also predicted to shift their distributions by a 100–500 m upslope in altitude and 1–581 km away from a species' current geographic centroid of their distribution.

There are, however, not only 'losers' but also 'winners' under climate change. For example, some species are expanding their geographic distribution. Modelling has revealed that an already widespread, arid-adapted Australian bee *Ceratina australensis* (Apidae: Xylocopinae) under Intergovernmental Panel on Climate Change (IPCC) climate change conditions predicted for 2070 will have an increased area of suitable habitat, including an increased range within protected areas (Dew et al., 2019).

2.3 Wrong place, wrong time? Temporal and spatial mismatches between plants and pollinators

2.3.1 Mismatches in phenology

Whilst species in tropical regions, and many eusocial bees, are perennially active, the majority of bee species have a limited flight season spanning only a few months (Prendergast, 2018). Similarly, most plants do not flower year-round and have a short blooming period. For specialist bees especially, if they emerge before or after their host plant blooms, they may suffer from insufficient resources. The cues that mediate the onset of blooming, and emergence of bees, may also be different for the two partners. If there are phenological mismatches, bees can be expected to suffer fitness losses, and whilst natural selection would select for adaptations to increase synchronisation, climate change may outpace the rates of adaptation, causing decoupling and eventually extinction.

On the other side of the pollination equation, if a plant loses its host species due to mismatches in phenology, it will suffer pollination limitation, leading to loss in yields for crops species and potential local extinction for wild plant species (Fig. 1). For plants, the situation is likely less dire than for bees, for whilst there are numerous oligolectic bee species, there are few plants visited by a single pollinator (Jordano et al., 2006). However, in a given location, a plant species may indeed depend upon one or few species for pollination, which may occur for crop or horticultural species grown outside of their native range, or when the plant exists in a habitat that only supports one or a few pollinators. There are also examples of specialised bee–flower interactions exist (e.g. Cane, 2018; Macior, 1968; Houston et al., 1993; Pauw, 2006), and pollination service declines associated with pollinators assemblage declines (Burkle et al., 2013). Modelling has revealed that under climate change, plant extinctions are likely to cause cascading extinctions throughout ecological networks (Schleuning et al., 2016).

The cues for plant flowering are better well known than cues for emergence for most bee species. A recent study into the phenology of bees over time using records of bee activity for 67 species collected over 9 years in the Rocky Mountains found that the main driver for phenology of bees was climate, with emergence advancing with earlier snowmelt timing, whilst later phenophases were best explained by functional traits, in particular overwintering stage and nest location (Stemkovski et al., 2020). Concerningly, looking at flower bloom phenology over the same period found that bee phenology was less sensitive to climate variation than flower phenology, pointing to the potential for reduced synchrony.

A number of studies have reported phenological mismatches between plants and their pollinators under climate change. A natural experimental study on two *Osmia* species (*O. cornuta* and *O. bicornis*) - a genus that is known to be an effective pollinator of stone fruits and apples - found that warming temperatures advanced flowering more strongly than bee emergences, and bee emergences were advanced in males, but not females (Kehrberger and Holzschuh, 2019). Moreover, this phenological mismatch was posited to lead to pollination limitations under climate change for the red-listed plant *Pulsatilla vulgaris*.

Another study has revealed that phenological mismatches due to climate change will impair fitness of both an important crop, *Cucurbita pepo*, and its main pollinator, the oligolectic *Eucera pruinosa*, with impacts already apparent at a northern part of the bee's range (Willis Chan and Raine, 2021). Less than half of the bee population emerged by the time the crop flowering window ended, and only 31.4% of the flowering period supported the period when both male and female bees were present to reproduce.

A large 4-year study of 132 plant species and 665 pollinating insect species in a Mediterranean plant-pollinator community also found that years of earlier flower bloom onset were associated with lower species richness, portending to losses in pollinators and pollination services under ongoing climate change causing phenological mismatches (Petanidou et al., 2014).

Experimental studies have found even short temporal mismatches can have severe fitness consequences. Using flight cages, mismatches in phenology in terms of earlier emergence of 3 solitary species of bees than flowers were simulated, in terms of (i) a mismatch of 3 days and (ii) a mismatch of 6 days and (iii) control: bees and flowers occur simultaneously (Schenk et al., 2018). All species suffered major fitness losses under the 6-day mismatch, with females showing strongly reduced activity and reproductive output. Under the 3-day mismatch, the late-spring specialist *Osmia brevicornis* produced fewer brood cells, and although both the early-spring species *O. cornuta* and the mid-spring species *O. bicornis* produced the same number of brood cells, *O. cornuta* decreased the number of female offspring, and *O. bicornis* spread the brood cells over fewer nests, which may increase offspring mortality due to parasitoids.

One of the largest scale studies to model impacts on climate-change-induced mismatches between plant and pollinator phenology revealed that up to half of pollinator species could suffer from resource shortages (Memmott et al., 2007). The study, using a highly resolved empirical network of interactions between 1420 pollinator species and 429 plant species, simulated how a doubling of atmospheric CO_2 could be expected to cause phenological shifts between plants and pollinators. Depending on model assumptions, mismatches between activity periods led to 17–50% of pollinators having reduced floral resources, with up to half of the activity periods of flower visitors falling completely outside the time of their floral host blooms, as well as causing decreased diet breadth of pollinators. There has, however, also been research reporting no mismatch between bees and flowering plants. Long-term datasets have found that it appears that plants and bees are undergoing almost parallel shifts in phenology (Bartomeus et al., 2011).

2.3.2 Mismatches in distribution

As noted earlier, species may experience shifts in their current distribution under climate change. Of course, for pollinators, they must rely on their host plant(s) shifting to a similar extent. When it comes to agriculture, current locations where a crop is grown may no longer support the populations of its pollinator, or, conversely, if growers need to grow their crops elsewhere because the climate no longer supports the effective growth of that crop, its pollinators may not migrate to that new location.

Whilst range contractions under climate change are predicted, there have also been studies that suggest plants and pollinators will not experience spatial mismatches. For example, using species distribution modelling for 11 cactus species from the southern Central Andes and their spatial (mis)match with their pollinators (18 species and 11 genera of animals, including bees, birds, wasps and moths) under 2 future scenarios (RCP 4.5 and RCP 8.5) for the years 2050 and 2070, an increase in the spatial match with their mutualists under future scenarios was actually predicted (Gorostiague et al., 2018).

Spatial mismatches are also predicted for plant species that provide human nutrition. Tomatoes are nutritious fruits high in lycopene, vitamin C and other nutrients (Canene-Adams et al., 2005). As a species that requires sonication (buzz pollination), the current main managed pollinator *Apis mellifera* cannot pollinate tomatoes, and thus this crop, as well as other plants with poricidal anthers, are entirely reliant on other alternative pollinators, and consequently are at risk of pollination deficits due to pollinator mismatches. Using baseline (1961–90) climate and future (2050s and 2080s) climatic projections in ecological niche models fitted with occurrences of 15 species of wild bees documented in the literature as pollinators of tomatoes, baseline and future potential distribution of suitable climatic conditions of targeted species were used to create maps of richness change across North America and how this corresponded with tomato growing regions (Carrasco et al., 2021). Not only did some species exhibit declines, but up to 11 of these species were projected to decline in the Eastern USA. Furthermore, this is anticipated to have impact on tomato pollination, as overlapping maps of species richness change with tomato crop regions, richness declines corresponded with areas with a high concentration of tomato crops.

2.4 Extreme events

Climate change is predicted to not only increase mean temperatures but also increase the frequency and intensity of extreme events, such as heat waves, droughts, floods, cyclones and fires (Banholzer et al., 2014). Compared with the more chronic aspects of climate change, these acute aspects which represent 'SHOCKs' (events that are a Sudden, High-magnitude Opportunity for a Catastrophic 'Kick') are less well-studied, and, by their nature, are more difficult to study being erratic and unpredictable (Erenler et al., 2020; Van De Pol et al., 2017). Bees appear to be particularly vulnerable to increased frequency of hurricanes, cyclones and typhoons; prolonged or extreme drought or floods; increased frequency or severity of wildfires, and temperature extremes (Klein et al., 2017). A meta-analysis involving 117 studies (74% being mensurative) on natural hazards to pollinators and pollination in natural and cultivated systems found evidence of community and population-level impacts to plants and pollinators from climatological (extreme heat, fire, drought), hydrological (flooding), meteorological (hurricanes) and geophysical (including tsunamis) extreme events (Nicholson and Egan, 2020). Whilst the majority of cases did report negative impacts, 19% found no significant effect (Nicholson and Egan, 2020).

The effect of heat waves has been experimentally studied on a number of *Bombus* species, and has revealed that responses are species-specific, with widespread species being relatively tolerant, whereas declining species are sensitive to hyperthermic stress (Zambra et al., 2020). Similar results of differential sensitivity of widespread, relatively 'warm-adapted' *Bombus* and declining, cold-adapted *Bombus* have been reported in terms of how simulated heat waves impact fertility and attractiveness of *Bombus* males. Here, heat shock (40°C) had minimal impacts on the warm-adapted *Bombus terrestris,* but negatively impacted sperm viability and DNA integrity, as well as the structure of cephalic labial glands and the production of pheromones, potentially making males less 'sexy' for cold-adapted declining *B. magnus* and *B. jonellus* (Martinet et al., 2021).

Whilst the impact of flooding on native bees has not been explicitly studied in a climate change context, extreme flooding has been found to devastate populations of the ground-nesting bee *Andrena vaga* (Fellendorf et al., 2004).

Even if studying impacts of these extreme 'unreplicable' events will fail to meet 'gold standards' of well-designed, well-replicated studies with controls, doing so nevertheless is highly valuable to better understand how populations may respond to extreme events, especially when they punctuate a long-term study, underscoring the need for ongoing, long-term monitoring of pollinator populations (Erenler et al., 2020).

2.4.1 Going up in flames

Fire is a natural part of many (but not all) landscapes; however, climate change is predicted to increase the severity, intensity and frequency of large, hot fires (Barbero et al., 2015, Aponte et al., 2016). Already, much of Australia and California (USA) have experienced widespread, catastrophic fires linked with climate change (Abram et al., 2021, Boer et al., 2020). The impact of fires on bees has been relatively understudied, especially in regions like Australia (Prendergast, in prep). Impacts are likely to range widely depending on the extent of the fire, the season, previous fire history, fire frequency and the dominant type of vegetation (Prendergast, in prep). In some regions, fire can benefit bees (Simanonok and Burkle, 2019), whereas in other contexts it can be a detriment to bees (Frankie et al., 1997). A very clear example of severe negative impacts of climate change-induced fires on a native bee species can be seen in the population of *Xylocopa (Lestis) aerata* on Kangaroo Island, off South Australia (Glatz et al., 2015). This species is strongly associated with *Banksia* trees as a nesting substrate and if fires are too frequent, there is no renewal of this re-seeding species, which moreover takes decades to reach a maturity stage suitable for the bees to nest in. Whilst they also nest in *Xanthorrhoea* stalks, these are only suitable 2–5 years after fire (Glatz et al., 2015). Large-scale bushfires in 2007 resulted in the population dwindling (Glatz et al., 2015). Whilst installation of artificial nesting substrates aided in population recovery, breeding success declined after the first year of installation due to climate change: the next couple of years were characterised by low spring and summer rain, such that although artificial nesting habitat was now available, the food that the bees needed for survival and reproducing the next generation was not (Hogendoorn et al., 2021). *X. aerata* then suffered further losses as extreme large-scale bushfires again decimated its habitat in 2019–20 (Hogendoorn et al., 2021). With climate change causing drier conditions, impacting flowering and leading to increased frequency and severity of extreme wildfire events, there is a real fear the Kangaroo Island population of *X. aerata* will become extinct (Hogendoorn et al., 2021). Nevertheless, in the absence of extreme fire events, using MaxEnt algorithms to model both the current distribution and effect of warming scenarios on the distribution of both *X. aerata* and its main host plants, suitable areas, whilst contracting towards the southeast, are not predicted to cause extinction (Santos et al., 2020). Although most other species of Australian bees have not been monitored to track how their populations respond to fires, using publicly available collection and GIS datasets, combined with life-history data, modelling extinction risk for 553 Australian native bee species from the 2019–20 catastrophic Australian wildfires revealed marked impacts (Dorey et al., 2021). Based on fire overlap, intensity, frequency and life-history traits, two species are now eligible for listing as Endangered and nine as Vulnerable under IUCN criteria (Dorey et al., 2021).

2.5 Climate change altering pollinator cues

2.5.1 Not so attractive

Climate change may indirectly alter plant–pollinator relationships by causing changes in plant morphology or signals that pollinators use as cues to forage. This has seldom been studied explicitly in a climate change context; however, we can predict that climate change, through water stress, can induce decreases in plant height, flower number, size, bloom duration and colour, and alter olfactory signals may cause disruptive impacts (Descamps et al., 2021b).

However, CO_2 enrichment may have beneficial impacts on bees by increasing investment in flower abundance (Jablonski et al., 2002).

2.5.2 Not so rewarding?

Climate-change-induced water stress and drought are known to impact flower blooming, leading to flower abortion. It can also further reduce food rewards offered by flowers by reducing nectar volume, concentration or composition (Descamps et al., 2021b). However not all plants respond uniformly, and nectar production or sugar amount or concentration may even increase under drought (Descamps et al., 2021b). Under elevated temperatures plants including crop species such as soybeans have been found to decrease pollen production (Scaven and Rafferty, 2013). Pollen viability has also been shown to decrease under drought conditions (Descamps et al., 2021b). Experimental research on a bee-pollinated plant, *Impatiens glandulifera*, revealed that both heat and water stress reduced flowering resources, pointing to potential negative impacts on bees: the number of flowers was reduced by 40–90% in response to drought, heat stress reduced flower longevity, and higher temperatures led to reductions in nectar volume and protein content (Descamps et al., 2021a).

2.6 Synergisms

Climate change and its multiple components (warming, extreme events, altered rainfall patterns, increased CO_2) do not occur in isolation of other environmental (biotic and abiotic) and anthropogenic variables, which can act synergistically to amplify or counteract the impacts of one variable of climate change acting in isolation.

2.6.1 Warming, CO_2 enrichment and nitrogen deposition

Climate change is associated with both global warming as well as high CO_2 concentrations. In addition, many parts of the globe are experiencing increased nitrogen deposition. The combined impacts of climate warming, CO_2 enrichment and N-deposition can alter plant chemistry, altering floral attractiveness and/or nutritional rewards for pollinators. The interacting effects of these global environmental drivers were studied in a model system using pumpkins and their *Bombus* pollinators (Hoover et al., 2012). It was revealed that they non-additively affect flower morphology, phenology, flower sex ratios and nectar chemistry (sugar and amino acids), which in turn altered the attractiveness of nectar to *Bombus* pollinators. This had fitness impacts, reducing *Bombus* worker longevity. Surprisingly, *Bombus* responded maladaptively, being more attracted to and consuming greater amounts of nectar that reduced their survival by 22%.

2.6.2 Climate change and land-use change

Land cover is also predicted to shift under climate change. As a result, species may not only experience shifts in distribution but also may be constrained by the availability and distribution of suitable habitat. A study simulating future land-cover changes and temperature changes on the distribution of bumblebee species endemic to East Asia revealed stark and sobering results, with 17–27 (59–93 %) of the species exhibiting range contractions, leading to losses of species richness in areas, and warranting the status of species to be upgraded according to IUCN threatened species criteria (Naeem et al., 2019). Considering both climate and land-use change together is far more realistic, and has been shown to greatly alter predictions to species' distributions compared with climate-only and static land-use models (Marshall et al., 2018). Modelling the combined impacts of dynamic land use and land-cover change along with climate change on distribution patterns of 48 species of European *Bombus* revealed that whilst there was considerable variation among species in their responses, the majority of species were predicted to lose considerable range, particularly under the extreme growth scenario (i.e. 'Growth Applied Strategy'(GRAS) – IPCC A1FI; mean projected temperature rise in Europe at 2100 is 5.6°C; a maximum change scenario driven by policies of deregulation and economic growth), with a number of local extinctions (Marshall et al., 2018).

2.6.3 Drought and competition

Loss in flowering resources due to drought means resources will become limiting, setting the stage for resource competition, resulting in fitness losses to the inferior competitor if two or more species share overlapping foraging niches. There is strong evidence that drought exacerbates competitive effect of introduced honeybees on wild bees. Ten years of data revealed that increased *A. mellifera* abundance corresponded with decreased *Bombus* use of the most abundant forage plant, reduced diet overlap and reduced *Bombus* abundances in the following year, with decreases in floral abundance from drought and reduced spring rainfall contributing to *Bombus* declines (Thomson, 2016, 2021).

2.6.4 Pesticides

Climate change may exacerbate adverse effects from exposure to pesticides, which already threatens pollinators, especially in agricultural landscapes (Stuligross and Williams, 2020). A semi-natural experiment with *B. terrestris* exploring the interaction between exposure to sublethal levels of the neonicotinoid insecticide, clothianidin and temperature found that bees exposed to the pesticide and higher temperatures exhibited decreased foraging bouts, whereas bees without exposure to the pesticide were unaffected by temperature (Kolano et al., 2021).

2.6.5 Pathogens and parasites

By altering the distribution of species, climate change may bring new species into contact with pathogens that they have not evolved tolerance to, making them highly susceptible to severe negative impacts of novel viruses, bacteria and parasites. Additionally, when organisms are under stress, such as thermal stress, they can have lowered immune defences, leading them more vulnerable to infection or adverse impacts from pathogens. The majority of research on interactions between climate and disease has, similar to most threatening processes, focused on the non-threatened, domesticated *A. mellifera*. Whether we can extrapolate to other bees is questionable given their largely different life-history strategies. Climate change may facilitate shifts in honeybee pest distribution, such as the small hive beetle (*Aethina tumida*), or cause changes in the developmental cycle leaving it vulnerable to *Tropileaps* mite (Le Conte and Navajas, 2008).

Considering inter-species interactions is particularly important, as research has found that whilst some pollinators may benefit from the direct effects of warmer temperatures, this is nullified by the detrimental impacts warming has, thereby favouring predators, pathogens or parasites. This was demonstrated in research conducted over 3 years on the subalpine solitary cavity-nesting oligolectic bee, *Osmia iridis*. Whilst increased ambient temperatures were correlated with increased availability of floral resources leading to increased rates of nest provisioning, this did not translate to increased reproductive output. This was because reproductive output was most strongly constrained by increased rates of brood parasitism by *Sapyga* wasps under increased temperatures (Forrest and Chisholm, 2017).

3 Advancing our understanding of climate change impacts on pollinators

To date, most information on the impacts of climate change has been biased towards *Bombus* species. Solitary bee species, despite comprising the majority of bee species, have been less well researched. There has also been a strong Northern Hemisphere bias.

Whilst a challenge, understanding how climate change impacts whole communities is needed as species do not exist in single plant-pollinator pairs, but rather interact within a network. Climate change impacts on plant-pollinator networks are relatively understudied, but studies looking at natural variation have found that during dry seasons, networks become more vulnerable to disturbance due to the smaller pool of floral hosts available to pollinators (Rabeling et al., 2019). To prevent cascading extinctions, identifying keystone species that maintain robustness in the face of drought and climatic extremes may help target conservation efforts.

As has been evident in my review here, species do not respond uniformly to climate change. Trait-based approaches have provided insights into which guilds of bees respond differentially to threatening processes or environmental variables (Prendergast et al., 2021, De Palma et al., 2015), and will be valuable in identifying 'winners' and 'losers', so as to target conservation at the most at-risk species or predict shifts in pollinator community composition.

Common to wild bee conservation as a whole, for many pollinator species predicting the population-level impacts of climate change remains an extreme challenge as we lack basic baseline information regarding what variables limit populations. Greater investment into natural history observations of species, such as the cues that trigger emergence, or the suite of floral resources a species uses would be helpful and a greater understanding of the distribution of species is required. Again, whilst this information is relatively well documented in Europe and the USA, it is extremely poorly documented in other regions such as Australia, South America, Asia and Africa. Long-term monitoring with appropriate methods, and investment in taxonomic expertise (Prendergast and Hogendoorn, 2021), is also necessary to accurately document changes in populations sizes or assemblage composition under climatic changes.

A review of impacts of climate change on insects concluded that what matters is not just the trends in climate parameters, such as average warming, but also the variance, as well as frequency and intensity of extreme events (Boggs, 2016). As such, more dynamic models may be required.

Climate-change-induced altered timing, distributions, population declines and changing interactions of flower-visiting taxa may impair pollination services. Of greatest concern are flora with restricted ranges, and highly specialised pollination mutualisms. This has been studied in the sexually deceptive orchid *Leporella fimbriata*, which relies on a single pollen vector, an ant *Myrmecia urens* (Kolanowska et al., 2021). Modelling the niches of both plant and pollinator under climate change, whilst the orchid's distribution will be relatively unaffected, that of the ant will be considerably reduced, leading to a decrease in spatial overlap. Given that there are no alternative pollinators to fill this niche, this could pose a major threat to the persistence of *L. fimbriata* and lead to local extinctions.

Economically important plant species may also suffer from reduced pollination services under climate change, with research again emphasising that predicting impacts of climate change on any given species must include its ecological partners. Combining ecological niche models with simulations of tree cover loss and dispersal constraints, the forecasted distribution of Brazil nut tree (*Bertholletia excels*) was expected to increase by up to 6% by 2090. However, the Brazil nut tree was projected to suffer from reductions in the co-occurrence of its pollinator taxa (16 wild bee species) by up to almost 80%, with these mutualists suffering a loss of almost 50% of suitable habitat (Sales et al., 2021). Pollinator richness was also projected to decrease by 20%, which may also cause losses in pollination services due to loss of redundancy and resilience on pollination networks.

Already climate change mismatches between plants and pollinators are having consequences for plant reproduction. Earlier snowmelt in Japan is advancing flower onset of the spring ephemeral *Corydalis ambigua* and the emergence of queen *Bombus hypocrita sapporoensis*, a major pollinator. However, the plant was responding to a greater extent than the bee; so the earlier the snow melts, the greater the phenological mismatch, whereby flowering of *C. ambigua* commenced up to 1 week earlier than bee emergence (Kudo and Cooper, 2019). This had direct impacts on plant reproduction, where seed set in natural populations was approximately 60% with a minimal mismatch, but dropped to approximately 30% with a week mismatch. Experimental manipulations revealed this phenological mismatch arose due to plants and pollinators responding to different cues: flowering phenology of *C. ambigua* was determined by snowmelt date and subsequent ambient temperature; in contrast, bee emergence responded to belowground temperature, and following emergence, foraging activity was influenced by air temperature. It can be anticipated this mismatch will not only detriment plant fitness but also pollinator fitness: although *Bombus* are polylectic, spring ephemerals like *C. ambigua* are important for queens emerging from hibernation and produce large volumes of nectar.

Species can adapt and evolve in unexpected ways to climate change, including altering their morphology. Whilst this may be adaptive for the pollinator or the plant, it may harm the other partner in the mutualism. These complex responses were demonstrated in a study looking at morphological changes in two long-tongued bees, *Bombus balteatus* and *Bombus sylvicola*, from specimens collected from 1966 to 1980 and from 2012 to 2014 in the central Rocky Mountains (Miller-Struttmann et al., 2015). Between these two periods, over the 40 years, tongue length decreased by 0.61% annually and 24.4% cumulatively. These decreases in tongue length were not correlated with decreasing body size, competition from subalpine invaders, nor, as could be anticipated, coevolution with floral traits. Instead, the reduction in tongue length was associated with decreased floral resources occurring under warmer summers, which favoured a more generalised foraging strategy, with the species expanding their foraging niches compared with historical data. This strategy may enable the *Bombus* species to cope with climate change; however, it may have implications for flowers in the community which have long corollas, as they may suffer reduced pollination services.

4 Conserving pollinators under climate change

There has been scant research into testing how management actions may aid in protecting pollinators under climate change. Research has emphasised that providing good foraging resources (in quantity and quality) may ameliorate the negative impacts of climate change (and other stressors in general). An experiment with *B. terrestris* colonies demonstrated that whilst heat stress reduced colony development, resulting in a lower investment in offspring production, especially for small colonies, when small colonies were fed with high-quality diets (dominated by *Salix*, demonstrated as an excellent resource for *B. terrestris* colony development, with high total amino acid content), the adverse effects of heat stress were much less pronounced (Vanderplanck et al., 2019).

Given that research has found climate change and other threats can interact synergistically, reducing other stressors such as exposure to toxic pesticides, crop monocultures, lack of suitable wildflowers and nutritional stress, and pathogens may allow bee populations to be more resilient to ongoing climate change pressures. Species can also evolve adaptively under the natural selection pressures climate change poses, but this has been little studied, and how evolution will occur among multiple, interacting partners is challenging to predict (Gilman et al., 2012). Nevertheless, so that genetic drift does not counteract positive natural selection, we can assist populations to adapt to climate change by ensuring they have higher effective population sizes (enhancing population abundance and ensuring populations do not become isolated and inbred) (Packer and Owen, 2001).

As outlined earlier, one major way that organisms can respond to climate change is to migrate to stay within their climatic envelope. Such adaptive shifts in distribution rely upon suitable, continuous habitat, which is becoming increasingly rare as humans fragment natural landscapes through anthropogenic land use - namely urban and agricultural development (Aizen and Feinsinger, 1994, 2003; Botsch et al., 2017; Brosi et al., 2008; Donaldson et al., 2002; Hung et al., 2017). To provide pollinators with the best chance to migrate, habitat needs to be preserved and restored. There have been no explicit climate change pollinator corridors established; however, there are increasing incentives in agricultural and urban areas to create pollinator corridors. Such corridors, or pollinator highways, are important for connecting high-quality habitats, and enabling species to move through otherwise inhospitable landscape matrixes (Griffin and Haddad, 2021, Gutiérrez-Chacón et al., 2019).

Shifts in species' distributions under climate change, and changes in habitat suitability, means we cannot be assured that protecting habitats where species currently occur will remain important habitats for them in the future. Increasing the overall area of legislatively protected habitat therefore is needed.

At present, efforts to predict the impacts of climate change have largely relied on retrospective studies looking at shifts in historical to contemporary distributions or phenology, lab-based manipulations or modelling approaches. Translocation approaches also have drawbacks, but experiments translocating plants and/or pollinators outside of their current ranges in space and time may allow proactive management, and empirically test model predictions on how populations will fare in novel niche spaces (Morton and Rafferty, 2017).

Furthermore, in the vertebrate and plant realm, assisted migration or assisted translocation has been advocated as a proactive measure to enable species to stay within their climate envelope (Gallagher et al., 2015). Such proposals are not without dissidents, especially when species are moved outside their historic range; however, they also hold great promise (Prendergast, 2020b). There have been no plans for assisted migration of wild pollinators, but provided risks of disease and disruption of plant-pollinator networks of the resident community are considered, this may be a vital tool to prevent climate-change-mediated extinction of vulnerable species. However, any assisted migration must be approached with scientific vigour and should not be undertaken without a careful appraisal of the potential risks (Prendergast, 2020b), such as introduced species becoming invasive and exerting negative impacts on the recipient community (Mueller and Hellmann, 2008).

Given known reductions in rainfall, a common call to action for conserving pollinators is to put out water. However, this may not accrue any major benefits since most bees obtain all the moisture they need from nectar, instead favouring the competitive honeybees, which need water to evaporatively cool their hive.

Conservation of biological diversity will provide insurance and resilience for pollination services under climate change (Senapathi et al., 2021, Bartomeus et al., 2013). Because species do not respond uniformly to the various components of climate change, it is imperative to conserve biological diversity, at the species level as well as intraspecific genetic diversity.

Climate change has been noted to exacerbate the impact of competition, and therefore minimising these adverse inter-species interactions for at-risk species should be a priority. *A. mellifera* has received much public, financial and media attention due to its economic importance, and the fact that it is a widespread, adaptable species (Wood et al., 2020, Iwasaki and Hogendoorn, 2021). However, this mass-managed species can outcompete wild bees and have disruptive impacts on plant-pollinator networks (Prendergast and Ollerton, 2021a, Prendergast et al., 2021, Geslin et al., 2017, Thomson, 2016,

Aizen and Feinsinger, 1994, Aizen et al., 2008). During 'crunch' times when resources become scarce due to drought, these competitive impacts may be particularly detrimental to native bee fitness. Consequently, regulating honeybee densities, and ensuring there are areas dedicated to native bee needs can help mitigate the potential for adverse impacts of competition from honeybees under climate change.

Of course, the greatest action that we can all take is addressing climate change itself by reducing humanity's carbon footprint. This means less land clearing, and planting more plants, especially trees which capture large amounts of carbon (Donkersley, 2019) e.g. through agro-forestry (Kay et al., 2020). This can be a win-win if trees are planted that are highly attractive to bees, such as *Eucalyptus* and *Corymbia* in Australia (Prendergast and Ollerton, 2021b), *Croton floribundus* in Brazil (Ramalho, 2004), or *Acer platanoides* in Germany (Hausmann et al., 2015). Reducing the demand for environmentally destructive practices is also important, and as consumers, we can all do this by reducing or eliminating meat consumption, given how livestock farming is a major driver of land clearing, especially meat production that results in replacement of forested habitats and wildflower fields with non-entomophilous grasslands and grain pastures, and livestock's major contribution to greenhouse gas emissions (Prendergast, 2020c, Poore and Nemecek, 2018).

Climate change may be the greatest threat jeopardising wild bees. Indeed, a recent study analysing a long-term dataset of wild bee occurrences across three states in Northeast USA revealed that climate factors emerged as being even more significant drivers of wild bee abundance and richness than landscape composition and quality (Kammerer et al., 2021). In particular, spring and summer/fall wild bee abundance and species richness were influenced to a greater extent by temperature and precipitation than landscape composition, quality or topography. Climate change predictions of warmer winters, a greater intensity of rainfall in winter and spring, and higher maximum temperatures are expected to decrease bee abundance and have variable effects on species richness. Not all taxa were equally sensitive, with *Osmia* (spring bees) being particularly associated with lower abundance under higher precipitation. For the summer/fall bee community, solitary bees were more susceptible than eusocial bees to decline in abundance following hot summers and to decrease in species richness under drought conditions.

5 Conclusion

Climate change is one of the greatest threats biodiversity is facing now and into the future. The impacts on pollinators have the potential to be dire, with flow-on effects for the wider ecosystem, including agroecosystems that humans depend on for nutrition. In the worst-case scenario, populations and species of pollinators will become extinct, which may result in pollination deficits, but most importantly, a loss of unique, irreplaceable components of the diversity that has evolved on earth (Prendergast, 2020a). There is, however, the potential for pollinators and pollination networks to be buffered by plastic or genetic responses, or for novel interactions to occur, and some species will even benefit from climate change.

6 Where to look for further information

The Intergovernmental Panel on Climate Change (IPCC) is the United Nations body for assessing the science related to climate change. The IPCC produces comprehensive Assessment Reports detailing the state of scientific, technical and socio-economic knowledge on climate change, its impacts and future risks as well as on strategies for mitigating climate change and reducing the rate of climate change. The IPCC also produces Special Reports on topics as well as Methodology Reports that provide guidelines for preparing greenhouse gas inventories: https://www.ipcc.ch/.

NASA also has an overview of climate change, its impacts and solutions: https://climate.nasa.gov/effects/.

To provide pollinators with suitable native forage to connect habitats to facilitate migration, there are a number of regional guides for suitable bee wildflowers:

- Australia: Wheen Bee has collaborated with scientists to produce 'Powerful Pollinators Planting Guides': https://www.wheenbeefoundation.org.au/our-work/projects/powerful-pollinators/.
- USA: The Pollinator Partnership has developed 'Ecoregional Planting Guides': https://www.pollinator.org/guides.
- UK: Guides on flowering plants for pollinators have been developed by The Royal Horticultural Society: https://www.rhs.org.uk/science/conservation-biodiversity/wildlife/plants-for-pollinators.

'Food and Climate Change: Healthy diets for a healthier planet' produced by the United Nations is a good summary on the impact of dietary choices on an individual's climate footprint: https://www.un.org/en/climatechange/science/climate-issues/food#:~:text=Plant%2Dbased%20foods%20%E2%80%93%20such%20as,intensities%20than%20animal%2Dbased%20foods.

7 References

Abram, N. J., Henley, B. J., Gupta, A. S., Lippmann, T. J., Clarke, H., Dowdy, A. J., Sharples, J. J., Nolan, R. H., Zhang, T. and Wooster, M. J. 2021. Connections of climate change and variability to large and extreme forest fires in southeast Australia. *Communications Earth & Environment* 2, 1-17.

Aizen, M. A. and Feinsinger, P. 2003. Bees not to be? Responses of insect pollinator faunas and flower pollination to habitat fragmentation. *How Landscapes Change*. Springer.

Aizen, M. A. and Feinsinger, P. 1994. Habitat fragmentation, native insect pollinators, and feral honey bees in Argentine'Chaco Serrano. *Ecological Applications* 4(2), 378-392.

Aizen, M. A., Morales, C. L. and Morales, J. M. 2008. Invasive mutualists erode native pollination webs. *PLoS Biology* 6(2), e31.

Aponte, C., De Groot, W. J. and Wotton, B. M. 2016. Forest fires and climate change: causes, consequences and management options. *International Journal of Wildland Fire* 25(8), i-ii.

Banholzer, S., Kossin, J. and Donner, S. 2014. The impact of climate change on natural disasters. In: Singh, A., Zommers, Z. (eds) *Reducing Disaster: Early Warning Systems For Climate Change*. Springer, Dordrecht. pp. 21-49. https://doi.org/10.1007/978-94-017-8598-3_2

Barbero, R., Abatzoglou, J. T., Larkin, N. K., Kolden, C. A. and Stocks, B. 2015. Climate change presents increased potential for very large fires in the contiguous United States. *International Journal of Wildland Fire* 24(7), 892-899.

Bartomeus, I., Ascher, J. S., Wagner, D., Danforth, B. N., Colla, S., Kornbluth, S. and Winfree, R. 2011. Climate-associated phenological advances in bee pollinators and bee-pollinated plants. *Proceedings of the National Academy of Sciences of the United States of America* 108(51), 20645-20649.

Bartomeus, I., Park, M. G., Gibbs, J., Danforth, B. N., Lakso, A. N. and Winfree, R. 2013. Biodiversity ensures plant-pollinator phenological synchrony against climate change. *Ecology Letters* 16(11), 1331-1338.

Batra, S. W. T. 1995. Bees and pollination in our changing environment. *Apidologie* 26(5), 361-370.

Battarbee, R. W., Cope, P., Freeman, C., Lawton, J., Mace, G., Mackay, A. W., Read, D. and Shepherd, J. 2008. *Biodiversity-Climate Interactions: Adaptation, Mitigation and Human Livelihoods*. London: The Royal Society.

Biesmeijer, J. C. 1997. *The Organisation of Foraging in Stingless Bees of the Genus Melipona: An Individual-Oriented Approach*. Utrecht: Utrecht University.

Biesmeijer, J. C., Richter, J. A. P., Smeets, M. A. J. P. and Sommeijer, M. J. 1999. Niche differentiation in nectar-collecting stingless bees: the influence of morphology, floral choice and interference competition. *Ecological Entomology* 24(4), 380-388.

Boer, M. M., De Dios, V. R. and Bradstock, R. A. 2020. Unprecedented burn area of Australian mega forest fires. *Nature Climate Change* 10(3), 171-172.

Boggs, C. L. 2016. The fingerprints of global climate change on insect populations. *Current Opinion in Insect Science* 17, 69-73.

Botsch, J. C., Walter, S. T., Karubian, J., González, N., Dobbs, E. K. and Brosi, B. J. 2017. Impacts of forest fragmentation on orchid bee (Hymenoptera: Apidae: Euglossini) communities in the Chocó biodiversity hotspot of northwest Ecuador. *Journal of Insect Conservation* 21(4), 633-643.

Brosi, B. J., Daily, G. C., Shih, T. M., Oviedo, F. and Durán, G. 2008. The effects of forest fragmentation on bee communities in tropical countryside. *Journal of Applied Ecology* 45(3), 773-783.

Burkle, L. A., Marlin, J. C. and Knight, T. M. 2013. Plant-pollinator interactions over 120 years: loss of species, co-occurrence, and function. *Science* 339(6127), 1611-1615.

Cane, J. H. 2018. Co-dependency between a specialist Andrena bee and its death camas host, Toxicoscordion paniculatum. *Arthropod-Plant Interactions* 12(5), 657-662.

Canene-Adams, K., Campbell, J. K., Zaripheh, S., Jeffery, E. H. and Erdman Jr., J. W. 2005. The tomato as a functional food. *The Journal of Nutrition* 135(5), 1226-1230.

Caradonna, P. J., Cunningham, J. L. and Iler, A. M. 2018. Experimental warming in the field delays phenology and reduces body mass, fat content and survival: implications for the persistence of a pollinator under climate change. *Functional Ecology* 32(10), 2345-2356.

Carrasco, L., Papeş, M., Lochner, E. N., Ruiz, B. C., Williams, A. G. and Wiggins, G. J. 2021. Potential regional declines in species richness of tomato pollinators in North America under climate change. *Ecological Applications* 31(3), e02259.

Dainese, M., Martin, E. A., Aizen, M. A., Albrecht, M., Bartomeus, I., Bommarco, R., Carvalheiro, L. G., Chaplin-Kramer, R., Gagic, V., Garibaldi, L. A., Ghazoul, J., Grab, H., Jonsson, M., Karp, D. S., Kennedy, C. M., Kleijn, D., Kremen, C., Landis, D. A., Letourneau, D. K., Marini, L., Poveda, K., Rader, R., Smith, H. G., Tscharntke, T., Andersson, G. K. S., Badenhausser, I., Baensch, S., Bezerra, A. D. M., Bianchi, F. J. J. A., Boreux, V., Bretagnolle, V., Caballero-Lopez, B., Cavigliasso, P., Ćetković, A., Chacoff, N. P., Classen, A., Cusser, S., Da Silva E Silva, F. D., De Groot, G. A., Dudenhöffer, J. H., Ekroos, J., Fijen, T., Franck, P., Freitas, B. M., Garratt, M. P. D., Gratton, C., Hipólito, J., Holzschuh, A., Hunt, L., Iverson, A. L., Jha, S., Keasar, T., Kim, T. N., Kishinevsky, M., Klatt, B. K., Klein, A. M., Krewenka, K. M., Krishnan, S., Larsen, A. E., Lavigne, C., Liere, H., Maas, B., Mallinger, R. E., Martinez Pachon, E., Martínez-Salinas, A., Meehan, T. D., Mitchell, M. G. E., Molina, G. A. R., Nesper, M., Nilsson, L., O'Rourke, M. E., Peters, M. K., Plećaš, M., Potts, S. G., Ramos, D. L., Rosenheim, J. A., Rundlöf, M., Rusch, A., Sáez, A., Scheper, J., Schleuning, M., Schmack, J. M., Sciligo, A. R., Seymour, C., Stanley, D. A., Stewart, R., Stout, J. C., Sutter, L., Takada, M. B., Taki, H., Tamburini, G., Tschumi, M., Viana, B. F., Westphal, C., Willcox, B. K., Wratten, S. D., Yoshioka, A., Zaragoza-Trello, C., Zhang, W., Zou, Y. and Steffan-Dewenter, I. 2019. A global synthesis reveals biodiversity-mediated benefits for crop production. *Science Advances* 5(10), eaax0121.

De Palma, A., Kuhlmann, M., Roberts, S. P., Potts, S. G., Börger, L., Hudson, L. N., Lysenko, I., Newbold, T. and Purvis, A. 2015. Ecological traits affect the sensitivity of bees to land-use pressures in European agricultural landscapes. *Journal of Applied Ecology* 52(6), 1567-1577.

Descamps, C., Boubnan, N., Jacquemart, A. L. and Quinet, M. 2021a. Growing and flowering in a changing climate: effects of higher temperatures and drought stress on the bee-pollinated species Impatiens glandulifera Royle. *Plants* 10(5), 988.

Descamps, C., Quinet, M. and Jacquemart, A.-L. 2021b. The effects of drought on plant-pollinator interactions: what to expect? *Environmental and Experimental Botany* 182, 104297.

Dew, R. M., Silva, D. P. and Rehan, S. M. 2019. Range expansion of an already widespread bee under climate change. *Global Ecology and Conservation* 17, e00584.

Donaldson, J., Nänni, I., Zachariades, C. and Kemper, J. 2002. Effects of habitat fragmentation on pollinator diversity and plant reproductive success in renosterveld shrublands of South Africa. *Conservation Biology* 16(5), 1267-1276.

Donkersley, P. 2019. Trees for bees. *Agriculture, Ecosystems and Environment* 270-271, 79-83.

Dorey, J. B., Fagan-Jeffries, E. P., Stevens, M. I., Schwarz, M. P., Dorey, J. B., Fagan-Jeffries, E. P., Stevens, M. I., Schwarz, M. P., Dorey, J. B., Fagan-Jeffries, E. P., Stevens, M. I., Schwarz, M. P., Dorey, J. B., Fagan-Jeffries, E. P., Stevens, M. I., Schwarz, M. P., Dorey, J. B., Fagan-Jeffries, E. P., Stevens, M. I., Schwarz, M. P., Dorey, J. B., Fagan-Jeffries, E. P., Stevens, M. I., Schwarz, M. P., Dorey, J. B., Fagan-Jeffries, E. P., Stevens, M. I., Schwarz, M. P., Dorey, J. B., Fagan-Jeffries, E. P., Stevens, M. I., Schwarz, M. P., Dorey, J. B., Fagan-Jeffries, E. P., Stevens, M. I., Schwarz, M. P., Dorey, J. B., Fagan-Jeffries, E. P., Stevens, M. I., Schwarz, M. P., Dorey, J. B., Fagan-Jeffries, E. P., Stevens, M. I., Schwarz, M. P., Dorey, J. B., Fagan-Jeffries, E. P., Stevens, M. I. and Schwarz, M. P. 2020. Morphometric comparisons and novel observations of diurnal and low-light-foraging bees. *Journal of Hymenoptera Research* 79, 117–144.

Dorey, J. B., Rebola, C. M., Davies, O. K., Prendergast, K. S., Parslow, B. A., Hogendoorn, K., Leijs, R., Hearn, L. R., Leitch, E. J., O'reilly, R. L., Marsh, J., Woinarski, J. C. Z. and Caddy-Retalic, S. 2021. Continental risk assessment for understudied taxa post-catastrophic wildfire indicates severe impacts on the Australian bee fauna. *Global Change Biology* 27(24), 6551–6567.

Erenler, H. E., Gillman, M. P. and Ollerton, J. 2020. Impact of extreme events on pollinator assemblages. *Current Opinion in Insect Science* 38, 34–39.

Fellendorf, M., Mohra, C. and Paxton, R. J. 2004. Devasting effects of river flooding to the ground-nesting bee, Andrena vaga (Hymenoptera: Andrenidae), and its associated fauna. *Journal of Insect Conservation* 8(4), 311–312.

Forrest, J. R. and Chisholm, S. P. 2017. Direct benefits and indirect costs of warm temperatures for high-elevation populations of a solitary bee. *Ecology* 98(2), 359–369.

Frankie, G. W., Vinson, S. B., Rizzardi, M. A., Griswold, T. L., O'keefe, S. and Snelling, R. R. 1997. Diversity and abundance of bees visiting a mass flowering tree species in disturbed seasonal dry forest, Costa Rica. *Journal of the Kansas Entomological Society*, 70(4), 281–296.

Fründ, J., Zieger, S. L. and Tscharntke, T. 2013. Response diversity of wild bees to overwintering temperatures. *Oecologia* 173(4), 1639–1648.

Gallagher, R. V., Makinson, R. O., Hogbin, P. M. and Hancock, N. 2015. Assisted colonization as a climate change adaptation tool. *Austral Ecology* 40(1), 12–20.

Garibaldi, L. A., Carvalheiro, L. G., Vaissière, B. E., Gemmill-Herren, B., Hipólito, J., Freitas, B. M., Ngo, H. T., Azzu, N., Sáez, A., Åström, J., An, J., Blochtein, B., Buchori, D., Chamorro García, F. J., Oliveira da Silva, F., Devkota, K., Ribeiro, Mde F., Freitas, L., Gaglianone, M. C., Goss, M., Irshad, M., Kasina, M., Pacheco Filho, A. J., Kiill, L. H., Kwapong, P., Parra, G. N., Pires, C., Pires, V., Rawal, R. S., Rizali, A., Saraiva, A. M., Veldtman, R., Viana, B. F., Witter, S. and Zhang, H. 2016. Mutually beneficial pollinator diversity and crop yield outcomes in small and large farms. *Science* 351(6271), 388–391.

Garibaldi, L. A., Steffan-Dewenter, I., Winfree, R., Aizen, M. A., Bommarco, R., Cunningham, S. A., Kremen, C., Carvalheiro, L. G., Harder, L. D., Afik, O., Bartomeus, I., Benjamin, F., Boreux, V., Cariveau, D., Chacoff, N. P., Dudenhöffer, J. H., Freitas, B. M., Ghazoul, J., Greenleaf, S., Hipólito, J., Holzschuh, A., Howlett, B., Isaacs, R., Javorek, S. K., Kennedy, C. M., Krewenka, K. M., Krishnan, S., Mandelik, Y., Mayfield, M. M., Motzke, I., Munyuli, T., Nault, B. A., Otieno, M., Petersen, J., Pisanty, G., Potts, S. G., Rader, R., Ricketts, T. H., Rundlöf, M., Seymour, C. L., Schüepp, C., Szentgyörgyi, H., Taki, H.,

Tscharntke, T., Vergara, C. H., Viana, B. F., Wanger, T. C., Westphal, C., Williams, N. and Klein, A. M. 2013. Wild pollinators enhance fruit set of crops regardless of honey bee abundance. *Science* 339(6127), 1608–1611.

Geslin, B., Gauzens, B., Baude, M., Dajoz, I., Fontaine, C., Henry, M., Ropars, L., Rollin, O., Thébault, E. and Vereecken, N. J. 2017. Massively introduced managed species and their consequences for plant-pollinator interactions. *Advances in Ecological Research*, 57, 147–199.

Gilman, R. T., Fabina, N. S., Abbott, K. C. and Rafferty, N. E. 2012. Evolution of plant-pollinator mutualisms in response to climate change. *Evolutionary Applications* 5(1), 2–16.

Glatz, R., Leijs, R. and Hogendoorn, K. 2015. *Biology, Distribution and Conservation of Green Carpenter Bee (Xylocopa aeratus: Apidae) on Kangaroo Island*. South Australia: South Australia Museum.

Gorostiague, P., Sajama, J. and Ortega-Baes, P. 2018. Will climate change cause spatial mismatch between plants and their pollinators? A test using Andean cactus species. *Biological Conservation* 226, 247–255.

Griffin, S. R. and Haddad, N. M. 2021. Connectivity and edge effects increase bee colonization in an experimentally fragmented landscape. *Ecography* 44(6), 919–927.

Gutiérrez-Chacón, C., Valderrama-A, C. and Klein, A. 2019. Biological corridors as important habitat structures for maintaining bees in a tropical fragmented landscape. *Journal of Insect Conservation* 24(1), 187–197.

Hausmann, S. L., Petermann, J. S. and Rolff, J. 2015. Wild bees as pollinators of city trees. *Insect Conservation and Diversity* 9(2), 97–107.

Heinrich, B. 2013. *The Hot-Blooded Insects: Strategies and Mechanisms of Thermoregulation*. Berlin: Springer Science & Business Media.

Hoffmann, A. A., Rymer, P. D., Byrne, M., Ruthrof, K. X., Whinam, J., Mcgeoch, M., Bergstrom, D. M., Guerin, G. R., Sparrow, B., Joseph, L., Hill, S. J., Andrew, N. R., Camac, J., Bell, N., Riegler, M., Gardner, J. L. and Williams, S. E. 2019. Impacts of recent climate change on terrestrial flora and fauna: some emerging Australian examples. *Austral Ecology* 44(1), 3–27.

Hogendoorn, K., Glatz, R. V. and Leijs, R. 2021. Conservation management of the green carpenter bee Xylocopa aerata (Hymenoptera: Apidae) through provision of artificial nesting substrate. *Austral Entomology* 60(1), 82–88.

Hoover, S. E., Ladley, J. J., Shchepetkina, A. A., Tisch, M., Gieseg, S. P. and Tylianakis, J. M. 2012. Warming, CO2, and nitrogen deposition interactively affect a plant-pollinator mutualism. *Ecology Letters* 15(3), 227–234.

Houston, T., Lamont, B., Radford, S. and Errington, S. 1993. Apparent mutualism between Verticordia nitens and V-Aurea (Myrtaceae) and their oil-ingesting bee pollinators (Hymenoptera, Colletidae). *Australian Journal of Botany* 41(3), 369–380.

Hung, K.-L. J., Ascher, J. S. and Holway, D. A. 2017. Urbanization-induced habitat fragmentation erodes multiple components of temporal diversity in a Southern California native bee assemblage. *PLoS ONE* 12(8), e0184136.

Iwasaki, J. M. and Hogendoorn, K. 2021. How protection of honey bees can help and hinder bee conservation. *Current Opinion in Insect Science* 46, 112–118.

Jablonski, L. M., Wang, X. and Curtis, P. S. 2002. Plant reproduction under elevated CO2 conditions: a meta-analysis of reports on 79 crop and wild species. *New Phytologist* 156(1), 9–26.

James, R. R. and Pitts-Singer, T. L. 2008. *Bee Pollination in Agricultural Ecosystems*. Oxford: Oxford University Press on Demand.

Jordano, P., Bascompte, J. and Olesen, J. M. 2006. The ecological consequences of complex topology and nested structure in pollination webs. *Plant-Pollinator Interactions: From Specialization to Generalization*, Chicago: University of Chicago Press, 173–199.

Kammerer, M., Goslee, S. C., Douglas, M. R., Tooker, J. F. and Grozinger, C. M. 2021. Wild bees as winners and losers: relative impacts of landscape composition, quality, and climate. *Global Change Biology* 27(6), 1250–1265.

Kay, S., Kühn, E., Albrecht, M., Sutter, L., Szerencsits, E. and Herzog, F. 2020. Agroforestry can enhance foraging and nesting resources for pollinators with focus on solitary bees at the landscape scale. *Agroforestry Systems* 94(2), 379–387.

Kehrberger, S. and Holzschuh, A. 2019. Warmer temperatures advance flowering in a spring plant more strongly than emergence of two solitary spring bee species. *PLoS ONE* 14(6), e0218824.

Kevan, P. G., Clark, E. A. and Thomas, V. G. 1990. Insect pollinators and sustainable agriculture. *American Journal of Alternative Agriculture* 5(1), 13–22.

Klein, S. B., Cabirol, A. B., Devaud, J. M., Barron, A. B. and Lihoreau, M. 2017. Why bees are so vulnerable to environmental stressors. *Trends in Ecology and Evolution* 32(4), 268–278.

Kolano, P. J., Røyset Aarønes, M., Borgå, K. and Nielsen, A. 2021. Temperature sensitive effects of the neonicotinoid clothianidin on bumblebee (Bombus terrestris) foraging behaviour. *Journal of Pollination Ecology* 28, 138–152.

Kolanowska, M., Michalska, E. and Konowalik, K. 2021. The impact of global warming on the niches and pollinator availability of sexually deceptive orchid with a single pollen vector. *Science of the Total Environment* 795, 148850.

Kudo, G. and Cooper, E. J. 2019. When spring ephemerals fail to meet pollinators: mechanism of phenological mismatch and its impact on plant reproduction. *Proceedings. Biological Sciences* 286(1904), 20190573.

Le Conte, Y. and Navajas, M. 2008. Climate change: impact on honey bee populations and diseases. *Revue Scientifique et Technique* 27(2), 499–510.

Lenoir, J., Hattab, T. and Pierre, G. 2017. Climatic microrefugia under anthropogenic climate change: implications for species redistribution. *Ecography* 40(2), 253–266.

Macior, L. W. 1968. Pollination adaptation in *Pedicularis groenlandica*. *American Journal of Botany* 55(8), 927–932.

Marshall, L., Biesmeijer, J. C., Rasmont, P., Vereecken, N. J., Dvorak, L., Fitzpatrick, U., Francis, F., Neumayer, J., Ødegaard, F., Paukkunen, J. P. T., Pawlikowski, T., Reemer, M., Roberts, S. P. M., Straka, J., Vray, S. and Dendoncker, N. 2018. The interplay of climate and land use change affects the distribution of Eu bumblebees. *Global Change Biology* 24(1), 101–116.

Martinet, B., Zambra, E., Przybyla, K., Lecocq, T., Anselmo, A., Nonclercq, D., Rasmont, P., Michez, D. and Hennebert, E. 2021. Mating under climate change: impact of simulated heatwaves on the reproduction of model pollinators. *Functional Ecology* 35(3), 739–752.

Martínez-López, O., Koch, J. B., Martínez-Morales, M. A., Navarrete-Gutiérrez, D., Enríquez, E. and Vandame, R. 2021. Reduction in the potential distribution of bumble bees (Apidae: Bombus) in Mesoamerica under different climate change scenarios: conservation implications. *Global Change Biology* 27(9), 1772–1787.

Mayr, A. V., Peters, M. K., Eardley, C. D., Renner, M. E., Röder, J. and Steffan-Dewenter, I. 2020. Climate and food resources shape species richness and trophic interactions of cavity-nesting Hymenoptera. *Journal of Biogeography* 47(4), 854–865.

Memmott, J., Craze, P. G., Waser, N. M. and Price, M. V. 2007. Global warming and the disruption of plant-pollinator interactions. *Ecology Letters* 10(8), 710–717.

Miller-Struttmann, N. E., Geib, J. C., Franklin, J. D., Kevan, P. G., Holdo, R. M., Ebert-May, D., Lynn, A. M., Kettenbach, J. A., Hedrick, E. and Galen, C. 2015. Functional mismatch in a bumble bee pollination mutualism under climate change. *Science* 349(6255), 1541–1544.

Morton, E. M. and Rafferty, N. E. 2017. Plant-pollinator interactions under climate change: the use of spatial and temporal transplants. *Applications in Plant Sciences* 5(6), 1600133.

Mueller, J. M. and Hellmann, J. J. 2008. An assessment of invasion risk from assisted migration. *Conservation Biology* 22(3), 562–567.

Naeem, M., Liu, M., Huang, J., Ding, G., Potapov, G., Jung, C. and An, J. 2019. Vulnerability of East Asian bumblebee species to future climate and land cover changes. *Agriculture, Ecosystems and Environment* 277, 11–20.

National Oceanic And Atmospheric Administration, N. C. F. E. I. 2020. State of the Climate: Global Climate Report for Annual 2020. *State of the Climate*. Uwa: National Oceanic and Atmospheric Administration.

Nicholson, C. C. and Egan, P. A. 2020. Natural hazard threats to pollinators and pollination. *Global Change Biology* 26(2), 380–391.

Packer, L. and Owen, R. 2001. Population genetic aspects of pollinator decline. *Conservation Ecology* 5(1). https://www.jstor.org/stable/26271799#metadata_info_tab_contents.

Pauw, A. 2006. Floral syndromes accurately predict pollination by a specialized oil-collecting bee (Rediviva peringueyi, Melittidae) in a guild of South African orchids (Coryciinae). *American Journal of Botany* 93(6), 917–926.

Pereboom, J. J. and Biesmeijer, J. C. 2003. Thermal constraints for stingless bee foragers: the importance of body size and coloration. *Oecologia* 137(1), 42–50.

Pérez-Méndez, N., Andersson, G. K. S., Requier, F., Hipólito, J., Aizen, M. A., Morales, C. L., García, N., Gennari, G. P. and Garibaldi, L. A. 2020. The economic cost of losing native pollinator species for orchard production. *Journal of Applied Ecology* 57(3), 599–608.

Petanidou, T., Kallimanis, A. S., Sgardelis, S. P., Mazaris, A. D., Pantis, J. D. and Waser, N. M. 2014. Variable flowering phenology and pollinator use in a community suggest future phenological mismatch. *Acta Oecologica* 59, 104–111.

Poore, J. and Nemecek, T. 2018. Reducing food's environmental impacts through producers and consumers. *Science* 360(6392), 987–992.

Potts, S. G., Imperatriz-Fonseca, V., Ngo, H. T., Aizen, M. A., Biesmeijer, J. C., Breeze, T. D., Dicks, L. V., Garibaldi, L. A., Hill, R., Settele, J. and Vanbergen, A. J. 2016. Safeguarding pollinators and their values to human well-being. *Nature* 540(7632), 220–229.

Prendergast, K. A. 2018. Nesting Biology of *Megachile ignita* Smith, 1853 (Hymenoptera: Megachilidae) in artificial nesting blocks in urbanised southwestern Australia. *Australian Entomologist* 45, 139–148.

Prendergast, K. S. and Hogendoorn, K. 2021. FORUM: methodological shortcomings and lack of taxonomic effort beleaguer Australian bee studies. *Austral Ecology* 46(5), 880–884.

Prendergast, K. S. 2020a. Beyond ecosystem services as justification for biodiversity conservation. *Austral Ecology* 45(2), 141–143.

Prendergast, K. S. 2020b. Critiquing the notion of a species natural range in an era of unprecedented change. *Austral Ecology* 45, 672–674.

Prendergast, K. S. 2020c. Defending biodiversity through our diets. *Austral Ecology* 45(5), 520–522.

Prendergast, K. S., Dixon, K. W. and Bateman, P. W. 2021. Interactions between the introduced European honey bee and native bees in urban areas varies by year, habitat type and native bee guild. *Biological Journal of the Linnean Society* 133(3), 725–743.

Prendergast, K. S. and Ollerton, J. 2021b. Plant-pollinator networks in Australian urban bushland remnants are not structurally equivalent to those in residential gardens. *Urban Ecosystems* 24(5), 973–987.

Rabeling, S. C., Lim, J. L., Tidon, R., Neff, J. L., Simpson, B. B. and Pawar, S. 2019. Seasonal variation of a plant-pollinator network in the Brazilian cerrado: implications for community structure and robustness. *PLoS ONE* 14(12), e0224997.

Ramalho, M. 2004. Stingless bees and mass flowering trees in the canopy of Atlantic Forest: a tight relationship. *Acta Botanica Brasilica* 18(1), 37–47.

Reid, H. and Huq, S. 2005. Climate change-biodiversity and livelihood impacts. *Tropical Forests and Adaptation to Climate Change* 57, 70.

Sales, L. P., Rodrigues, L. and Masiero, R. 2021. Climate change drives spatial mismatch and threatens the biotic interactions of the Brazil nut. *Global Ecology and Biogeography* 30(1), 117–127.

Santos, A. A., Leijs, R., Picanço, M. C., Glatz, R. and Hogendoorn, K. 2020. Modelling the climate suitability of green carpenter bee (Xylocopa aerata) and its nesting hosts under current and future scenarios to guide conservation efforts. *Austral Ecology* 45(3), 271–282.

Scaven, V. L. and Rafferty, N. E. 2013. Physiological effects of climate warming on flowering plants and insect pollinators and potential consequences for their interactions. *Current Zoology* 59(3), 418–426.

Schenk, M., Krauss, J. and Holzschuh, A. 2018. Desynchronizations in bee-plant interactions cause severe fitness losses in solitary bees. *Journal of Animal Ecology* 87(1), 139–149.

Schleuning, M., Fründ, J., Schweiger, O., Welk, E., Albrecht, J., Albrecht, M., Beil, M., Benadi, G., Blüthgen, N., Bruelheide, H., Böhning-Gaese, K., Dehling, D. M., Dormann, C. F., Exeler, N., Farwig, N., Harpke, A., Hickler, T., Kratochwil, A., Kuhlmann, M., Kühn, I., Michez, D., Mudri-Stojnić, S., Plein, M., Rasmont, P., Schwabe, A., Settele, J., Vujić, A., Weiner, C. N., Wiemers, M. and Hof, C. 2016. Ecological networks are more sensitive to plant than to animal extinction under climate change. *Nature Communications* 7, 13965.

Senapathi, D., Fründ, J., Albrecht, M., Garratt, M. P. D., Kleijn, D., Pickles, B. J., Potts, S. G., An, J., Andersson, G. K. S., Bänsch, S., Basu, P., Benjamin, F., Bezerra, A. D. M., Bhattacharya, R., Biesmeijer, J. C., Blaauw, B., Blitzer, E. J., Brittain, C. A., Carvalheiro, L. G., Cariveau, D. P., Chakraborty, P., Chatterjee, A., Chatterjee, S., Cusser, S., Danforth, B. N., Degani, E., Freitas, B. M., Garibaldi, L. A., Geslin, B., de Groot, G. A., Harrison, T., Howlett, B., Isaacs, R., Jha, S., Klatt, B. K., Krewenka, K., Leigh, S., Lindström, S. A. M., Mandelik, Y., McKerchar, M., Park, M., Pisanty, G., Rader, R., Reemer, M., Rundlöf, M., Smith, B., Smith, H. G., Silva, P. N., Steffan-Dewenter, I., Tscharntke, T., Webber, S., Westbury, D. B., Westphal, C., Wickens, J. B., Wickens, V. J., Winfree, R., Zhang, H. and

Klein, A. M. 2021. Wild insect diversity increases inter-annual stability in global crop pollinator communities. *Proceedings. Biological Sciences* 288(1947), 20210212.

Simanonok, M. P. and Burkle, L. A. 2019. Nesting success of wood-cavity-nesting bees declines with increasing time since wildfire. *Ecology and Evolution* 9(22), 12436-12445.

Slominski, A. H. and Burkle, L. A. 2019. Solitary bee life history traits and sex mediate responses to manipulated seasonal temperatures and season length. *Frontiers in Ecology and Evolution* 7. https://www.frontiersin.org/articles/10.3389/fevo.2019.00314/full.

Somerville, D. C. 2020. Management of bee nutrition in drought and fire affected areas [Online]. *Australia: AgriFutures Australia*. Available at: https://www.agrifutures.com.au/wp-content/uploads/2020/04/20-004.pdf [Accessed 26/07/2021].

Soroye, P., Newbold, T. and Kerr, J. 2020. Climate change contributes to widespread declines among bumble bees across continents. *Science* 367(6478), 685-688.

Stemkovski, M., Pearse, W. D., Griffin, S. R., Pardee, G. L., Gibbs, J., Griswold, T., Neff, J. L., Oram, R., Rightmyer, M. G., Sheffield, C. S., Wright, K., Inouye, B. D., Inouye, D. W. and Irwin, R. E. 2020. Bee phenology is predicted by climatic variation and functional traits. *Ecology Letters* 23(11), 1589-1598.

Stuligross, C. and Williams, N. M. 2020. Pesticide and resource stressors additively impair wild bee reproduction. *Proceedings. Biological Sciences* 287(1935), 20201390.

Theobald, J. C., Coates, M. M., Wcislo, W. T. and Warrant, E. J. 2007. Flight performance in night-flying sweat bees suffers at low light levels. *Journal of Experimental Biology* 210(22), 4034-4042.

Thomson, D. M. 2016. Local bumble bee decline linked to recovery of honey bees, drought effects on floral resources. *Ecology Letters* 19(10), 1247-1255.

Thomson, D. M. 2021. Novel data support model linking floral resources and honey bee competition with bumble bee abundances in coastal scrub. *Journal of Pollination Ecology* 27, 47-56.

Van De Pol, M., Jenouvrier, S., Cornelissen, J. H. C. and Visser, M. E. 2017. Behavioural, ecological and evolutionary responses to extreme climatic events: challenges and directions. *Philosophical Transactions of the Royal Society of London. Series B, Biological Sciences* 372(1723), 20160134.

Vanderplanck, M., Martinet, B., Carvalheiro, L. G., Rasmont, P., Barraud, A., Renaudeau, C. and Michez, D. 2019. Ensuring access to high-quality resources reduces the impacts of heat stress on bees. *Scientific Reports* 9(1), 12596.

Willis Chan, D. S. and Raine, N. E. 2021. Phenological synchrony between the hoary squash bee (Eucera pruinosa) and cultivated acorn squash (Cucurbita pepo) flowering is imperfect at a northern site. *Current Research in Insect Science* 1, 100022.

Wood, T. J., Michez, D., Paxton, R. J., Drossart, M., Neumann, P., Gérard, M., Vanderplanck, M., Barraud, A., Martinet, B., Leclercq, N. and Vereecken, N. J. 2020. Managed honey bees as a radar for wild bee decline? *Apidologie* 51(6), 1100-1116.

Zambra, E., Martinet, B., Brasero, N., Michez, D. and Rasmont, P. 2020. Hyperthermic stress resistance of bumblebee males: test case of Belgian species. *Apidologie* 51(5), 911-920.

Chapter 5

Assessing the impact of disease on pollinators

Patrycja Pluta and Robert J. Paxton, Martin Luther University Halle-Wittenberg, Germany

1 Introduction

Spurred by the report of a sensational 75% loss of insect biomass over a period of 27 years around the city of Krefeld in northwest Germany (Hallmann et al., 2017), more recent analyses of insect trends strongly support a worldwide decline in terrestrial insect biodiversity as well as biomass (van Klink et al., 2020; Wagner et al., 2021). As insects, particularly bees, are the most important pollinators of the majority of crops requiring pollination (Klein et al., 2007) and as there is an ongoing agricultural expansion of pollinator-dependent crops (Aizen et al., 2019), the decline of insects has obvious negative implications for pollination and crop yields in agricultural settings.

Insect decline is often seen as a multifactorial problem, itself associated with agricultural expansion (Raven and Wagner, 2021). Chief among the factors causing the decline of insect pollinators are thought to be (i) lack of habitat and associated essential resources (flowers providing pollen and nectar, suitable nesting sites), (ii) pesticides and (iii) disease organisms (Brown and Paxton, 2009; Vanbergen and the Insect Pollinators Initiative, 2013; Goulson et al., 2015; Potts et al., 2016). The relative importance of these three likely varies from

http://dx.doi.org/10.19103/AS.2022.0111.13

pollinator species to pollinator species, though pests and pathogens feature prominently (Dicks et al., 2021).

In agricultural settings, where managed bees such as the ubiquitous western honey bee (*Apis mellifera* L.) are typically employed for pollination (McGregor, 1976; Kevan et al., 1990; Free, 1993; Delaplane and Mayer, 2000), two factors may promote the impact of pathogens: host density and host dispersal. Firstly, elevated densities can facilitate pathogen transmission and theoretically result in higher disease prevalence (Anderson and May, 1981). Pollinators are often promoted in or near crops requiring pollination (Free, 1993), and their increased density clearly represents a threat in terms of enhanced disease transmission to conspecifics and, through pathogen spillover, to heterospecifics (Gisder and Genersch, 2017). Secondly, translocation has led to the dispersal and subsequent emergence of many pathogens in plants, animals and humans (Daszak et al., 2000). Pollinators are not exempt. The transport of managed pollinators, particularly *Apis mellifera* and the bumble bee *Bombus terrestris*, is likely to blame for the dispersal and emergence of infectious diseases that threaten conspecifics and heterospecifics (Cameron et al., 2011; Schmid-Hempel et al., 2014; Wilfert et al., 2016) alike, with knock-on consequences for the supply of pollination services to crops.

In this chapter we focus on diseases and their impact on insect pollinators; the direct impacts of habitat (degradation and loss) and pesticides are dealt with elsewhere in this book. We furthermore focus on bee pests and pathogens (disease-causing organisms) because of the importance of bees in crop pollination (Klein et al., 2007) and because most is known about their pests and pathogens.

2 A bestiary of honey bee diseases

Apis mellifera is arguably the best monitored of insects and numerically the most important commercial pollinator worldwide. The world stock of beehives (colonies) has grown consistently over the past 6 decades to around 90 million in the last 4 years (2016–2019; see Fig. 1a), likely driven by the agricultural expansion in pollinator-dependent crops (Aizen et al., 2008; Aizen and Harder, 2009). The logic of the argument is that farmers growing more, or a greater area of, pollinator-dependent crops like almonds and apples may pay for pollination services, encouraging beekeeping as an economic activity and thereby boosting the number of managed honey bee colonies.

Although these data might suggest that we should not be concerned about the world's population of honey bees, they belie the high annual colony losses suffered by beekeepers over the past 2 decades, particularly in Northern temperate regions of the world (Osterman et al., 2021). The massive die-off of US honey bees in the winter of 2006/2007 brought the issue of colony losses to

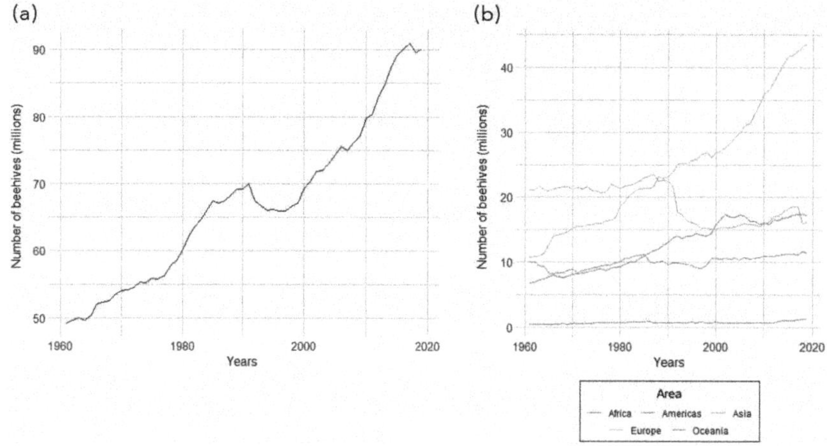

Figure 1 (a) The world stock of honey bee colonies (hives) from 1961 to 2019; (b) the number of colonies split by geographic regions (data from faostat: https://www.fao.org/faostat/en/#home).

world media and political attention (Oldroyd, 2007). Since 2007, North America has reported annual (overwinter) colony losses of 26%, including >40% losses in the last 3 years (2017/2018, 2018/2019, 2019/2020; https://beeinformed.org/citizen-science/loss-and-management-survey/) whilst those in Europe since 2007 have been around 16%, all well above 11% elsewhere in the world (Osterman et al., 2021) or the 'normal' expected mortality of ca. 10% (Chauzat et al., 2016). Part of the apparent discrepancy between the upward trend in the standing stock of honey bee colonies and high annual losses is because colony gains are in subtropical or tropical regions of the world (China, South America; Fig. 1b), whereas losses are in Northern temperate regions (Osterman et al., 2021), part is because beekeepers make up for colony losses by multiplying their surviving colonies.

Regardless of these arguments over global colony dynamics, there is widespread consensus that disease organisms (pests and pathogens) are a major cause of elevated honey bee colony mortality (Cox-Foster et al., 2007; Genersch et al., 2010; Johnson, 2010; Staveley et al., 2014; Barron, 2015; Chauzat et al., 2016). We now describe the major disease organisms of honey bees and their impact on hosts (for summary, see Table 1). Our list is not comprehensive; e.g. we have not addressed the havoc currently wreaked on honey bees by Asian hornet (*Vespa velutina*) predation in NW Europe (Keeling et al., 2017). Rather, we focus on pests and pathogens with a broad distribution or with the potential to cause considerable harm in the near future.

Table 1 Major parasites and pests of the western honey bee (*Apis mellifera*)

Disease/common name	Causative agent
Mites	
Varroosis	*Varroa destructor*
Tropilaelaps mite	*Tropilaelaps* spp.
Acarapisosis/tracheal mite	*Acarapis woodi*
Viruses	
Deformed wing virus	Deformed wing virus
Acute bee paralysis virus	Acute bee paralysis virus
Israeli acute paralysis virus	Israeli acute paralysis virus
Kashmir bee virus	Kashmir bee virus
Slow bee paralysis virus	Slow bee paralysis virus
Chronic bee paralysis virus	Chronic bee paralysis virus
Sacbrood virus	Sacbrood virus
Black queen cell virus	Black queen cell virus
Fungi	
Nosemosis type A	*Nosema apis*
Nosemosis type C	*Nosema ceranae*
Chalkbrood	*Ascophaera apis*
Bacteria	
American foulbrood	*Paenibacillus larvae*
European foulbrood	*Melisococcus plutonius*
Trypanosomatids	
Crithidia	*Crithidia mellificae*
Lotmaria	*Lotmaria passim*
Coleoptera	
Small hive beetle	*Aethina tumida*
Lepidoptera: Pyralidae	
Greater wax moth	*Galleria mellonella*
Lesser wax moth	*Achroia grisella*

2.1 Ectoparasitic mites

Social insects have long been known to harbour a diversity of parasitic mites, and the honey bee is no exception (Eickwort, 1990). We briefly detail three ectoparasitic mites of honey bees that are particularly well studied, in part because of their very serious impact on honey bee health.

2.1.1 Varroa destructor

The exotic mite *Varroa destructor* is a well-known ectoparasite of honey bees, present in almost every apiary worldwide, with the exception of Australia (Traynor et al., 2020). Successfully tolerated by its original host, the eastern honey bee *Apis cerana*, after jumping over to *Apis mellifera* the mite has been held responsible - directly or indirectly - for the elevated colony losses of managed honey bees (Cox-Foster et al., 2007; Genersch et al., 2010; Rosenkranz et al., 2010; Nazzi and Le Conte, 2016).

A direct, detrimental effect of the mite is caused by its feeding behaviour. The parasite invades host pupae in brood cells as well as adult honey bees and feeds on their haemolymph and fat bodies (Ramsey et al., 2019), resulting in lowered body weight and water content. Heavy infestation of the brood cells may result in the death of pupae. Low body weight is associated with decreased sperm production in drones, which hampers their reproductive performance (Duay et al., 2002) and subsequently queen reproductivity. Feeding by mites on adult workers results in flight impairment of the host (Duay et al., 2002) and loss of orientation, thus severely affecting homing (Kralj and Fuchs, 2006). Besides the effects inflicted on individuals, varroa mites compromise the colony; the decreased number of emerging honey bees, coupled with the lower performance of foragers and reproductives (queens and drones), inexorably leads to colony decline (Neumann and Carreck, 2010).

Although varroa mites alone can potentially impair colony fitness, the major threat they pose arises from the pathogens - particularly viruses - they transmit whilst feeding on hosts (Yang and Cox-Foster, 2005; Neumann and Carreck, 2010). To date, a plethora of viruses is thought to be vectored by varroa mites (Levin et al., 2019; Beaurepaire et al., 2020; Yañez et al., 2020), some of which elicit severe disease. The presence of *V. destructor* has therefore shaped the current distribution of mite-transmitted viruses across the world, facilitating viral spread and, through a change in the route of viral transmission, leading to increased severity of viral disease (virulence) (Martin et al., 2012; Mondet et al., 2014).

Three other *Varroa* species, *jacobsoni*, *rindereri* and *underwoodi*, are known from south and east Asian honey bee species, though they are rarely associated with *Apis mellifera* (Chantawannakul et al., 2016). The exception is *V. jacobsoni* that parasitises *Apis mellifera* colonies in Papua New Guinea. Two *Euvarroa* species, *sinhai* and *wongsirii*, are hosted by Asian honey bees and occasionally by *Apis mellifera* imported into the region, though the biology of these mites is less well known (Chantawannakul et al., 2016) and their impacts on hosts are likely modest.

2.1.2 *Tropilaelaps* spp.

Tropilaelaps is a genus of ectoparasitic mite whose primary hosts are honey bee species native to tropical and subtropical south and east Asia, but which can now also be found on *Apis mellifera* imported into the region (de Guzman et al., 2017). Despite their high rates of reproduction than can exceed those of varroa mites, they have not (yet) been recorded widely outside of their native range. Of four known species in the genus *Tropilaelaps*, namely *mercedesae*, *clerae*, *thaii* and *koenigerum*, the first two are the most prevalent and have successfully jumped over to *Apis mellifera* after its introduction to South and East Asia (Chantawannakul et al., 2016). The primary host of *T. mercedesae* is *Apis dorsata*, a widely distributed honey bee species of South and East Asia (Smith, 1991). Contrastingly, *T. clerae* and its primary hosts, *Apis dorsata binghami* and *Apis breviligula*, are restricted in distribution, and thus the mite is less prevalent than *T. mercedesae*. *Tropilaelaps thaii* and *T. koenigerum* are less well known (for further details, see Chantawannakul et al., 2016).

Tropilaelaps spp. and *V. destructor* share similarities in the life cycle, feeding behaviour and negative effects on honey bee hosts. An important distinction between these two mite genera is the inability of *Tropilaelaps* spp. to feed on adult honey bees due to their soft chelicerae (mouthparts) (Koeniger et al., 1988). This drastically shortens mite survivorship outside brood cells. This difference in biology means that *Tropilaelaps* spp. may not be able to spread to temperate regions of the world where honey bees enter a broodless period during the winter months.

Colonies infested with *Tropilaelaps* spp. mites suffer from a decreased number of emerging workers; those that do emerge are often crippled and weakened. The immunity of individual hosts is diminished due to direct damage by mite feeding and the lack of proteins that are harvested by the mite from pupal haemolymph (Khongphinitbunjong et al., 2015). The health status of a single, parasitised honey bee projects onto the general fitness of the colony and heavy infestation can lead to colony collapse. *Tropilaelaps* spp. mites, such as varroa mites, vector honey bee viruses (Dainat et al., 2009; Forsgren et al., 2009).

2.1.3 Acarapis woodi

Acarapis woodi is a microscopic parasitic mite inhabiting the tracheal tubes of young honey bees and the base of the wings of older honey bees, whose stiff tracheal hairs are too hard for the mite to penetrate into the host's tracheae. Initially discovered in the United Kingdom in the early 1900s, the mite has been reported across all continents (Matheson, 1993), although in many countries *Acarapis woodi* outbreaks occur only sporadically. Currently, it is not widely reported (Traynor et al., 2016), possibly because of widespread miticide use to treat colonies against varroa mites. The life cycle of *Acarapis woodi* is limited (exclusively for males and almost exclusively for females) to tracheal tubes, where mites pierce the tracheal tubes to feed on host haemolymph, with a short outside phoretic period reserved for post-mating host switching by females. Acarapisosis usually proceeds covertly and elicits symptoms only after heavy infestation. Affected honey bees are unable to fly and can be seen crawling around the hive floor, the wings may be displaced and the abdomens are elongated.

2.2 Viruses

Honey bees host a rich diversity of viruses. To date, over 70 have been associated with them (McMenamin and Genersch, 2015; Grozinger and Flenniken, 2019; Beaurepaire et al., 2020), though only a small fraction of them has been studied in any depth. Most honey bee viruses are short (ca. 10 kb), positive sense, single-stranded RNA or (+)ssRNA viruses that produce a negative RNA strand using the machinery of the host cell to replicate. Because of their small genome size, they lack a good mechanism of RNA repair during replication and therefore have high mutation rates and likely a high capacity to evolve and adapt, e.g. to a change in transmission pathway or to novel hosts (Holmes, 2009).

2.2.1 Deformed wing virus

The best-studied (Martin and Brettell, 2019) and arguably the most important virus in terms of its impact, deformed wing virus (DWV), has long been reported in honey bees, even before varroa infestation (Bailey and Ball, 1991). The mite's presence changed DWV's main route of transmission from (presumed) faecal-oral to indirect via the varroa mite as a vector (Yañez et al., 2020). This change precipitated a dramatic increase in DWV's prevalence (number of infected colonies) and pathogen load (amount of virus per honey bee) (Martin et al., 2012; Mondet et al., 2014) as well as virulence (harm caused to the host) (Bowen-Walker et al., 1999; Nazzi et al., 2012; Di Prisco et al., 2016). DWV is notoriously found in collapsing colonies and is associated with high overwinter mortality rates in honey bees (Dainat et al., 2012; Francis et al., 2013; Natsopoulou et al., 2017). DWV and its transmission by varroa mites are therefore the leading causes of elevated honey bee colony mortality in the past 2 decades, at least in temperate regions of the world.

DWV, a (+)ssRNA virus, belongs to the family Iflaviridae (Picornavirales) and comprises three variants or genotypes: the first detected genotype A (Bailey and Ball, 1991), genotype B, previously known as *Varroa destructor virus 1* (VDV-1) (Ongus et al., 2004), and genotype C, the most recently discovered although rarely reported (Mordecai et al., 2016). Current opinion is that DWV-genotype A (DWV-A) is merely mechanically transmitted to honey bees, whereas DWV-genotype B (DWV-B) can also replicate inside *V. destructor* and therefore the mite acts as a biological vector for it (Gisder and Genersch, 2021). These subtleties aside, it is important to bear in mind that both DWV-A and DWV-B are virulent when vectored by varroa mites and can lead to colony collapse (McMahon et al., 2016).

In line with the nature of other RNA viruses, DWV is characterised by high rates of mutation and recombination, which potentially result in the development of new variants and recombinants (Moore et al., 2011; Ryabov et al., 2014). The comparative virulence among DWV genotypes is currently unclear. Whilst laboratory experiments demonstrate higher virulence of DWV-B over that of genotype A in adults (McMahon et al., 2016; for field experiments, see also Norton et al., 2021), there is no obvious fitness or mortality differences between genotypes in pupae (Tehel et al., 2019). Initial mild and covert infections transmitted via the oral route are aggravated when the virus is injected directly into the haemolymph/haemocoel of honey bee pupae or adults by the mite. The classic symptom of DWV is the presence of wing deformities in emerging honey bees when these have been infected as pupae (by varroa feeding), although the pathogenesis of DWV remains unclear; high viral titres can also be found in freshly emerging honey bees with normal wings (Tehel et al., 2019; Gusachenko et al., 2020). Honey bees infected as adults clearly cannot exhibit deformed wings because their wings were already fully formed during pupation. The lack of penetrance of the trait 'deformed wings' in freshly emerging honey bees and the lack of other clear DWV symptoms make it difficult to recognise that a colony of honey bees is suffering from a DWV infection. DWV prevalence is typically very high (Genersch et al., 2010; Traynor et al., 2016); potentially all colonies are infected when infested by varroa mites.

2.2.2 Acute bee paralysis virus, Israeli acute paralysis virus and Kashmir bee virus

Acute bee paralysis virus (ABPV), Israeli acute paralysis virus (IAPV) and Kashmir bee virus (KBV) are a group of very closely related (+)ssRNA viruses in the family Dicistroviridae (Bailey et al., 1963; for review: de Miranda et al., 2010). All three may cause covert infections and maintain low viral titres. However, all three, like DWV, can be transmitted by varroa mites (Chen et al., 2004; Ball and Allen, 1988); varroa mite transmission serves as an activator of viral replication, leading to acute onset with a dramatic increase in mortality and colony loss (Ball and Allen, 1988).

Typical symptoms of infection by these Dicistroviridae include impaired flight, trembling progressing to paralysis and loss of hair, which gives honey bees a dark appearance. These characteristics, however, can be difficult to notice at the colony level due to the rapid death of diseased individuals. ABPV has been suggested as a cause of high honey bee mortality since the host shift of varroa mites to *Apis mellifera* (Ball and Allen, 1988; Bakonyi et al., 2002). Moreover, IAPV was one of the main suspects of so-called Colony Collapse Disorder (the mass die-off of colonies) that took North American beekeepers and scientists by surprise with a sudden wave of colony depopulation in winter 2006/2007 (vanEngelsdorp et al., 2009). However, these three viruses are so virulent when transmitted by varroa mites that they kill pupae in the pupal stage, where they die un-opened. In the short term, then, these viruses can lead to colony loss. But, in the long term, they are so virulent when vectored by varroa mites that they are not transmitted by infected pupae because pupae die in their sealed brood cells. In effect, these viruses eliminate themselves from a honey bee population (Martin, 2001; Sumpter and Martin, 2004; McMahon et al., 2018) and their prevalence is typically very low (Genersch et al., 2010; Traynor et al., 2016).

2.2.3 Slow bee paralysis virus

Slow bee paralysis virus (SBPV) is a (+)ssRNA virus of the family Iflavirus infecting a variety of bees, managed and wild (de Miranda et al., 2010). Transmission by varroa mites results in high viral titres and subsequent high colony mortality (Carreck et al., 2010). Other routes of transmission result in only covert infections. Affected honey bees exhibit paralysis in two anterior leg pairs 10 days after infection. After a spike in mortality closely following varroa introduction (Carreck, 2005; Carreck et al., 2010), SBPV is now considered a rare virus (McMahon et al., 2015) and has not been detected in a multi-year survey of honey bees in the United States (Traynor et al., 2016).

2.2.4 Chronic bee paralysis virus

Chronic bee paralysis virus (CBPV) is an unclassified (+)ssRNA virus with wide distribution and apparently increasing prevalence (Traynor et al., 2016). It has a distinct set of symptoms comprised of two syndromes that may occur simultaneously in the colony; type 1 syndrome is characterised by trembling, paralysis, distended abdomen and loss of flight ability (crawling). Type 2 syndrome results in loss of hair on the abdomen, which gives honey bees a black, shiny appearance (Dittes et al., 2020a). Moreover, affected honey bees are refused entry to the hive and can even be attacked by healthy hive-mates, which may create an illusion to beekeepers of robbing. The symptomatic similarity to ABPV lessens with the length of host survival; CBPV-affected honey bees can live for several days, whereas honey bees infected with ABPV die already after 1–2 days. Despite its noticeable symptoms, CBPV is rarely thought to result in the collapse of a colony but can impair its performance due to a lack of healthy, capable worker honey bees.

2.2.5 Sacbrood virus

Sacbrood virus (SBV) is another varroa-transmitted (+)ssRNA virus in the family Iflaviridae (Shen et al., 2005). Its name derives from the typical appearance of infected honey bee larvae which fail to pupate, accumulating fluid within the unshed skin, which resembles a sac. Subsequently, the dead larvae dry out and darken, acquiring a gondola-like shape and forming a scale at the base of the brood cell. Loss of brood at the point of pupation places the colony in a disadvantageous position; not enough new worker honey bees emerge to provide care and supplies (pollen and nectar) for the colony. Covert infections in adult honey bees facilitate the spread of the virus to other larvae through feeding (Beaurepaire et al., 2020). Similar to DWV, SBV is prone to mutation and recombination, which results in the emergence of different variants with altered virulence (Huang et al., 2021b). Despite its global distribution (Beaurepaire et al., 2020) and the damage it inflicts on brood, SBV is not considered a major factor in the loss of *Apis mellifera* colonies.

2.2.6 Black queen cell virus

Black queen cell virus (BQCV) is a (+)ssRNA virus in the family Dicistrovirus that is widely distributed (Beaurepaire et al., 2020) and at high prevalence in honey bees (e.g. McMahon et al., 2015; Traynor et al., 2016). It has a primarily faecal-oral transmission route and is apparently not transmitted by varroa mites. As suggested by the name, BQCV poses a threat to the queen-rearing industry, with high titres responsible for elevated mortality of queen larvae. Following their death, remnants of queen larvae acquire a black colour that stains the queen brood cell.

Although BQCV rarely results in symptomatic infections in managed hives, experimental infection of worker larvae results in high host mortality (Doublet et al., 2015), suggesting that the virus might exert a significant impact on colony health. BQCV, like the other Dicistroviridae ABPV, IAPV and KBV, is highly virulent when experimentally injected into host pupae and adults (Remnant et al., 2019; Al Naggar and Paxton, 2020). Research suggests a connection between infection of BQCV and the Microsporidian *Nosema ceranae* in adult honey bees; mixed infections result in higher mortality of adult workers (Doublet et al., 2015).

2.2.7 Novel viruses

Due to advances in molecular biological methods, and especially next-generation sequencing technologies, myriad novel viruses have been detected in honey bees (Beaurepaire et al., 2020). Although distinguishing between viruses causing active infections and viruses passively present on or within honey bees is problematic, recent discoveries shed light on how abundant and diverse is the world of bee viruses.

2.3 Fungi: Nosema spp. and chalkbrood

Nosema spp. are Microsporidia, highly derived spore-forming fungi that are intracellular parasites of animal hosts. The genus contains two highly honey bee-pathogenic species: *Nosema apis*, whose primary host is *Apis mellifera*, and *Nosema ceranae*, whose primary host is *Apis cerana* of south and east Asia but which has jumped host to *Apis mellifera* and is nowadays found across the world (Klee et al., 2007). Note that *N. apis* and *N. ceranae* have recently been reclassified to the genus *Vairimorpha* (Tokarev et al., 2020). Of many possible colony collapse culprits, *Nosema* spp. seconds the varroa mite and its associated viruses in the number of reported cases (e.g. Higes et al., 2008).

Nosema ceranae in *Apis mellifera* originally appeared better adapted to higher temperatures and to dominate in warm climates whilst *N. apis* retained its reign in cool temperate climates (Natsopoulou et al., 2015). More recent evidence suggests that *N. ceranae* may even dominate *N. apis* as a parasite of *Apis mellifera* in Canada's cooler, northern climate (Emsen et al., 2016). Both species are characterised by seasonality, with higher incidence and spore loads reported in the spring. Both parasitise the adult host's ventriculus (part of the gut), resulting in dysentery and within-hive defaecation, though experimental evidence suggests that *N. ceranae* may also infect larvae (Urbieta-Magro et al., 2019). The disease may also progress furtively, with a sudden collapse of the colony. Symptoms such as a distended abdomen and loss of flight ability, which results in crawling behaviour, are often found in colonies heavily infected with *N. ceranae*. The comparative virulence of these two *Nosema* species in *Apis mellifera* varies across studies from slight to moderately higher virulence of *N. ceranae* (Milbrath et al., 2015; Natsopoulou et al., 2016). Nevertheless, both species may have deleterious effects on colony fitness and ultimately lead to its collapse (Higes et al., 2008).

A more recognisable fungal pathogen of honey bees is the spore-forming *Ascosphaera apis*, the causative agent of chalkbrood (Aronstein and Murray, 2010). If spores are ingested by a honey bee larva, it then succumbs to the developing fungus, drying out to form a 'mummy' that seems to be covered in a white or mottled black-and-white 'chalk'. *Ascosphaera apis* is likely omnipresent in *Apis mellifera* colonies, at least in temperate regions, and is thought to break out only under conditions stressful for larvae, particularly if they are chilled below 35°C during development.

2.4 Bacteria: American and European Foulbrood

The two most economically important bacterial diseases of honey bees, American Foulbrood (AFB) and European Foulbrood (EFB), are exclusively diseases of brood (larvae).

AFB, caused by the Gram-positive, rod-shaped bacterium *Paenibacillus larvae*, is a worldwide disease of *Apis mellifera* known for its detrimental effects on colonies during, usually sudden, outbreaks (Hansen and Brødsgaard, 1999; Genersch, 2010). It consists of four genotypes (ERIC I-IV), of which the first two are the most prevalent. Spores, the infectious form of *P. larvae*, are extremely resilient and capable of surviving for 60 or more years under harsh environmental conditions. AFB is highly contagious; less than a dozen spores are required to elicit an infection in a honey bee larva. Upon ingestion, the pathogen germinates and massively proliferates in the midgut. Typical symptoms are observed in capped brood, in which the cap sealing the cell is dark and concave. Excreted proteases decompose the larval body within, which transforms into a brown, colloid-like mass known as the ropey stage (with a fishy or foul smell, hence the name 'foulbrood'). The remnants of the larva then dry, forming a scale that adheres to the floor of the brood cell, called the foulbrood scale. The classic symptoms are often obscured during infection with highly virulent strains (ERIC II-IV) by the rapid death of infected larvae before capping. Dead larvae are removed by worker honey bees and do not form foulbrood scales, producing an irregular brood pattern (so-called 'shotgun' pattern), though this pattern can have a number of other causes, too (e.g. queen infertility). All four genotypes of AFB lead to almost 100% mortality in honey bee larvae, with the time of death varying from 7 to 12 days post-infection. Infection can profoundly affect colony strength and lead to complete colony collapse.

EFB is caused by the Gram-positive bacterium *Melissococcus plutonius*, frequently accompanied by infection with other bacteria: *Enterococcus faecalis*, *P. alvei* and *Achromobacter Euridice*, the latter also commonly occurring in healthy honey bees (White, 1912; Forsgren, 2010). Affected honey bee larvae may die before capping and be removed from the hive or die post-capping, where the contagious material is preserved (and, when fresh, has a foul smell). The disease does not inflict 100% mortality; honey bee pupation, although delayed, may occur and result in the emergence of small, asymptomatic adults which serve as a bacterial reservoir. Sudden outbreaks of EFB disease are often followed by equally sudden recovery. The onset of symptoms may mirror a period of unfavourable weather conditions or lack of food resources. Similar to ABF, if EFB-infected larvae die, they first form a dark colloid-like mass which then dries into a scale. Co-infection with other bacteria may alter the characteristics of the disease, most notably changing the scent of decomposing larvae.

2.5 Trypansomes: Crithidia mellificae *and* Lotmaria passim

Despite growing interest in the Trypanosomatidae of honey bees, the pathogenicity of the two most prevalent members, *Crithidia mellificae* and *Lotmaria passim,* remains elusive. The former seems well adapted to the warm and acidic gut of adult honey bees, whereas the latter may spill over from other hosts into honey bees (Palmer-Young et al., 2021). These gut-infesting kinetoplasts occur in two forms: flagellated, motile choanomastigote and non-flagellated amastigote. Studies report contradictory results regarding their impact on infected honey bees, although the outcome of disease might be highly dependent on the dose and the phase of the pathogen's growth, which influences its infectivity (Gómez-Moracho et al., 2020; Liu et al., 2020). Few studies have linked trypanosomatids to colony collapse, although this needs further investigation (Cornman et al., 2012; Ravoet et al., 2013). *Crithidia mellificae* and *L. passim* are often found in colonies infected with *N. ceranae* (Runckel et al., 2011), yet the nature of the interaction between these pathogens remains unclear.

2.6 Common pests: the small hive beetle and wax moths

The small hive beetle (*Aethina tumida*), a minor pest of African *Apis mellifera* (for review: Neumann and Ellis, 2008), came into the spotlight after its discovery in US honey bee hives in 1996. Although African honey bees seem to be resistant to infestation, European *Apis mellifera* in temperate climates seem to lack vital behavioural hygienic traits that can either prevent the beetle from entering the hive or facilitate its eradication (Ellis et al., 2004).

The life cycle of the small hive beetle is divided into the within-hive phase, when reproducing adults and wandering larvae feed on honey bee resources (honey and pollen stores), and a phase outside the hive in the soil, where larvae undergo pupation. The beetle can also survive and reproduce outside the hive by feeding on fruit. Even the strongest European *Apis mellifera* colonies seem to be powerless against the voracious beetle, with reports of colony collapse only 2 weeks after infestation.

Two species of wax moth (Lepidoptera and Pyralidae; see Table 1) infest old wax combs, which their larvae consume. They cause considerable damage to combs stored by beekeepers but do not usually represent a problem inside a hive because honey bee workers can effectively defend their nest against moths.

2.7 Measuring disease impact on honey bee colonies (and pollination)

A healthy colony can grow in size, and a large colony with a numerous worker force is likely to have a considerable number of forager honey bees that can act as pollinators. The size of a colony can be measured directly by visually quantifying the number of honey bees, brood, pollen and honey reserves, e.g. using the so-called Liebefeld method (Dainat et al., 2020), which can be semi-automated through digital photography linked to AI-informed recognition software. Numerous fully automated devices that can operate remotely are nowadays also available on the market that record the weight of a colony and send the information in real time over the internet to a beekeeper's digital device to report on colony health (Rafael Braga et al., 2020). Poor colony growth might suggest disease impact, which can be investigated through veterinary inspection (Dittes et al., 2020b), incorporating the use of molecular methods for pathogen detection so that the cause of poor colony health can be correctly diagnosed and an appropriate remedy deployed.

The regular inspection of honey bee colonies is an obligatory part of beekeeping, all the more so for commercial colonies managed for the pollination of crops. During the swarm season, which usually coincides with the spring/early summer flowering of many crop plants in temperate regions, hive inspections every 9 days are recommended so as to ensure that a colony does not swarm before the next inspection (Hooper, 1976). A hive inspection should include not only a check (i) on whether the colony is about to swarm (e.g. building of queen cells) but also (ii) for the presence of the queen (iii) that there are sufficient reserves of honey and (iv) sufficient space to store additional honey as well as (v) for diseases, i.e. through visual inspection of adult honey bees (for symptoms of viral infection - discoloration, trembling and deformed wings - and for varroa mites), of the brood (for foulbrood, chalkbrood, other fungal and viral diseases), of the stored honey and pollen (for the small hive beetle) and of hive debris on the hive's bottom board. If disease is suspected, a full veterinary inspection can be undertaken (Dittes et al., 2020b).

Across temperate parts of its native range, most colonies of *Apis mellifera* are managed and can be inspected as described above. Tropical parts of *A. mellifera*'s native range maintain large populations of wild colonies (Jaffé et al., 2010). This is also likely true of feral Africanized *Apis mellifera* in Latin America, where the feral population may exceed that of managed colonies even in localities such as the Yucatan Peninsula of Mexico, with one of the world's highest densities of managed *Apis mellifera* colonies (Moritz et al., 2013). It is very challenging to monitor the health of wild or feral colonies because the colony's brood is often protected within a tree trunk or rock face. In such cases, disease presence can be determined by the molecular quantification of pathogens in honey bees collected from the hive entrance or its drones where they assemble for mating, high in the sky (Yañez et al., 2012; Forfert et al., 2016). But measuring disease impact on colony size or performance *per se* (as in the number of exiting forager honey bees) can only be evaluated indirectly by quantifying the flow of honey bees into and out of a colony.

3 The poorly known wild bee diseases

Concern over the decline of pollinators (and pollination) was initially focussed on wild species of insect and vertebrates such as wild bee species, humming birds and bats, and the wild plants they pollinate (Buchmann and Nabhan, 1996). In-depth analysis of museum records subsequently revealed that wild bees have declined in diversity in the Netherlands and the United Kingdom (Biesmeijer et al., 2006). Garibaldi et al.'s (2013) landmark paper, highlighting the important role of wild bee (and other insects) species in the pollination of crops, has only served to draw more attention to the fate of wild bees and their conservation. A recent analysis of records held by the Global Biodiversity Information Facility (GBIF) has revealed that the decline in wild bee species diversity is a worldwide phenomenon (Zattara and Aizen, 2021).

Bumble bees (*Bombus* spp.) are a social subset of ca. 260 of the world's total ca. 20 000 wild bee species that are rather charismatic because of their large size and characteristic, sometimes species-specific colour patterns. Of all wild bee taxa, they have therefore been relatively well studied. Monitoring data reveal that many *Bombus* spp. have declined in range over the past decades (Williams and Osborne, 2009; Cameron and Sadd, 2020), both in North America (Cameron et al., 2011) and in Europe (Nieto et al., 2014).

Another large group of wild bees, the eusocial (colony-forming) and pantropical stingless bees, are also considered important pollinators of numerous crop plants (Heard, 1999). There is a long history of management of some of these species back into antiquity; e.g. archaeological evidence suggests the Maya peoples have managed the Neotropical stingless bee *Melipona beecheii* for well over 2 millennia, and this and many other stingless bee species continue to be managed nowadays (Osterman et al., 2021). Stingless bee populations are thought to be in decline largely through habitat deterioration and destruction (Freitas et al., 2009); their pests and pathogens are less well known and deserve greater scrutiny. Because of the paucity of data on their diseases, we do not address them further in this chapter.

To what extent are pests and pathogens to blame for the decline of wild bees? In contrast to the relatively good knowledge of honey bee pests and pathogens, the disease organisms of wild bees are poorly researched, and there is little knowledge of the impact they may have on wild bee populations (Brown and Paxton, 2009). Exceptions are the few wild bee species that are managed for use in crop pollination: currently eight solitary bee species and nine *Bombus* taxa (Osterman et al., 2021). The scrutiny to which *Bombus* spp. have been subject to basic ecological and evolutionary research has also provided deeper insight into their pests and pathogens, comprehensively reviewed in Schmid-Hempel (1998). Evidence points to the important role of pathogens in wild bee decline (Dicks et al., 2021), particularly in the decline of bumble bees (Cameron et al., 2011; Arbetman et al., 2017; Cameron and Sadd, 2020).

We now overview the major known pests and pathogens of (i) bumble bees (Table 2) and (ii) other managed and wild bee species (Table 3), which we collectively term solitary bees.

3.1 Bumble bees

Bombus comprises a genus of large-bodied social (or socially parasitic) bee species that are frequent visitors to flowers in temperate and cooler regions of the world (e.g. the Andes of South America), where they are considered important pollinators (Goulson, 2009). In addition, their large size and often

Table 2 Major parasites and pests of bumble bees (*Bombus* spp.)

Disease/common name	Causative agent
Viruses	
Deformed wing virus	Deformed wing virus
Acute bee paralysis virus	Acute bee paralysis virus
Israeli acute paralysis virus	Israeli acute paralysis virus
Slow bee paralysis virus	Slow bee paralysis virus
Black queen cell virus	Black queen cell virus
Fungi	
Nosemosis	*Nosema bombi*
Nosemosis type C	*Nosema ceranae*
Trypanosomatids	
Crithidia	*Crithidia bombi*
Gregarines	
Apicystis	*Apicystis bombi*
Mites	
Locustacaris tracheal mite	*Locustacaris buchneri*
Nematodes	
Sphaerularia nematode	*Sphaerularia bombi*
Hymenoptera	
Parasitoid wasp	*Syntretus* sp.
Cuckoo bumble bees	*Psithyrus* spp.
Lepidoptera: Pyralidae	
Wax moth	*Aphomia sociella*
Diptera	
Conopid flies	Conopidae spp.

Table 3 Some examples of widespread parasites and pests of wild bees

Disease/common name	Causative agent
Fungi	
Nosemosis	*Nosema thomsoni*
Nosemosis type C	*Nosema ceranae*
Chalkbrood	*Ascophaera* spp.
Trypanosomatids	
Crithidia	*Crithidia* spp.
Gregarines	
Apicystis	*Apicystis bombi*
Diptera	
Conopid flies	Conopidae
Bee flies	Bombuyliidae
Satellite flies	Anthomyiidae
Strepsiptera	
Twisted-wing insects	*Stylops* spp.
Hymenoptera	
Nomad bees	*Nomada* spp.
Pests	
Oil and blister beetles	Meloidae

colourful adult patterns have attracted the attention of lay and scientists alike, contributing to the knowledge of their pests and parasites. Major diseases of bumble bees and their respective causative agents are given in Table 2.

3.1.1 The Microsporidia: Nosema bombi *and* Nosema ceranae

Bumble bees are parasitised by a species-specific Microsporidian, *Nosema bombi*, which can be responsible for a drastic reduction in fertility (Otti and Schmid-Hempel, 2007). Its route of infection is faecal-oral, and ingested *N. bombi* spores can colonise the adult host's intestine, Malpighian tubes and reproductive organs, causing diarrhoea, limiting sperm production in males and elevating mortality (Otti and Schmid-Hempel, 2007). Heavily diseased gynes (queen-destined females) and males have scarcely a chance of successful mating.

The honey bee Microsporidian *Nosema ceranae* is known to infect bumble bees, e.g. the common and widespread *Bombus terrestris* (Fürst et al., 2014), although consensus has not been reached yet on whether the detection of *N. ceranae* is mere contamination (the host may act as a mechanical vector) or reflects an actual infection (Gisder et al., 2020).

3.1.2 *Widespread Protozoa:* Crithidia bombi *and* Apicystis bombi

The trypanosomatid *Crithidia bombi* is a gut parasite of adult *Bombus* spp., transmitted by the faecal-oral route and often rising to high prevalence (Ruiz-González et al., 2012). Although it does not inflict direct negative effects on the reproductive system of its hosts, infection leads to higher mortality of gynes, and their success rates are lower (Rutrecht and Brown, 2008). Furthermore, *C. bombi* can impair the cognitive function of worker bees and reduce foraging effectiveness. *C. bombi* often occurs in combination with other parasites, which further aggravates the effects of infection.

Apicystis bombi is a gregarine parasite of various bee species, with bumble bees considered to be its primary host. Oral ingestion of spores results in colonisation of the gut and fat body. Parasites reproduce in fat cells, causing the organ to lose its cream colour and become hypertrophic (swollen). It is considered a serious disease organism of adult bumble bees; affected gynes have a shortened lifespan, which hinders successful mating and thus colony establishment (Schoonvaere et al., 2020).

3.1.3 Viruses: Deformed wing virus, black queen cell virus and others

In recent years, increasing attention has been given to emerging infectious diseases in wild pollinators, including bumble bees (Fürst et al., 2014; Nanetti et al., 2021). Advanced molecular techniques have led to the discovery of typical honey bee viruses in bumble bees and other wild bees (Table 2), as well as the presence of a range of novel viruses that are seemingly associated with *Bombus* spp. (Pascall et al., 2019). Of the various honey bee-associated viral pathogens, two appear to be the most prevalent in *Bombus* spp., namely DWV and BQCV (Fürst et al., 2014; McMahon et al., 2015). Although not apparently as pathogenic as in their reservoir host *Apis mellifera*, DWV inflicts negative effects on *Bombus* spp. and has been reported to cause overt infection with characteristic 'deformed wing' symptoms (Genersch et al., 2006).

3.1.4 Other Eukaryote pests and parasites

The nematode *Spherularia bombi* is a serious parasite of *Bombus* spp. queens; it is not known to infect worker bumble bees. Queens are exposed to *S. bombi* during winter hibernation in the soil, when juvenile nematodes enter overwintering females and colonise the host's haemocoel, draining host resources during and after hibernation. Infection results in impaired development of the host's corpora allata and ovaries, usually terminating a queen's reproduction. Affected queens do not find nests after winter diapause but seek new places to overwinter, where they shed large numbers of juvenile-stage *S. bombi* into the soil (Poinar and van der Laan, 1972). Infection with *S. bombi* is common in many *Bombus* spp. queens.

Bumble bees are associated with many mite species that vary in their supposed impact from parasitic (i.e. detrimental to host health) to beneficial (Alford, 1975). One relatively well-characterised mite is *Locustacaris buchneri* (Table 2), a widely prevalent parasite of multiple bumble bee species which is found in the tracheal tubes of both queens and worker bees, where it feeds on the haemolymph of the host (Rutrecht and Brown, 2008). Yet the scarcity of research about *L. buchneri* means that little is known about its true impact on colony fitness and reproduction. This is true for many of the numerous mite species associated with bumble bees.

The parasitic wasp *Syntretus* sp. is considered a rare parasite of bumble bees. Females lay eggs within *Bombus* individuals, thereby developing offspring feed on the tissues of the host, quickly leading to its death. *Syntretus* spp. may infest queens as well as workers, shortening their lifespan, reducing colony performance and potentially resulting in colony failure (Rutrecht and Brown, 2008).

The wax moth *Aphomia sociella* is considered a pest of bumble bee nests, especially those managed in artificial boxes. The ravenous larvae of *A. sociella* feed on various resources of the colony, from pollen and wax to bee eggs and larvae. Heavy infestation can cause the death of the colony due to a lack of stored resources (pollen and honey) and workers.

Conopidae are a group of true flies (Diptera) that lay their eggs inside their hosts, usually other insects; the conopid egg hatches and consumes the host from inside, eventually killing it. Some conopid species specialise in parasitising *Bombus* spp., manipulating their host so that it digs underground to die, where the conopid can more successfully hibernate (Müller, 1994). They clearly exert a cost on host populations which can be quite profound in temperate regions as summer progresses, with up to 47% of host workers infected in one study (Schmid-Hempel et al., 1990).

3.1.5 Social parasites (cuckoo bumble bees of the subgenus* Psithyrus*)

Social parasitic bumble bee species (genus *Bombus*, subgenus *Psithyrus*; also known as cuckoo or inquiline bumble bees) usurp nests of other *Bombus* species (termed 'true' bumble bees), often killing the host queen; host workers are thereby enslaved and provision the sexual offspring of the social parasite. Though socially parasitic *Bombus (Psithyrus)* spp. could theoretically regulate host populations, evidence from multiannual records in Great Britain suggests little temporal fluctuation in social parasitism (Antonovics and Edwards, 2011). In New Zealand, where only true bumble bee species (but not their social parasites) have been introduced, spring foundress queens are encouraged to nest around crop fields requiring pollination by offering them underground nesting chambers (Donovan, 2007). The efficacy of this method of boosting *Bombus* colonies for crop pollination may be substantively reduced elsewhere in the world where cuckoo bumble bees are found, e.g. northern temperate regions, because of frequent nest usurpation by cuckoo queens.

3.2 Solitary bees

Under this heading, we include the managed solitary bees, of which 22 species are or have been trialled for pollination management (eight species are currently in use; Osterman et al., 2021), and all other wild bee species are not subsumed within honey bees and bumble bees. This includes ca. 20 000 species of bee, 10% of which are not solitary but possess some degree of sociality and between 10% and 30% of which are cleptoparasitic (brood parasitic) on other wild bee species (Wcislo, 1987; Danforth et al., 2019).

Though the recent review of solitary bees by Danforth et al. (2019) is very informative on solitary bee pests and parasites, we still largely lack detailed information on their impact on host populations, with the most detailed knowledge derived from the foremost commercial solitary bee pollinator, *Megachile rotundata* (Pitts-Singer and Cane, 2011). Other knowledge on solitary bee parasites is largely derived from autecological studies of a single host species at a single field site (e.g. Paxton et al., 1996). Major diseases of solitary bees and their respective causative agents are given in Table 3.

3.2.1 Fungi: Microsporidia and chalkbrood

Microsporidia have been detected in various solitary bees (Grupe and Alisha Quant, 2020; Martínez-López et al., 2021), although their effects on fitness and mortality are poorly researched. Artificial infection of *Osmia bicornis* larvae with *N. ceranae* resulted in poor rates of infection, but treated larvae suffered apparent survival costs (Bramke et al., 2019). The Lepidopteran parasite *Nosema thomsoni* has been found in solitary bees (Ravoet et al., 2014; Schoonvaere et al., 2018), but we lack studies providing an in-depth analysis of its course of infection. Genetic examination of museum specimens of the hoary squash bee (*Eucera pruinosa*) revealed the presence of *Nosema* sp. in over a decade-old pinned specimens (Vaudo et al., 2018). The North European *Andrena scotica* (synonym *carantonica*) harbours the Microsporidian *Antonospora scoticae* (Fries et al., 1999); there is likely a diversity of yet-to-be-discovered Microsporidia within other solitary bee species.

A prominent fungal disease of wild bees is chalkbrood, caused by *Ascosphaera* spp. and closely related to *A. apis* that infects honey bees. Wild bees, especially those of the family Megachilid, are known to be susceptible to *Ascosphaera* spp. infection. Chalkbrood spores are ingested by a larva with stored brood food (the 'bee bread' deposited by its mother bee in an offspring's natal cell). Spore germination, penetration and subsequent growth of the fungus inside the larval host's body cavity cause lethal mechanical and enzymatic damage to the larva that becomes swollen and which, after drying out, resembles a 'mummy' of chalk-like consistency. The high pathogenicity of *Ascosphaera* spp. can lead to the death of 60% of brood. It is therefore a serious pathogen of commercial rearing facilities of *Megachile rotundata* (Richards, 1984).

3.2.2 Other microparasites (those not visible to the naked eye) of wild bees

Two *Crithidia* spp. have been found in solitary bees: *Crithidia mellificae* (reservoir host: honey bees) in *Osmia cornuta* and *Crithidia bombi* (reservoir host: bumble bees) in both *Osmia lignaria* and *M. rotundata*. Similar to their impact on their reservoir hosts, trypanosomatid infections tend to be mild and only subtly affect the mortality of solitary bee hosts (Figueroa et al., 2021).

Infection by *Apicystis bombi* has been documented in *Osmia bicornis*, considered a model solitary bee species; infected individuals, both females and males, show significantly reduced rates of survival, and sublethal effects might include compromised mating and reproductive success (Tian et al., 2018).

We know less about other solitary bee microparasites, such as bacteria, and therefore of their role as drivers of host decline. In some instances, for example, the microorganisms associated with the bee bread provisions of larvae, these microorganisms may play a positive and important role in protecting the provisions from decay and in supporting larval nutrition (Dharampal et al., 2019; Steffan et al., 2019). Increasing awareness of the viruses of honey bees has, though, prompted investigation of the role of these same viruses in the decline of solitary bees, revealing the presence of many of them across the community of flower visitors (Nanetti et al., 2021). We take up this topic more fully below (section: disease transmission, spillover and spillback), though note here that definitive evidence that these 'honey bee' viruses are parasitic (i.e. reduce host fitness) is in many instances lacking.

3.2.3 Macroparasites (those visible to the naked eye)

Close investigation of a solitary bee species often reveals a rich diversity of symbiotic organisms, some parasitic, some potentially beneficial. A case study of one solitary bee, *Andrena scotica*, revealed adults to be parasitised by conopid flies (Diptera, Conopidae), Strepsiptera, generalist parasitic mermithid nematodes and diplogasterid potentially beneficial nematodes in intersegmental glands whilst brood cells were parasitised by bee flies (Diptera, Bombyliidae), satellite flies (Diptera, Anthomyiidae), oil or blister beetles (Coleoptera, Meloidae) and cleptoparasitic nomad bees (*Nomada marshamella*; Hymenoptera, Anthophila) (Paxton et al., 1996). This rich assemblage of pests and parasites associated with this one solitary bee species, many with highly specialised life histories tied to just one or a few host bee species, is by no means unusual (Danforth et al., 2019). Though up to 40% of adults were parasitised by deadly conopid flies and a further 10% of brood cells (offspring) had been replaced by nomad cleptoparasitic bees (Paxton et al., 1996), it is unclear whether these parasites in any way regulated host *A. scotica* population dynamics (and therefore provision of pollination). A recent review of the literature (Danforth et al., 2019) comes to the same conclusion; i.e. we lack the data to know whether pests and parasites cause the decline of solitary bees.

3.3 Assessing impacts of disease on bumble bees and other wild bee species

Honey bees are eusocial, living in perennial colonies that are usually maintained in a beehive and managed by a beekeeper, features facilitating their inspection and assessment of the impact of pests and pathogens on them. The same may be said for managed (commercial) bumble bee colonies. For wild bees, most of which are solitary, with a single generation per year (Michener, 1974), and nesting in hidden cavities above or below ground, assessment of the impact on their populations of pests and diseases is all the more complicated.

One option to assess the impact on them of disease agents is to undertake long-term monitoring of their populations coupled with an independent assessment of their pests and parasites, including the use of molecular techniques to assay microparasites (as for honey bees: Dittes et al., 2020b). Multi-generation (multiannual) monitoring of wild bees through destructive sampling, e.g. using pan traps/bowls, though costly, is a standard technique for monitoring wild bee species (Westphal et al., 2008) that, from a cost-benefit perspective, pays its way (Breeze et al., 2021), though is not without its critics (Tepedino and Portman, 2021). A downside is that this approach is merely correlational; it can inform about changes in the population size of wild bees and even about the association of a change in population size with a pathogen, but it cannot demonstrate the cause of a decline.

A case study of *Andrena scotica* and its novel Microsporidian *Antonospora scoticae* reveals some of the problems (for details of the study, see Paxton et al., 1997, supplemented by personal observation, RJP). At one nesting aggregation at Törnbottens Stugby on Öland in South Sweden, over 85% of adults emerging in 1995 were infected with *A. scoticae*, some with over 10^9 spores concentrated in the bees' fat bodies. Nesting females were generally uninfected or with a low spore count, suggesting that *A. scoticae* exerts a heavy fitness cost on hosts. Though the population of this wild bee species at Törnbottens Stugby was studied by RJP over 5 intensive years of fieldwork (1993–1997), formal estimate of its size by mark-recapture was only undertaken in 1994 and 1995, when the population comprised ca. 5000 females; however, by 1997 the field site was devoid of *A. scotica* (RJP, unpublished data). The Microsporidian could have caused the death of the Törnbottens Stugby population of *Andrena scotica*, but it is not possible to exclude other factors such as inclement weather during a critical phase of the life cycle, either alone or in combination with the Microsporidian; correct diagnosis of a decline is necessary to devise effective remedies to restore a population. As the Microsporidian was at the time an undescribed species infecting a little researched host that nests underground, it was difficult to conceive of control measures that could have rescued the population from extirpation. Moreover, the Microsporidian was seemingly host-specific; monitoring of all wild bees by pan-trapping might not have alerted to a decline of *A. scotica*, or its decline might have been ignored as an idiosyncrasy of the species (which is most likely the case).

Cutting the Gordian knot of demonstrating causation requires experimental studies, but these are difficult with non-model species such as the vast majority of wild bees. In lieu of experimentation, next-generation sequencing (NGS) approaches might provide a high-resolution means to infer causation (Grozinger and Zayed, 2020); for example, Tsvetkov et al. (2021) have recently demonstrated that the declining North American *Bombus terricola* often harbours the parasites *Nosema ceranae, Crithidia bombi, Lotmaria passim* as well as the viruses BQCV and SBV, whereupon bees were also associated with an up-regulation of genes employed in defence against these parasites.

4 Disease transmission, spillover and spillback

Many routes of transmission exist for pathogens and pests of the honey bee that are possible because of the host's eusocial colony organisation and perennial lifecycle (Schmid-Hempel, 1998). Intraspecific transmission can occur horizontally through within-hive interactions such as trophallaxis, grooming, cannibalism and direct contact augmented by the constricted space within a hive. Transmission between colonies occurs via hive robbing (honey bees from one colony removing resources from another, usually adjacent, colony), by bee drifting (a returning forager entering the wrong colony), at shared flower resources as well as by human-induced contamination of beekeeping tools, clothing, hive frames, honey or beeswax. Introduction of an infected queen to a colony may result in the horizontal spread of a disease agent among other colony inhabitants and its vertical spread to offspring, with the route of transmission of considerable significance for disease epidemiology (Schmid-Hempel, 1998; Fries and Camazine, 2001). Though differences in apiary size (number of honey bee colonies per apiary) have been shown theoretically to have little impact on disease prevalence and epidemiology (Bartlett et al., 2019), empirical evidence suggests that disease prevalence increases with colony density, presumably because of greater opportunities for between-colony (horizontal) transmission (Forfert et al., 2016). Maintaining colonies in hives may itself lead to persistent infection within the colony (Bartlett et al., 2021).

Less is known about the transmission and epidemiology of the pests and parasites of wild bees. An exception is *Crithidia bombi*, which parasitises multiple bumble bee species (Ruiz-González et al., 2012) and has been demonstrated to be transmitted through shared use of the same flower (Durrer and Schmid-Hempel, 1994). Even hover flies may pick up *C. bombi* from infected flowers and transport it onwards, acting as mechanical vectors (Davis et al., 2021). This example highlights the important role that flowers play as transmission hubs for a range of parasites across a diversity of pollinator (flower-visiting) insect species (McArt et al., 2014; Graystock et al., 2015), in which one insect deposits (by defecation, by regurgitation, by other forms of excretion or by mere physical proximity) a pathogen propagule onto a flower and from which a subsequent insect visitor to the same flower acquires the pathogen (Fig. 2). It also brings to

Figure 2 Two or more bee species may sequentially visit the same flower, offering ample opportunity for pathogen transmission (spillover).

the fore the question of whether a transmitted pathogen leads to an infection in a recipient host, with the host potentially acting as a biological vector (in which case the host is defined as being competent), or whether the transmitted pathogen merely resides in or on a host without replicating, with the host acting either as a form of horizontal vector or as a dead-end host. These differences are important because they impact the parasite's epidemiology and evolutionary trajectory as well as host population fitness (McMahon et al., 2018).

The global distribution of managed honey bees, even in otherwise non-native regions (Hung et al., 2018), their increasing abundance (Osterman et al., 2021) and the ubiquity of many of their parasites (see Section 3.1.3) have consequences for wild pollinator species. Pathogen spillover (interspecific disease transmission) from honey bees as reservoir hosts to wild bee species as recipient hosts has been increasingly researched because the potential negative effects of honey bee pathogens could undermine the already fragile health status of wild bee species (Fig. 3).

Spillover has been documented for several honey bee pathogens (e.g. Ravoet et al., 2014; Graystock et al., 2020), with increasing attention now being given to viral spillover. Experimental evidence using flight cages first demonstrated that IAPV can be transmitted from honey bees to bumble bees (Singh et al., 2010). Correlational evidence strongly supports the ongoing viral transmission of DWV and BQCV from honey bees to bumble bees in

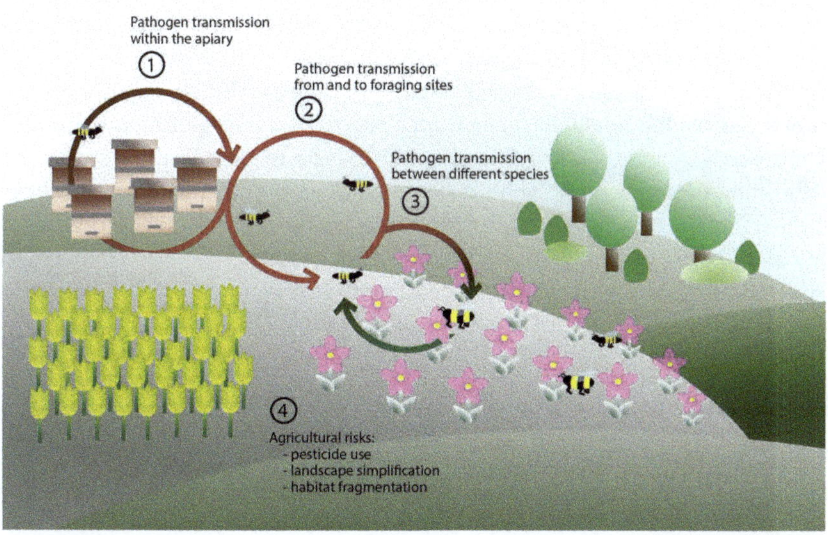

Figure 3 Schematic landscape describing pathogen spillover and spillback among honey bees and wild bee species at shared flowers, a phenomenon that may be promoted by agricultural practices and other global change pressures.

Europe (Fürst et al., 2014; McMahon et al., 2015; Manley et al., 2019), where the honey bee is native, as well as in North America (Alger et al., 2019; Pritchard et al., 2021), where the honey bee is not native. Honey bee viruses have been detected in a large number of wild bee species (Nanetti et al., 2021), suggesting that viral spillover may be a very common phenomenon. Though evidence for the impact of DWV on bumble bees is equivocal (cf. Fürst et al., 2014; Tehel et al., 2020), this and other RNA viruses have high adaptive potential (Holmes, 2009), thereby posing a serious threat to wild pollinator health. Viruses that have spilled over from a reservoir to a novel host may subsequently spill back to the original host with altered virulence due to adaptation to the novel host environment, though we currently lack evidence for this phenomenon.

Spillover is not solely attributable to the honey bee as a reservoir host. *Apicystis bombi*, a pathogen of bumble bees, has been increasingly found in managed honey bee colonies (Plischuk et al., 2011; Schulz et al., 2019), though the consequences of the spillover of *A. bombi* for pathogen epidemiology and recipient honey bee host fitness remain open conjectures. Pathogens may also spill over from managed bumble bees used for commercial pollination to wild bumble bees, both in North America (reservoir host *Bombus impatiens*: Colla et al., 2006) and Europe (reservoir host *Bombus terrestris*: Murray et al., 2013). In Chile and Argentina, the deployment of commercial Eurasian *B. terrestris* in the past 3 decades is associated with the introduction of Eurasian variants of bumble bee parasites (*A. bombi* and *C. bombi*), which have likely caused population collapse of the native giant bumble bee of South America, *Bombus dahlbombi* (Arbetman et al., 2013; Schmid-Hempel et al., 2014).

Hopefully, it is clear from this section that pathogen spillover is a major determinant of the health of pollinator populations and therefore the provision of the ecosystem service of pollination. It needs to be seriously considered in the management of pollinators. Before transporting a pollinator species, whether native or not, to a crop requiring pollination, one should evaluate the risk of pathogen spillover. Encouragement of local pollinators that already exist in or around a crop ought to be favoured in the first instance as a more sustainable approach to pollination service provision.

5 Defence mechanisms of bees

Though challenged by a wide diversity of pests and pathogens, large and small, bees have evolved an impressive array of response mechanisms to fight their foes. We divide these into 'individual immunity', mechanisms possessed by each individual and which function equally in social and solitary insects, and 'social immunity', describing the behavioural and physiological traits employed by a group of two or more social bees to fight pests and pathogens (Cremer et al., 2007).

5.1 Individual immunity

Despite their vulnerability to infectious diseases, bees have developed well-functioning pathogen defence responses, comprising mechanical, behavioural and physiological (immune) mechanisms (Evans and Spivak, 2010). A hard exocuticle with a waxy epithelial covering provides insects with a barrier against mechanical damage and limits entry points for pathogens. Bees employ a range of behaviours to defend themselves against their macroparasites. A well-studied case in point is the interaction between bumble bees and conopid fly parasitoids (Müller and Schmid-Hempel, 1993). A worker bumble bee normally spends the night in its warm hive. If a concopid fly has laid an egg within its abdomen, it is more likely to spend the night outdoors at a lower temperature, which arrests the development of the conopid egg/larva and may even kill it, thus extending the bumble bee's lifespan. In a wicked twist of coevolution, if the conopid larva does succeed in developing inside its bumble bee host by eating it from the inside out, it manipulates its host to dig itself underground just before host death, where the conopid can pupate with a higher probability of survival than if the bumble bee had died in the open.

Although bees (like other invertebrates) do not possess an adaptive immune response as found in all vertebrates including ourselves, their innate immune mechanisms provide a wide range of physiological and other molecular defences that protect them from bacteria, fungi and viruses. For example, melanisation, encapsulation, nodulation and phagocytosis are performed by haemocytes in response to septic or aseptic trauma. These cellular responses are rapid, in contrast to the humoral mechanisms which deploy the production of antimicrobial peptides (AMPs). In detail, canonical innate immune defence employs pathogen recognition and signalling cascades (e.g. Toll, Imd, JAT/STAT and JNK pathways), leading to the production of AMPs to fight against bacteria and fungi by altering the structure of a pathogen's cell membrane and leading to its destruction. Haemocytes and AMPs are produced by the bee's fat body, a factory for various immune-related compounds (Evans et al., 2006).

For fighting viruses, the best-described dimension of the honey bee's innate immune system is the RNAi (RNA interference) pathway (McMenamin et al., 2018), which is also found in bumble bees (Barribeau et al., 2015) and likely all bee species. In essence, the host cell recognises and then degrades dsRNA (double-stranded RNA), thereafter degrading (knocking down or silencing) other intracellular RNA with the very same sequence, including RNA viruses.

The honey bee's gastrointestinal tract harbours a rich diversity of commensal microorganisms dominated by five major bacterial phylotypes that are very consistent within and between hives (Kwong and Moran, 2016). Their social life enables all individuals in a hive to become colonised by the same gut flora, which are thought to have a major beneficial impact on the health and fitness of the individual by maintaining an optimal gut environment, enabling proper nutrition, detoxification and stimulation of immune pathways (Emery et al., 2017) as well as defending the bee against pathogens (Forsgren et al., 2010). Solitary bees also possess a rich gut microbiota, though it is much more variable in composition compared to that of social bees (Martinson et al., 2011).

5.2 Social immunity

Many behaviours of social bees are considered to be acts of collective defence against pathogens. Through grooming, honey bees can remove external parasites from either their own body or other colony members' bodies (Pritchard, 2016). Collective 'balling' of predatory wasps allows the Eastern honey bee (*A. cerana*) to kill (through heat exhaustion) yellowjackets and hornets much larger than an individual honey bee (Ono et al., 1995). Antimicrobial excretions can be passed to other honey bees together with food via trophallaxis to protect all colony members. This practice is a double-edged sword as pathogens that have escaped the effects of AMPs may also spread rapidly within the colony (Naug, 2008). Hygienic behaviour deploying removal of parasitised brood from the hive can help with parasite management, especially in reducing *Varroa destructor* invasion (Wilson-Rich et al., 2009). When faced with severe conditions (shortage of food and extreme temperatures), bees can cannibalise their brood to provide nutrition for the colony and limit the spread of pathogens (Schmickl and Crailsheim, 2001). Reducing pathogen dispersal among brood can also be achieved through social fever, when adult bees raise the temperature of the brood above the tolerance limit of pathogens, e.g. *Ascosphaera apis* (Starks et al., 2000). Finally, infected individuals may abandon the colony in the act of self-sacrifice to prevent the transmission of pathogens to other nestmates (Schmid-Hempel, 1998).

6 Synergies with other risk factors

The interaction between a host bee and its pests and pathogens is, in many instances, context-dependent, being modulated by, for example, host resource supply, human activities such as agricultural intensification and the associated use of pesticides and climate. Global change pressures may therefore drive altered pest and pathogen epidemiology and virulence (Proesmans et al., 2021).

Nutrition represents an important factor that can alter the host-parasite relationship. Honey bee hosts with adequate resources or an intact microbiome are better able to tolerate viral pathogens than hosts with a depleted microbiome (Dosch et al., 2021). Under a scenario of agricultural intensification, floral scarcity in farmland threatens the provision of an adequate nutritional supply (Jha and Kremen, 2013), which is needed for the maintenance of a well-functioning immune system (Smith, 2007). A mass-flowering crop that increases floral abundance may alleviate nutritional shortage within the agricultural landscape, but it may also lure in pollinators, creating a perfect opportunity for horizontal pathogen transmission (spillover), increasing the prevalence of pathogens among the community of bees (Piot et al., 2019). Alternatively, a superabundance of flowers may dilute pathogens, reducing transmission and prevalence (Graystock et al., 2020).

Pesticides impair a honey bee's detoxification capacities and immune response, increasing its susceptibility to pathogen infection (di Prisco et al., 2013). In the laboratory, synergy is seen between *Nosema* spp. and neonicotinoid insecticides used in farming, aggravating the negative effects of both agents on honey bees (Doublet et al., 2015). A recent, comprehensive review of the impacts of pesticides on bees (Siviter et al., 2021) highlights synergistic interactions between two or more pesticides on host health; the data on pest/pathogen-pesticide interactions on bee health, whilst generally suggesting an additive or synergistic impacts, currently do not permit a definitive statement. This is an area that deserves great research attention across a range of managed and wild bee species.

Climate change, similar to agricultural intensification, may lead to the loss of natural habitats for wild and managed pollinators, depriving them of food resources and niches (Walter, 2020), with knock-on consequences for the impact of pests and pathogens on bees. Climate warming may allow for the expansion of exotic pathogens into new regions, e.g. *Nosema ceranae* (Natsopoulou et al., 2015), which may be especially dangerous due to the lack of adaptation of naïve species to the new threat. Heat *per se* has an ambivalent role in pathogen susceptibility, on the one hand, potentially lowering host immune response (Laughton et al., 2017), on the other, lowering the survivorship of some pathogens (Zaragoza-Trello et al., 2021).

7 Prevention of diseases

Disease prevention in managed pollinators poses many challenges. One is their unrestricted movement from one crop to another for pollination. For example, the transport of ca. 2 million honey bee colonies from across the United States to California every February for almond pollination risks rapid transmission of pests and pathogens within the stock of US honey bees (Cavigli et al., 2016).

For honey bees and potentially also other bee species, drifting of workers from one colony to the next or the robbing by one colony's members of another, diseased colony's resources represent very efficient routes of transmission. Another challenge is the shared use of flowers such that pathogens can easily be transmitted within and between bee species. A honey bee forager visits hundreds to thousands of flowers every day, which creates ample opportunities for pathogen transmission. Since it is not possible or sensible to restrict foraging, disease prevention must be limited to the management or treatment of the hive itself. For social species such as honey bees, a third challenge is their size and dynamism; it is difficult to treat effectively all individuals of a colony that may comprise over 40 000 individual worker honey bees, each replaced on a 3- to 6-week schedule.

Typical disease prevention for honey bees therefore aims at certain hygienic beekeeper standards (for one beekeeping hygiene manual, see: Tyl et al., 2014). In addition, low honey bee density (Forfert et al., 2016), small hive size and periodic brood breaks may further aid in disease control and diminish disease spread, though we note that maintaining a small hive size (and a small number of workers) likely diminishes the pollination potential of a honey bee colony and its likelihood of surviving the winter in northern climates like experienced in Canada.

Given the risk of disease spillover from managed to wild bumble bees (Colla et al., 2006; Murray et al., 2013), some countries or regions have imposed restrictions on the importation of non-native species or subspecies of *Bombus*, e.g. the United Kingdom and the Canary Isles. It is unclear whether these retrospective measures have prevented disease spread, though evidence from South America clearly points to the introduction of the non-native *Bombus terrestris* as the cause of disease spread among native bumble bees (Arbetman et al., 2013; Schmid-Hempel et al., 2014).

Every year, US alfalfa farmers import cocoons of *M. rotundata* from further north (Canada), where *Ascosphaera* spp. are absent. The widespread proliferation of chalkbrood in US facilities does not permit a sustainable US *M. rotundata* industry. This is not so much 'prevention' as 'mitigation' by the annual replacement of diseased bees.

Next, we address additional topics in disease prevention, though note that they largely relate to the honey bee and its diseases.

7.1 *Varroa control*

Given the importance of the varroa-virus nexus for honey bee health, it is understandable that the most commonly applied measures to control honey bee diseases are associated with the reduction in varroa mite infestation. Since *V. destructor* is a vector and 'virulence amplifier' of multiple viruses, managing varroa infection in the hive contributes greatly to maintaining the vitality of the colony. Purchasing varroa-free colonies is not a long-term solution as honey bees will eventually acquire the mites from other hives through robbing and drifting.

Chemicals (miticides) are frequently used to control varroa and are often unavoidable as the infestation progresses and may lead to a collapse of the colony if untreated after 3–4 years (e.g. McMahon et al., 2016). The choice of chemical substances used to control mite infestation is vast, ranging from essential oils, propolis extracts and algal preparations to synthetic miticides (Box 1). Beekeeping management may also help reduce the number of varroa mites in a hive; examples include artificial swarming and the removal of capped drone brood. Varroa control is a vast topic comprehensively described by Jack and Ellis (2021).

Box 1. Differences between 'hard' and 'soft' chemicals used to control varroa mites in honey bee colonies

Beekeepers often resort to chemical control of *Varroa destructor* mites in their honey bee (*Apis mellifera*) colonies because it is often difficult to otherwise reduce mite numbers. If not controlled, varroa mites (and the viruses they transmit, especially the deformed wing virus) lead to the collapse and death of the colony. Two classes of chemicals ('hard' and 'soft') are often employed in varroa control, each with their merits and demerits, which we bullet point here.

'Hard' chemicals:

- Synthetic compounds
- Examples include amitraz, coumaphos, flumethrin, tau-fluvalinate
- Advantages
    - Well-established efficacy
    - Convenience to use
    - Low in cost
    - Climate-independent efficiency
- Disadvantages
    - Lead to mite resistance of a chemical
    - Negative effects on bee reproduction and memory
    - Accumulation in bee products, potential to enter the human food chain

'Soft' chemicals:

- Compounds found naturally in the environment, even in the hive
- Thymol, formic acid, oxalic acid, lactic acid

- Advantages
  - No resistance yet reported
  - Impact varroa on adult bees but also in capped brood cells (formic acid)
  - Minimal accumulation in bee products
- Disadvantages
  - Limited use in temperate regions (e.g. formic acid at lower temperatures)
  - Use limited to broodless conditions (e.g. lactic and oxalic acids)
  - Lower and poorly quantified efficiency compared to 'hard' chemicals

See Jack and Ellis (2021) for a complete review of methods of varroa control in the beehive.

7.2 Viruses

The prevention of viruses in honey bees, at least those transmitted by varroa mites, is best achieved by controlling varroa mites (see above). For viral diseases of honey bee brood, namely those caused by SBV and CBPV, a break in brood rearing, e.g. by generating an artificial swarm or by re-queening the colony, is an effective means of reducing viral titres and relieving or eliminating overt disease symptoms (Dittes et al., 2020b). For viruses in wild bees (bumble bees and solitary bees) that spill over from honey bees, two plausible options are to decrease honey bee density (by relocating honey bee hives) and increase flower density, which is likely to diminish the force of infection (the probability that a honey bee and a wild bee visit the same flower); the latter effect has been inferred for protozoan parasites (Graystock et al., 2020). However, we lack direct tests of their effectiveness.

Given that bees, like other invertebrates, lack an adaptive (acquired) immune system, they cannot be 'vaccinated' in the way that humans can so as to fight viral disease. As described in Section 5, they do, though, possess an innate immune system, which includes the RNAi (RNA interference) pathway. By administering dsRNA with a specific sequence, one can efficiently knock down a gene's expression or, in the case of RNA viruses, induce host cells to destroy viral RNA. This approach has been used with apparent success to control a range of pathogens in honey bees, including varroa mites and DWV (Desai et al., 2012; Leonard et al., 2020), though successful dsRNA administration requires refinement and its efficacy in the field has not yet been demonstrated (Paxton, 2020). RNAi does, though, offer a breakthrough in the control of pernicious viral diseases such as DWV.

Chemical control of bee viruses such as DWV is an additional approach (e.g. Tang et al., 2021), though contamination of honey (for honey bees) and efficacy in the field remain open questions.

7.3 Breeding

Some honey bee colonies express higher tolerance to varroa mites or viruses, withstanding invasion without treatment (Locke, 2016). Resistance or tolerance characteristics of honey bees can be maintained or enhanced by targeted breeding (Rinderer et al., 2010). Promoting colonies that display resistance against or tolerance to parasites and express social immunity traits, such as hygienic behaviour and grooming that are known to be genetically determined, is highly desirable in order to improve the intrinsic defence mechanisms of honey bees (Rothenbuhler, 1964; Arechavaleta-Velasco et al., 2012).

7.4 Novel methods of diagnosis and treatment

With rapid progress in molecular biology, research on bee diseases has advanced quickly, giving rise to novel methods of disease prevention and treatment. One possible option for varroa control is the use of gene drives (Faber et al., 2021). Promotion of genes reducing mite reproduction, survival rates or resistance to miticides would help immensely with effective varroa management, although successfully establishing gene drive in varroa population poses multiple obstacles, such as mite inbreeding and haplodiploidy, to be first overcome.

Extensive research into the honey bee microbiome has uncovered the positive effects of lactic acid bacteria (LAB) on honey bee health (Royan, 2019). Additionally, LAB has been reported to inhibit the growth of *P. larvae*, the causative agent of AFB (Forsgren et al., 2010); their inclusion in honey bee nutritional supplements may be a powerful means of enhancing honey bee immune defence.

Bee disease diagnosis poses another challenge; two major problems are the great number of asymptomatic individuals in an affected honey bee colony and the very circumstantial switch to overt infection that precludes proper visual detection of a disease. Moreover, due to the small size of individuals, bees often need to be sacrificed in order to identify efficiently an internal pathogen. To circumvent this latter problem, several means of non-lethal collection and detection of infectious agents from individual honey bees have been proposed, with notable successes in haemolymph collection (in which disease agents can be detected by molecular approaches) from antennae (Borsuk et al., 2017) and via a needle (Huang et al., 2021a).

Insight into disease pathogenesis is crucial in order to establish methods of treatment and prevention. A big step for honey bee viral research was the recent construction of DWV tagged with a green fluorescent protein (GFP) (Evans et al., 2021). Bees infected with GFP-DWV can be non-lethally screened in real time via camera recording. Fluorescence-marked viruses open new opportunities for research on this pathogen, such as DWV propagation, tissue specificity, host species specificity, interference with other pathogens and many others.

8 Future trends

The charm of bees and their various, often peculiar, methods of immune defence does not negate the fact that they suffer considerable annual losses from pathogens. The most important disease agents of honey bees are varroa mites and the viruses they transmit. Both are at high prevalence and act synergistically (Nazzi et al., 2012), often resulting in the death of affected colonies. Despite the lower number of known wild bee parasites, wild bees also face grave consequences from well-established pathogens and others spilling over from heterospecifics.

Notable progress in the detection of bee pathogens has brought to light the desperate need for new and effective methods for disease control, e.g. to treat honey bee viruses, especially DWV. Well-known disease-causing agents and newly emerging pests and pathogens, in interaction with climate change and agricultural practices, reduce pollinator survival. On a more upbeat note, it is important to emphasise that the initial dramatic effect of an emerging exotic pathogen may lessen with time as the host adapts and evolves greater resistance or tolerance towards the disease agent.

Human interference in honey bee colony life (disease treatment, food supplementation, swarm prevention and breeding focussed on production traits) may impede the process of natural selection and leaves the managed hive vulnerable to infestation and disease outbreak. Moreover, pathogens do not remain evolutionarily stagnant, undergoing change through mutation and selection, not only adapting to new hosts but also to new vectors. This scenario may be tempered, however, by human practices aimed at appropriate conservation efforts and bee management that aim at reducing pathogen virulence.

Together with changes in hosts and parasites, future research and monitoring techniques may be better able to capture data on honey bee diseases in real time (Rafael Braga et al., 2020). The use of NGS (genomics and metagenomics) marks a new standard in bee molecular research and may prove invaluable in elucidating the genetic diversity of bees, their parasites and their interactions as drivers of their diversity (Tsvetkov et al., 2021). The obscure pathogenesis of many honey bee diseases is now being challenged by means of reverse genetics (Gusachenko et al., 2020). Novel methods of disease treatment arising from recent research, such as the use of LAB, RNAi and gene drive, promise hope for limiting pathogen spread and mitigating bee decline.

9 Where to look for further information

9.1 Honey bees

COLOSS (https://coloss.org) is a worldwide network of researchers interested in the health and well-being of honey bees, with numerous working groups ('task forces') addressing pressing issues in honey bee health. Membership is open to all. As well as organising two conferences per year, they have co-published with the International Bee Research Association (https://ibra.org.uk) the *BeeBook*, a compendium of (open access) chapters addressing all matters related to honey bees and bee research; volume II, titled *Standard Methods for* Apis mellifera *Pest and Pathogen Research*, contains chapters on each major taxon of a honey bee disease-causing organism (https://coloss.org/beebook/volume-2/).

9.2 Wild bees

Schmid-Hempel's (1998) *Parasites in Social Insects* is a tour de force of the major pests and pathogens of social insects, with much emphasis on bumble bees, whilst Richards' (1984) pamphlet on managing the alfalfa leafcutter bee (*Megachile rotundata*) reviews the diseases and control measures for this species, the most important managed solitary bee. The literature on the pests and pathogens of other wild bees is scattered, though Danforth et al. (2019) provide an excellent overview in their book *The Solitary Bees*.

The International Union for the Conservation of Nature (IUCN) has a wild bee specialist group (WBSG) tasked with the development of best practice guidelines for the implementation of conservation measures for wild bees (https://www.iucn.org/ssc-groups/invertebrates/wild-bee/aims-wbsg). Initiated in 2021 and set to run till 2025, at the time of writing this chapter, WBSG's website promises to provide details on how best to conserve wild bees.

10 References

Aizen, M. A., Aguiar, S., Biesmeijer, J. C., Garibaldi, L. A., Inouye, D. W., Jung, C., Martins, D. J., Medel, R., Morales, C. L., Ngo, H., Pauw, A., Paxton, R. J., Sáez, A. and Seymour, C. L. (2019). Global agricultural productivity is threatened by increasing pollinator dependence without a parallel increase in crop diversification, *Global Change Biology* 25(10), 3516–3527. doi: 10.1111/gcb.14736.

Aizen, M. A., Garibaldi, L. A., Cunningham, S. A. and Klein, A. M. (2008). Long-term global trends in crop yield and production reveal no current pollination shortage but increasing pollinator dependency, *Current Biology: CB* 18(20), 1572–1575. doi: 10.1016/j.cub.2008.08.066.

Aizen, M. A. and Harder, L. D. (2009). The global stock of domesticated honey bees is growing slower than agricultural demand for pollination, *Current Biology: CB* 19(11), 915–918. doi: 10.1016/j.cub.2009.03.071.

Al Naggar, Y. and Paxton, R. J. (2020). Mode of transmission determines the virulence of *Black queen cell virus* in adult honey bees, posing a future threat to bees and apiculture, *Viruses* 12(5), 535. doi: 10.3390/v12050535.

Alford, D. V. (1975). *Bumblebees*, London, UK: Davis-Poynter.

Alger, S. A., Burnham, P. A., Boncristiani, H. F. and Brody, A. K. (2019). RNA virus spillover from managed honeybees (*Apis mellifera*) to wild bumblebees (*Bombus* spp.), *PLoS ONE* 14(6), e0217822. doi: 10.1371/journal.pone.0217822.

Anderson, R. M. and May, R. M. (1981). The population dynamics of microparasites and their invertebrate hosts, *Philosophical Transactions of the Royal Society of London. B, Biological Sciences* 291(1054), 451–524. doi: 10.1098/rstb.1981.0005.

Antonovics, J. and Edwards, M. (2011). Spatio-temporal dynamics of bumblebee nest parasites (*Bombus* subgenus *Psythirus* ssp.) and their hosts (*Bombus* spp.), *Journal of Animal Ecology* 80(5), 999–1011. doi: 10.1111/j.1365-2656.2011.01846.x.

Arbetman, M. P., Gleiser, G., Morales, C. L., Williams, P. and Aizen, M. A. (2017). Global decline of bumblebees is phylogenetically structured and inversely related to species range size and pathogen incidence, *Proceedings. Biological Sciences* 284(1859), 20170204. doi: 10.1098/rspb.2017.0204.

Arbetman, M. P., Meeus, I., Morales, C. L., Aizen, M. A. and Smagghe, G. (2013). Alien parasite hitchhikes to Patagonia on invasive bumblebee, *Biological Invasions* 15(3), 489–494. doi: 10.1007/s10530-012-0311-0.

Arechavaleta-Velasco, M. E., Alcala-Escamilla, K., Robles-Rios, C., Tsuruda, J. M. and Hunt, G. J. (2012). Fine-scale linkage mapping reveals a small set of candidate genes influencing honey bee grooming behavior in response to varroa mites, *PLoS ONE* 7(11), e47269. doi: 10.1371/journal.pone.0047269.

Aronstein, K. A. and Murray, K. D. (2010). Chalkbrood disease in honey bees, *Journal of Invertebrate Pathology* 103 (Suppl. 1), S20–S29. doi: 10.1016/j.jip.2009.06.018.

Bailey, L. and Ball, B. V. (1991). *Honey Bee Pathology*, London, UK: Academic Press. doi: org/10.1016/B978-0-12-073481-8.50002-3.

Bailey, L., Gibbs, A. J. and Woods, R. D. (1963). Two viruses from adult honey bees (*Apis mellifera* Linnaeus), *Virology* 21(3), 390–395. doi: 10.1016/0042-6822(63)90200-9.

Bakonyi, T., Farkas, R., Szendroi, A., Dobos-Kovács, M. and Rusvai, M. (2002). Detection of acute bee paralysis virus by RT-PCR in honey bee and *Varroa destructor* field samples: rapid screening of representative Hungarian apiaries, *Apidologie* 33(1), 63–74. doi: 10.1051/APIDO:2001004.

Ball, B. V. and Allen, M. F. (1988). The prevalence of pathogens in honey bee (*Apis mellifera*) colonies infested with the parasitic mite *Varroa jacobsoni*, *Annals of Applied Biology* 113(2), 237–244. doi: 10.1111/j.1744-7348.1988.tb03300.x.

Barribeau, S. M., Sadd, B. M., du Plessis, L., Brown, M. J., Buechel, S. D., Cappelle, K., Carolan, J. C., Christiaens, O., Colgan, T. J., Erler, S., Evans, J., Helbing, S., Karaus, E., Lattorff, H. M., Marxer, M., Meeus, I., Näpflin, K., Niu, J., Schmid-Hempel, R., Smagghe, G., Waterhouse, R. M., Yu, N., Zdobnov, E. M. and Schmid-Hempel, P. (2015). A depauperate immune repertoire precedes evolution of sociality in bees, *Genome Biology* 16(1), 83. doi: 10.1186/s13059-015-0628-y.

Barron, A. B. (2015). Death of the bee hive: understanding the failure of an insect society, *Current Opinion in Insect Science* 10(0), 45–50. doi: 10.1016/j.cois.2015.04.004.

Bartlett, L. J., Boots, M., Brosi, B. J., de Roode, J. C., Delaplane, K. S., Hernandez, C. A. and Wilfert, L. (2021). Persistent effects of management history on honeybee colony virus abundances, *Journal of Invertebrate Pathology* 179, 107520. doi: 10.1016/j.jip.2020.107520.

Bartlett, L. J., Rozins, C., Brosi, B. J., Delaplane, K. S., de Roode, J. C., White, A., Wilfert, L. and Boots, M. (2019). Industrial bees: the impact of apicultural intensification on local disease prevalence, *Journal of Applied Ecology* 56(9), 2195–2205. doi: 10.1111/1365-2664.13461.

Beaurepaire, A., Piot, N., Doublet, V., Antunez, K., Campbell, E., Chantawannakul, P., Chejanovsky, N., Gajda, A., Heerman, M., Panziera, D., Smagghe, G., Yañez, O., de Miranda, J. R. and Dalmon, A. (2020). Diversity and global distribution of viruses of the western honey bee, *Apis mellifera*, *Insects* 11(4), 239. doi: 10.3390/insects11040239.

Biesmeijer, J. C., Roberts, S. P., Reemer, M., Ohlemüller, R., Edwards, M., Peeters, T., Schaffers, A. P., Potts, S. G., Kleukers, R., Thomas, C. D., Settele, J. and Kunin, W. E. (2006). Parallel declines in pollinators and insect-pollinated plants in Britain and the Netherlands, *Science* 313(5785), 351–354. doi: 10.1126/science.1127863.

Borsuk, G., Ptaszyńska, A. A., Olszewski, K., Domaciuk, M., Krutmuang, P. and Paleolog, J. (2017). A new method for quick and easy hemolymph collection from Apidae adults, *PLoS ONE* 12(1), e0170487. doi: 10.1371/journal.pone.0170487.

Bowen-Walker, P. L., Martin, S. J. and Gunn, A. (1999). The transmission of deformed wing virus between honeybees (*Apis mellifera* L.) by the ectoparasitic mite *Varroa jacobsoni* Oud, *Journal of Invertebrate Pathology* 73(1), 101–106. doi: 10.1006/jipa.1998.4807.

Bramke, K., Müller, U., McMahon, D. P. and Rolff, J. (2019). Exposure of larvae of the solitary bee *Osmia bicornis* to the honey bee pathogen *Nosema ceranae* affects life history, *Insects* 10(11), 380. doi: 10.3390/insects10110380.

Breeze, T. D., Bailey, A. P., Balcombe, K. G., Brereton, T., Comont, R., Edwards, M., Garratt, M. P., Harvey, M., Hawes, C., Isaac, N., Jitlal, M., Jones, C. M., Kunin, W. E., Lee, P., Morris, R. K. A., Musgrove, A., O'Connor, R. S., Peyton, J., Potts, S. G., Roberts, S. P. M., Roy, D. B., Roy, H. E., Tang, C. Q., Vanbergen, A. J. and Carvell, C. (2021). Pollinator monitoring more than pays for itself, *Journal of Applied Ecology* 58(1), 44–57. doi: 10.1111/1365-2664.13755.

Brown, M. J. F. and Paxton, R. J. (2009). The conservation of bees: a global perspective, *Apidologie* 40(3), 410–416. doi: 10.1051/apido/2009019.

Buchmann, S. L. and Nabhan, G. P. (1996). *The Forgotten Pollinators*, Washington, DC: Island Press.

Cameron, S. A., Lozier, J. D., Strange, J. P., Koch, J. B., Cordes, N., Solter, L. F. and Griswold, T. L. (2011). Patterns of widespread decline in North American bumble bees, *Proceedings of the National Academy of Sciences of the United States of America* 108(2), 662–667. doi: 10.1073/pnas.1014743108.

Cameron, S. A. and Sadd, B. M. (2020). Global trends in bumble bee health, *Annual Review of Entomology* 65(1), 209–232. doi: 10.1146/annurev-ento-011118-111847.

Carreck, N. (2005). The epidemiology of slow paralysis virus in honey bee colonies infested by *Varroa destructor* in the UK. In: Conference: 39th International Apicultural Congress. Dublin, Ireland. doi: 10.13140/2.1.1056.2562.

Carreck, N. L., Ball, B. V. and Martin, S. J. (2010). Honey bee colony collapse and changes in viral prevalence associated with *Varroa destructor*, *Journal of Apicultural Research* 49(1), 93–94. doi: 10.3896/IBRA.1.49.1.13.

Cavigli, I., Daughenbaugh, K. F., Martin, M., Lerch, M., Banner, K., Garcia, E., Brutscher, L. M. and Flenniken, M. L. (2016). Pathogen prevalence and abundance in honey bee colonies involved in almond pollination, *Apidologie* 47(2), 251–266. doi: 10.1007/s13592-015-0395-5.

Chantawannakul, P., de Guzman, L. I., Li, J. and Williams, G. R. (2016). Parasites, pathogens, and pests of honeybees in Asia, *Apidologie* 47(3), 301–324. doi: 10.1007/s13592-015-0407-5.

Chauzat, M.-P., Jacques, A., Laurent, M., Bougeard, S., Hendrikx, P. and Ribière-Chabert, M. (2016). Risk indicators affecting honeybee colony survival in Europe: one year of surveillance, *Apidologie* 47(3), 348–378. doi: 10.1007/s13592-016-0440-z.

Chen, Y., Zhao, Y., Hammond, J., Hsu, H. T., Evans, J. and Feldlaufer, M. (2004). Multiple virus infections in the honey bee and genome divergence of honey bee viruses, *Journal of Invertebrate Pathology* 87(2–3), 84–93. doi: 10.1016/j.jip.2004.07.005.

Colla, S. R., Otterstatter, M. C., Gegear, R. J. and Thomson, J. D. (2006). Plight of the bumble bee: pathogen spillover from commercial to wild populations, *Biological Conservation* 129(4), 461–467. doi: 10.1016/j.biocon.2005.11.013.

Cornman, R. S., Tarpy, D. R., Chen, Y., Jeffreys, L., Lopez, D., Pettis, J. S., vanEngelsdorp, D. and Evans, J. D. (2012). Pathogen webs in collapsing honey bee colonies, *PLoS ONE* 7(8), e43562. doi: 10.1371/journal.pone.0043562.

Cox-Foster, D. L., Conlan, S., Holmes, E. C., Palacios, G., Evans, J. D., Moran, N. A., Quan, P. L., Briese, T., Hornig, M., Geiser, D. M., Martinson, V., vanEngelsdorp, D., Kalkstein, A. L., Drysdale, A., Hui, J., Zhai, J., Cui, L., Hutchison, S. K., Simons, J. F., Egholm, M., Pettis, J. S. and Lipkin, W. I. (2007). A metagenomic survey of microbes in honey bee colony collapse disorder, *Science* 318(5848), 283–287. doi: 10.1126/SCIENCE.1146498.

Cremer, S., Armitage, S. A. O. and Schmid-Hempel, P. (2007). Social immunity, *Current Biology: CB* 17(16), R693–R702. doi: 10.1016/j.cub.2007.06.008.

Dainat, B., Dietemann, V., Imdorf, A. and Charrière, J. (2020). A scientific note on the "Liebefeld Method" to estimate honey bee colony strength: its history, use, and translation, *Apidologie* 51(3), 422–427. doi: 10.1007/s13592-019-00728-2.

Dainat, B., Evans, J. D., Chen, Y. P., Gauthier, L. and Neumann, P. (2012). Dead or alive: deformed wing virus and *Varroa destructor* reduce the life span of winter honeybees, *Applied and Environmental Microbiology* 78(4), 981–987. doi: 10.1128/AEM.06537-11.

Dainat, B., Ken, T., Berthoud, H. and Neumann, P. (2009). The ectoparasitic mite *Tropilaelaps mercedesae* (Acari, Laelapidae) as a vector of honeybee viruses, *Insectes Sociaux* 56(1), 40–43. doi: 10.1007/s00040-008-1030-5.

Danforth, B. N., Minckley, R. L. and Neff, J. L. (2019). *The Solitary Bees: Biology, Evolution, Conservation*, Princeton, NJ: Princeton University Press.

Daszak, P., Cunningham, A. A. and Hyatt, A. D. (2000). Emerging infectious diseases of wildlife–threats to biodiversity and human health, *Science* 287(5452), 443–449. doi: 10.1126/science.287.5452.443.

Davis, A. E., Deutsch, K. R., Torres, A. M., Mata Loya, M. J., Cody, L. V., Harte, E., Sossa, D., Muñiz, P. A., Ng, W. H. and McArt, S. H. (2021). *Eristalis* flower flies can be mechanical vectors of the common trypanosome bee parasite, *Crithidia bombi*, *Scientific Reports* 11(1), 15852. doi: 10.1038/s41598-021-95323-w.

de Guzman, L. I., Williams, G. R., Khongphinitbunjong, K. and Chantawannakul, P. (2017). Ecology, life history, and management of *Tropilaelaps* mites, *Journal of Economic Entomology* 110(2), 319–332. doi: 10.1093/jee/tow304.

de Miranda, J. R., Cordoni, G. and Budge, G. (2010). The Acute bee paralysis virus–Kashmir bee virus–Israeli acute paralysis virus complex, *Journal of Invertebrate Pathology* 103 (Suppl. 1), S30–S47. doi: 10.1016/j.jip.2009.06.014.

Delaplane, K. S. and Mayer, D. F. (2000). *Crop Pollination by Bees*, Wallingford, UK: CABI Publishing. doi: 10.1079/9780851994482.0000.

Desai, S. D., Eu, Y. J., Whyard, S. and Currie, R. W. (2012). Reduction in deformed wing virus infection in larval and adult honey bees (*Apis mellifera* L.) by double-stranded RNA ingestion, *Insect Molecular Biology* 21(4), 446–455. doi: 10.1111/j.1365-2583.2012.01150.x.

Dharampal, P. S., Carlson, C., Currie, C. R. and Steffan, S. A. (2019). Pollen-borne microbes shape bee fitness, *Proceedings. Biological Sciences* 286(1904), 20182894. doi: 10.1098/rspb.2018.2894.

Di Prisco, G., Annoscia, D., Margiotta, M., Ferrara, R., Varricchio, P., Zanni, V., Caprio, E., Nazzi, F. and Pennacchio, F. (2016). A mutualistic symbiosis between a parasitic mite and a pathogenic virus undermines honey bee immunity and health, *Proceedings of the National Academy of Sciences of the United States of America* 113(12), 3203–3208. doi: 10.1073/pnas.1523515113.

Di Prisco, G., Cavaliere, V., Annoscia, D., Varricchio, P., Caprio, E., Nazzi, F., Gargiulo, G. and Pennacchio, F. (2013). Neonicotinoid clothianidin adversely affects insect immunity and promotes replication of a viral pathogen in honey bees, *Proceedings of the National Academy of Sciences of the United States of America* 110(46), 18466–18471. doi: 10.1073/pnas.1314923110.

Dicks, L. V., Breeze, T. D., Ngo, H. T., Senapathi, D., An, J., Aizen, M. A., Basu, P., Buchori, D., Galetto, L., Garibaldi, L. A., Gemmill-Herren, B., Howlett, B. G., Imperatriz-Fonseca, V. L., Johnson, S. D., Kovács-Hostyánszki, A., Kwon, Y. J., Lattorff, H. M. G., Lungharwo, T., Seymour, C. L., Vanbergen, A. J. and Potts, S. G. (2021). A global-scale expert assessment of drivers and risks associated with pollinator decline, *Nature Ecology and Evolution* 5(10), 1453–1461. doi: 10.1038/s41559-021-01534-9.

Dittes, J., Schäfer, M. O., Aupperle-Lellbach, H., Mülling, C. K. W. and Emmerich, I. U. (2020a). Overt infection with Chronic bee paralysis virus (CBPV) in two honey bee colonies, *Veterinary Sciences* 7(3), 142. doi: 10.3390/vetsci7030142.

Dittes, J., Aupperle-Lellbach, H., Schäfer, M. O., Mülling, C. K. W. and Emmerich, I. U. (2020b). Veterinary diagnostic approach of common virus diseases in adult honeybees, *Veterinary Sciences* 7(4), 159. doi: 10.3390/vetsci7040159.

Donovan, B. J. (2007). *Apoidea (Insecta: Hymenoptera), Fauna of New Zealand* (vol. 57), doi: 10.7931/J2/FNZ.57.

Dosch, C., Manigk, A., Streicher, T., Tehel, A., Paxton, R. J. and Tragust, S. (2021). The gut microbiota can provide viral tolerance in the honey bee, *Microorganisms* 9(4), 871. doi: 10.3390/microorganisms9040871.

Doublet, V., Labarussias, M., de Miranda, J. R., Moritz, R. F. and Paxton, R. J. (2015). Bees under stress: sublethal doses of a neonicotinoid pesticide and pathogens interact to elevate honey bee mortality across the life cycle, *Environmental Microbiology* 17(4), 969–983. doi: 10.1111/1462-2920.12426.

Duay, P., De Jong, D. and Engels, W. (2002). Decreased flight performance and sperm production in drones of the honeybee (*Apis mellifera*) slightly infested by *Varroa destructor* mites during pupal development, *Genetics and Molecular Research: GMR* 1(3), 227–232. Available at: http://www.ncbi.nlm.nih.gov/pubmed/14963829.

Durrer, S. and Schmid-Hempel, P. (1994). Shared use of flowers leads to horizontal pathogen transmission, *Proceedings of the Royal Society of London. Series B: Biological Sciences* 258(1353), 299–302. doi: 10.1098/rspb.1994.0176.

Eickwort, G. C. (1990). Associations of mites with social insects, *Annual Review of Entomology* 35(1), 469–488. doi: 10.1146/annurev.en.35.010190.002345.

Ellis, J. D., Delaplane, K. S., Richards, C. S., Hepburn, R., Berry, J. A. and Elzen, P. J. (2004). Hygienic behavior of Cape and European *Apis mellifera* (Hymenoptera: Apidae) toward *Aethina tumida* (Coleoptera: Nitidulidae) eggs oviposited in sealed bee brood, *Annals of the Entomological Society of America* 97(4), 860–864. doi: 10.1603/0013-8746(2004)097[0860:HBOCAE]2.0.CO;2.

Emery, O., Schmidt, K. and Engel, P. (2017). Immune system stimulation by the gut symbiont *Frischella perrara* in the honey bee (*Apis mellifera*), *Molecular Ecology* 26(9), 2576–2590. doi: 10.1111/mec.14058.

Emsen, B., Guzman-Novoa, E., Hamiduzzaman, M. M., Eccles, L., Lacey, B., Ruiz-Pérez, R. A. and Nasr, M. (2016). Higher prevalence and levels of *Nosema ceranae* than *Nosema apis* infections in Canadian honey bee colonies, *Parasitology Research* 115(1), 175–181. doi: 10.1007/s00436-015-4733-3.

Evans, J. D., Aronstein, K., Chen, Y. P., Hetru, C., Imler, J. L., Jiang, H., Kanost, M., Thompson, G. J., Zou, Z. and Hultmark, D. (2006). Immune pathways and defence mechanisms in honey bees *Apis mellifera*, *Insect Molecular Biology* 15(5), 645–656. doi: 10.1111/j.1365-2583.2006.00682.x.

Evans, J. D., Banmeke, O., Palmer-Young, E. C., Chen, Y. and Ryabov, E. V. (2021). Beeporter: tools for high-throughput analyses of pollinator-virus infections, *Molecular Ecology Resources* 22(3), 978–987. doi: 10.22541/au.161884293.36210275/v1.

Evans, J. D. and Spivak, M. (2010). Socialized medicine: individual and communal disease barriers in honey bees, *Journal of Invertebrate Pathology* 103 (Suppl. 1), S62–S72. doi: 10.1016/j.jip.2009.06.019.

Faber, N. R., Meiborg, A. B., Mcfarlane, G. R., Gorjanc, G. and Harpur, B. A. (2021). A gene drive does not spread easily in populations of the honey bee parasite *Varroa destructor*, *Apidologie* 52(6), 1112–1127, doi: 10.1007/s13592-021-00891-5.

Figueroa, L. L., Compton, S., Grab, H. and McArt, S. H. (2021). Functional traits linked to pathogen prevalence in wild bee communities, *Scientific Reports* 11(1), 7529. doi: 10.1038/s41598-021-87103-3.

Forfert, N., Natsopoulou, M. E., Paxton, R. J. and Moritz, R. F. A. (2016). Viral prevalence increases with regional colony abundance in honey bee drones (*Apis mellifera* L), *Infection, Genetics and Evolution: Journal of Molecular Epidemiology and Evolutionary Genetics in Infectious Diseases* 44, 549–554. doi: 10.1016/j.meegid.2016.07.017.

Forsgren, E. (2010). European foulbrood in honey bees, *Journal of Invertebrate Pathology* 103 (Suppl. 1), S5–S9. doi: 10.1016/j.jip.2009.06.016.

Forsgren, E., de Miranda, J. R., Isaksson, M., Wei, S. and Fries, I. (2009). Deformed wing virus associated with *Tropilaelaps mercedesae* infesting European honey bees (*Apis mellifera*), *Experimental and Applied Acarology* 47(2), 87–97. doi: 10.1007/s10493-008-9204-4.

Forsgren, E., Olofsson, T. C., Vásquez, A. and Fries, I. (2010). Novel lactic acid bacteria inhibiting *Paenibacillus larvae* in honey bee larvae, *Apidologie* 41(1), 99–108. doi: 10.1051/apido/2009065.

Francis, R. M., Nielsen, S. L. and Kryger, P. (2013). Varroa-virus interaction in collapsing honey bee colonies, *PLoS ONE* 8(3), e57540. doi: 10.1371/journal.pone.0057540.

Free, J. B. (1993). Insect Pollination of Crops. 2nd. edn. London, UK: Academic Press.

Freitas, B. M., Imperatriz-Fonseca, V. L., Medina, L. M., Kleinert, AdM. P., Galetto, L., Nates-Parra, G. and Quezada-Euán, J. J. G. (2009). Diversity, threats and conservation of native bees in the Neotropics, *Apidologie* 40(3), 332–346. doi: 10.1051/apido/2009012.

Fries, I. and Camazine, S. (2001). Implications of horizontal and vertical pathogen transmission for honey bee epidemiology, *Apidologie* 32(3), 199–214. doi: 10.1051/apido:2001122.

Fries, I., Paxton, R. J., Tengö, J., Slemenda, S. B., da Silva, A. J. and Pieniazek, N. J. (1999). Morphological and molecular characterization of *Antonospora scoticae n.* gen., n. sp. (Protozoa, microsporidia) a parasite of the communal bee, Andrena scotica Perkins, 1916 (Hymenoptera, andrenidae), *European Journal of Protistology* 35(2), 183–193. doi: 10.1016/S0932-4739(99)80036-4.

Fürst, M. A., McMahon, D. P., Osborne, J. L., Paxton, R. J. and Brown, M. J. (2014). Disease associations between honeybees and bumblebees as a threat to wild pollinators, *Nature* 506(7488), 364–366. doi: 10.1038/nature12977.

Garibaldi, L. A., Steffan-Dewenter, I., Winfree, R., Aizen, M. A., Bommarco, R., Cunningham, S. A., Kremen, C., Carvalheiro, L. G., Harder, L. D., Afik, O., Bartomeus, I., Benjamin, F., Boreux, V., Cariveau, D., Chacoff, N. P., Dudenhöffer, J. H., Freitas, B. M., Ghazoul, J., Greenleaf, S., Hipólito, J., Holzschuh, A., Howlett, B., Isaacs, R., Javorek, S. K., Kennedy, C. M., Krewenka, K. M., Krishnan, S., Mandelik, Y., Mayfield, M. M., Motzke, I., Munyuli, T., Nault, B. A., Otieno, M., Petersen, J., Pisanty, G., Potts, S. G., Rader, R., Ricketts, T. H., Rundlöf, M., Seymour, C. L., Schüepp, C., Szentgyörgyi, H., Taki, H., Tscharntke, T., Vergara, C. H., Viana, B. F., Wanger, T. C., Westphal, C., Williams, N. and Klein, A. M. (2013). Wild pollinators enhance fruit set of crops regardless of honey bee abundance, *Science* 339(6127), 1608–1611. doi: 10.1126/science.1230200.

Genersch, E. (2010). American Foulbrood in honeybees and its causative agent, *Paenibacillus larvae*, *Journal of Invertebrate Pathology* 103 (Suppl. 1), S10–S19. doi: 10.1016/j.jip.2009.06.015.

Genersch, E., von der Ohe, W., Kaatz, H., Schroeder, A., Otten, C., Büchler, R., Berg, S., Ritter, W., Mühlen, W., Gisder, S., Meixner, M., Liebig, G. and Rosenkranz, P. (2010). The German bee monitoring project: a long term study to understand periodically high winter losses of honey bee colonies, *Apidologie* 41(3), 332–352. doi: 10.1051/apido/2010014.

Genersch, E., Yue, C., Fries, I. and de Miranda, J. R. (2006). Detection of deformed wing virus, a honey bee viral pathogen, in bumble bees (*Bombus terrestris* and *Bombus* pascuorum) with wing deformities, *Journal of Invertebrate Pathology* 91(1), 61–63. doi: 10.1016/j.jip.2005.10.002.

Gisder, S. and Genersch, E. (2017). Viruses of commercialized insect pollinators, *Journal of Invertebrate Pathology* 147, 51–59. doi: 10.1016/j.jip.2016.07.010.

Gisder, S. and Genersch, E. (2021). Direct evidence for infection of *Varroa destructor* mites with the bee-pathogenic deformed wing virus variant B, but not variant A, via fluorescence *in situ* hybridization analysis, *Journal of Virology* 95(5), e01786-20, / jvi/95/5/JVI.01786-20.atom. doi: 10.1128/JVI.01786-20.

Gisder, S., Horchler, L., Pieper, F., Schüler, V., Šima, P. and Genersch, E. (2020). Rapid gastrointestinal passage may protect *Bombus terrestris* from becoming a true host for *Nosema ceranae*, *Applied and Environmental Microbiology* 86(12), e00629-20. doi: 10.1128/AEM.00629-20.

Gómez-Moracho, T., Buendía-Abad, M., Benito, M., García-Palencia, P., Barrios, L., Bartolomé, C., Maside, X., Meana, A., Jiménez-Antón, M. D., Olías-Molero, A. I., Alunda, J. M., Martín-Hernández, R. and Higes, M. (2020). Experimental evidence of harmful effects of *Crithidia* mellificae and *Lotmaria passim* on honey bees, *International Journal for Parasitology* 50(13), 1117–1124. doi: 10.1016/j.ijpara.2020.06.009.

Goulson, D. (2009). *Bumblebees. Behaviour, Ecology and Conservation*, Oxford, UK: Oxford University Press.

Goulson, D., Nicholls, E., Botías, C. and Rotheray, E. L. (2015). Bee declines driven by combined stress from parasites, pesticides, and lack of flowers, *Science* 347(6229), 1255957. doi: 10.1126/science.1255957.

Graystock, P., Goulson, D. and Hughes, W. O. H. (2015). Parasites in bloom: flowers aid dispersal and transmission of pollinator parasites within and between bee species, *Proceedings. Biological Sciences* 282(1813), 20151371. doi: 10.1098/rspb.2015.1371.

Graystock, P., Ng, W. H., Parks, K., Tripodi, A. D., Muñiz, P. A., Fersch, A. A., Myers, C. R., McFrederick, Q. S. and McArt, S. H. (2020). Dominant bee species and floral abundance drive parasite temporal dynamics in plant-pollinator communities, *Nature Ecology and Evolution* 4(10), 1358–1367. doi: 10.1038/s41559-020-1247-x.

Grozinger, C. M. and Flenniken, M. L. (2019). Bee viruses: ecology, pathogenicity, and impacts, *Annual Review of Entomology* 64(1), 205–226. doi: 10.1146/annurev-ento-011118-111942.

Grozinger, C. M. and Zayed, A. (2020). Improving bee health through genomics, *Nature Reviews. Genetics* 21(5), 277–291. doi: 10.1038/s41576-020-0216-1.

Grupe, A. C. and Alisha Quandt, C. A. (2020). A growing pandemic: a review of *Nosema* parasites in globally distributed domesticated and native bees, *PLoS Pathogens* 16(6), e1008580. doi: 10.1371/journal.ppat.1008580.

Gusachenko, O. N., Woodford, L., Balbirnie-Cumming, K., Campbell, E. M., Christie, C. R., Bowman, A. S. and Evans, D. J. (2020). Green bees: reverse genetic analysis of deformed wing virus transmission, replication, and tropism, *Viruses* 12(5), 532. doi: 10.3390/v12050532.

Hallmann, C. A., Sorg, M., Jongejans, E., Siepel, H., Hofland, N., Schwan, H., Stenmans, W., Müller, A., Sumser, H., Hörren, T., Goulson, D. and de Kroon, H. (2017). More than 75 percent decline over 27 years in total flying insect biomass in protected areas, *PLoS ONE* 12(10), e0185809. doi: 10.1371/journal.pone.0185809.

Hansen, H. and Brødsgaard, C. J. (1999). American foulbrood: a review of its biology, diagnosis and control, *Bee World* 80(1), 5-23. doi: 10.1080/0005772X.1999.11099415.

Heard, T. A. (1999). The role of stingless bees in crop pollination, *Annual Review of Entomology* 44, 183–206. doi: 10.1146/annurev.ento.44.1.183.

Higes, M., Martín-Hernández, R., Botías, C., Bailón, E. G., González-Porto, A. V., Barrios, L., Del Nozal, M. J., Bernal, J. L., Jiménez, J. J., Palencia, P. G. and Meana, A. (2008). How natural infection by *Nosema ceranae* causes honeybee colony collapse, *Environmental Microbiology* 10(10), 2659–2669. doi: 10.1111/j.1462-2920.2008.01687.x.

Holmes, E. C. (2009). *The Evolution and Emergence of RNA Viruses*, Oxford, UK: Oxford University Press. doi: 10.3201/eid1605.100164.

Hooper, T. (1976). *Guide to Bees and Honey*, Poole, UK: Blandford Press.

Huang, S., Li, J., Zhang, Y., Li, Z., Evans, J. D., Rose, R., Gilligan, T. M., LeBrun, A., He, N., Zheng, T., Zhang, T., Hamilton, M. and Chen, Y. P. (2021a). A novel method for the detection and diagnosis of virus infections in honey bees, *Journal of Virological Methods* 293, 114163. doi: 10.1016/j.jviromet.2021.114163.

Huang, W. F., Zhang, Y., Mehmood, S., Wang, Z., Hou, C. and Li, Z. (2021b). Updating Sacbrood virus quantification PCR method using a TaqMan-MGB probe, *Veterinary Sciences* 8(4), 63. doi: 10.3390/vetsci8040063.

Hung, K. J., Kingston, J. M., Albrecht, M., Holway, D. A. and Kohn, J. R. (2018). The worldwide importance of honey bees as pollinators in natural habitats, *Proceedings. Biological Sciences* 285(1870), 20172140. doi: 10.1098/rspb.2017.2140.

Jack, C. J. and Ellis, J. D. (2021). Integrated pest management control of *Varroa destructor* (Acari: Varroidae), the most damaging pest of (*Apis mellifera* L. (Hymenoptera: Apidae)) colonies, *Journal of Insect Science* 21(5), 6. doi: 10.1093/jisesa/ieab058.

Jaffé, R., Dietemann, V., Allsopp, M. H., Costa, C., Crewe, R. M., Dall'olio, R., DE LA Rúa, P., El-Niweiri, M. A., Fries, I., Kezic, N., Meusel, M. S., Paxton, R. J., Shaibi, T., Stolle, E. and Moritz, R. F. (2010). Estimating the density of honeybee colonies across their natural range to fill the gap in pollinator decline censuses, *Conservation Biology: The Journal of the Society for Conservation Biology* 24(2), 583–593. doi: 10.1111/j.1523-1739.2009.01331.x.

Jha, S. and Kremen, C. (2013). Resource diversity and landscape-level homogeneity drive native bee foraging, *Proceedings of the National Academy of Sciences of the United States of America* 110(2), 555–558. doi: 10.1073/pnas.1208682110.

Johnson, R. (2010). Honey bee colony collapse disorder. *Honey Bees and Colony Collapse Disorder: Select Analyses*, Congress of the USA, pp. 69–92.

Keeling, M. J., Franklin, D. N., Datta, S., Brown, M. A. and Budge, G. E. (2017). Predicting the spread of the Asian hornet (*Vespa Velutina*) following its incursion into Great Britain, *Scientific Reports* 7(1), 6240. doi: 10.1038/s41598-017-06212-0.

Kevan, P. G., Clark, E. A. and Thomas, V. G. (1990). Insect pollinators and sustainable agriculture, *American Journal of Alternative Agriculture* 5(1), 13–22. doi: 10.1017/S0889189300003179.

Khongphinitbunjong, K., de Guzman, L. I., Tarver, M. R., Rinderer, T. E. and Chantawannakul, P. (2015). Interactions of *Tropilaelaps mercedesae*, honey bee viruses and immune response in *Apis mellifera*, *Journal of Apicultural Research* 54(1), 40–47. doi: 10.1080/00218839.2015.1041311.

Klee, J., Besana, A. M., Genersch, E., Gisder, S., Nanetti, A., Tam, D. Q., Chinh, T. X., Puerta, F., Ruz, J. M., Kryger, P., Message, D., Hatjina, F., Korpela, S., Fries, I. and Paxton, R. J. (2007). Widespread dispersal of the microsporidian *Nosema ceranae*, an emergent pathogen of the western honey bee, *Apis mellifera*, *Journal of Invertebrate Pathology* 96(1), 1–10. doi: 10.1016/j.jip.2007.02.014.

Klein, A. M., Vaissière, B. E., Cane, J. H., Steffan-Dewenter, I., Cunningham, S. A., Kremen, C. and Tscharntke, T. (2007). Importance of pollinators in changing landscapes for world crops, *Proceedings. Biological Sciences* 274(1608), 303–313. doi: 10.1098/rspb.2006.3721.

Koeniger, N., Muzaffar, N. and Koeniger, N. (1988). Lifespan of the parasitic honeybee mite, *Tropilaelaps clareae*, on Apis Cerana, Dorsata and Mellifera, *Journal of Apicultural Research* 27(4), 207–212. doi: 10.1080/00218839.1988.11100804.

Kralj, J. and Fuchs, S. (2006). Parasitic *Varroa destructor* mites influence flight duration and homing ability of infested *Apis mellifera* foragers, *Apidologie* 37(5), 577–587. doi: 10.1051/apido:2006040.

Kwong, W. K. and Moran, N. A. (2016). Gut microbial communities of social bees, *Nature Reviews. Microbiology* 14(6), 374–384. doi: 10.1038/nrmicro.2016.43.

Laughton, A. M., O'Connor, C. O. and Knell, R. J. (2017). Responses to a warming world: integrating life history, immune investment, and pathogen resistance in a model insect species, *Ecology and Evolution* 7(22), 9699–9710. doi: 10.1002/ece3.3506.

Leonard, S. P., Powell, J. E., Perutka, J., Geng, P., Heckmann, L. C., Horak, R. D., Davies, B. W., Ellington, A. D., Barrick, J. E. and Moran, N. A. (2020). Engineered symbionts activate honey bee immunity and limit pathogens, *Science* 367(6477), 573–576. doi: 10.1126/science.aax9039.

Levin, S., Sela, N., Erez, T., Nestel, D., Pettis, J., Neumann, P. and Chejanovsky, N. (2019). New viruses from the ectoparasite mite *Varroa destructor* infesting *Apis mellifera* and *Apis* cerana, *Viruses* 11(2), 94. doi: 10.3390/v11020094.

Liu, Q., Lei, J., Darby, A. C. and Kadowaki, T. (2020). Trypanosomatid parasite dynamically changes the transcriptome during infection and modifies honey bee physiology, *Communications Biology* 3(1), 51. doi: 10.1038/s42003-020-0775-x.

Locke, B. (2016). Natural varroa mite-surviving *Apis mellifera* honeybee populations, *Apidologie* 47(3), 467–482. doi: 10.1007/s13592-015-0412-8.

Manley, R., Temperton, B., Doyle, T., Gates, D., Hedges, S., Boots, M. and Wilfert, L. (2019). Knock-on community impacts of a novel vector: spillover of emerging DWV-B from varroa-infested honeybees to wild bumblebees, *Ecology Letters* 22(8), 1306–1315. doi: 10.1111/ele.13323.

Martin, S. J. (2001). The role of varroa and viral pathogens in the collapse of honeybee colonies: a modelling approach, *Journal of Applied Ecology* 38(5), 1082–1093. doi: 10.1046/j.1365-2664.2001.00662.x.

Martin, S. J. and Brettell, L. E. (2019). Deformed wing virus in honeybees and other insects, *Annual Review of Virology* 6(1), 49–69. doi: 10.1146/annurev-virology-092818-015700.

Martin, S. J., Highfield, A. C., Brettell, L., Villalobos, E. M., Budge, G. E., Powell, M., Nikaido, S. and Schroeder, D. C. (2012). Global honey bee viral landscape altered by a parasitic mite, *Science* 336(6086), 1304–1306. doi: 10.1126/science.1220941.

Martínez-López, V., Ruiz, C., Muñoz, I., Ornosa, C., Higes, M., Martín-Hernández, R. and De la Rúa, P. (2021). Detection of Microsporidia in pollinator communities of a Mediterranean biodiversity hotspot for wild bees, *Microbial Ecology*. doi: 10.1007/s00248-021-01854-0.

Martinson, V. G., Danforth, B. N., Minckley, R. L., Rueppell, O., Tingek, S. and Moran, N. A. (2011). A simple and distinctive microbiota associated with honey bees and bumble bees, *Molecular Ecology* 20(3), 619–628. doi: 10.1111/j.1365-294X.2010.04959.x.

Matheson, A. (1993). World bee health report, *Bee World* 74(4), 176–212. doi: 10.1080/0005772X.1993.11099183.

McArt, S. H., Koch, H., Irwin, R. E. and Adler, L. S. (2014). Arranging the bouquet of disease: floral traits and the transmission of plant and animal pathogens, *Ecology Letters* 17(5), 624–636. doi: 10.1111/ele.12257.

McGregor, S. E. (1976). *Insect Pollination of Cultivated Crop Plants*, Washington, DC: United States Department of Agriculture, Agricultural Research Services.

McMahon, D. P., Fürst, M. A., Caspar, J., Theodorou, P., Brown, M. J. F. and Paxton, R. J. (2015). A sting in the spit: widespread cross-infection of multiple RNA viruses across wild and managed bees, *Journal of Animal Ecology* 84(3), 615–624. doi: 10.1111/1365-2656.12345.

McMahon, D. P., Natsopoulou, M. E., Doublet, V., Fürst, M., Weging, S., Brown, M. J., Gogol-Döring, A. and Paxton, R. J. (2016). Elevated virulence of an emerging viral genotype as a driver of honeybee loss, *Proceedings. Biological Sciences* 283(1833), 20160811. doi: 10.1098/rspb.2016.0811.

McMahon, D. P., Wilfert, L., Paxton, R. J. and Brown, M. J. F. (2018). Emerging viruses in bees: from molecules to ecology, *Advances in Virus Research* 101, 251–291. doi: 10.1016/bs.aivir.2018.02.008.

McMenamin, A. J., Daughenbaugh, K. F., Parekh, F., Pizzorno, M. C. and Flenniken, M. L. (2018). Honey bee and bumble bee antiviral defense, *Viruses* 10(8), 395. doi: 10.3390/v10080395.

McMenamin, A. J. and Genersch, E. (2015). Honey bee colony losses and associated viruses, *Current Opinion in Insect Science* 8, 121–129. doi: 10.1016/j.cois.2015.01.015.

Michener, C. D. (1974). *The Social Behavior of the Bees. A Comparative Study*, Cambridge, MA: Belknap Press of Harvard University Press.

Milbrath, M. O., van Tran, T., Huang, W. F., Solter, L. F., Tarpy, D. R., Lawrence, F. and Huang, Z. Y. (2015). Comparative virulence and competition between *Nosema apis* and *Nosema ceranae* in honey bees (*Apis mellifera*), *Journal of Invertebrate Pathology* 125, 9–15. doi: 10.1016/j.jip.2014.12.006.

Mondet, F., de Miranda, J. R., Kretzschmar, A., Le Conte, Y. and Mercer, A. R. (2014). On the front line: quantitative virus dynamics in honeybee (*Apis mellifera* L.) colonies along a new expansion front of the parasite *Varroa destructor*, *PLoS Pathogens* 10(8), e1004323. doi: 10.1371/journal.ppat.1004323.

Moore, J., Jironkin, A., Chandler, D., Burroughs, N., Evans, D. J. and Ryabov, E. V. (2011). Recombinants between deformed wing virus and *Varroa destructor* virus-1 may prevail in *Varroa destructor*-infested honeybee colonies, *Journal of General Virology* 92(1), 156–161. doi: 10.1099/vir.0.025965-0.

Mordecai, G. J., Wilfert, L., Martin, S. J., Jones, I. M. and Schroeder, D. C. (2016). Diversity in a honey bee pathogen: first report of a third master variant of the deformed wing virus quasispecies, *ISME Journal* 10(5), 1264–1273. doi: 10.1038/ismej.2015.178.

Moritz, R. F. A., Kraus, F. B., Huth-Schwarz, A., Wolf, S., Carrillo, C. A. C., Paxton, R. J. and Vandame, R. (2013). Number of honeybee colonies in areas with high and low beekeeping activity in Southern Mexico, *Apidologie* 44(1), 113–120. doi: 10.1007/s13592-012-0163-8.

Müller, C. B. (1994). Parasite-induced digging behaviour in bumble bee workers, *Animal Behaviour* 48(4), 961–966.

Müller, C. B. and Schmid-Hempel, P. (1993). Exploitation of cold temperature as defence against parasitoids in bumblebees, *Nature* 363(6424), 65–67. doi: 10.1038/363065a0.

Murray, T. E., Coffey, M. F., Kehoe, E. and Horgan, F. G. (2013). Pathogen prevalence in commercially reared bumble bees and evidence of spillover in conspecific populations, *Biological Conservation* 159, 269–276. doi: 10.1016/j.biocon.2012.10.021.

Nanetti, A., Bortolotti, L. and Cilia, G. (2021). Pathogens spillover from honey bees to other arthropods, *Pathogens* 10(8), 1044. doi: 10.3390/pathogens10081044.

Natsopoulou, M. E., McMahon, D. P., Doublet, V., Bryden, J. and Paxton, R. J. (2015). Interspecific competition in honeybee intracellular gut parasites is asymmetric and favours the spread of an emerging infectious disease, *Proceedings. Biological Sciences* 282(1798), 20141896. doi: 10.1098/rspb.2014.1896.

Natsopoulou, M. E., McMahon, D. P., Doublet, V., Frey, E., Rosenkranz, P. and Paxton, R. J. (2017). The virulent, emerging genotype B of deformed wing virus is closely linked to overwinter honeybee worker loss, *Scientific Reports* 7(1), 5242. doi: 10.1038/s41598-017-05596-3.

Natsopoulou, M. E., Doublet, V. and Paxton, R. J. (2016). European isolates of the Microsporidia *Nosema apis* and *Nosema ceranae* have similar virulence in laboratory tests on European worker honey bees, *Apidologie* 47(1), 57–65. doi: 10.1007/s13592-015-0375-9.

Naug, D. (2008). Structure of the social network and its influence on transmission dynamics in a honeybee colony, *Behavioral Ecology and Sociobiology* 62(11), 1719–1725. doi: 10.1007/s00265-008-0600-x.

Nazzi, F., Brown, S. P., Annoscia, D., Del Piccolo, F., Di Prisco, G., Varricchio, P., Della Vedova, G., Cattonaro, F., Caprio, E. and Pennacchio, F. (2012). Synergistic parasite-pathogen interactions mediated by host immunity can drive the collapse of honeybee colonies, *PLoS Pathogens* 8(6), e1002735. doi: 10.1371/journal.ppat.1002735.

Nazzi, F. and Le Conte, Y. (2016). Ecology of *Varroa destructor*, the major ectoparasite of the western honey bee, *Apis mellifera*, *Annual Review of Entomology* 61(1), 417–432. doi: 10.1146/annurev-ento-010715-023731.

Neumann, P. and Carreck, N. L. (2010). Honey bee colony losses. *Journal of Apicultural Research* 49, 1–6. doi: 10.3896/IBRA.1.49.1.01.

Neumann, P. and Ellis, J. D. (2008). The small hive beetle (Aethina tumida Murray, Coleoptera: Nitidulidae): distribution, biology and control of an invasive species. *Journal of Apicultural Research* 47, 181–183. doi: 10.1080/00218839.2008.11101453.

Nieto, A., et al. (2014). European red list of bees. In: Luxembourg: International Union for Conservation of Nature and Natural Resources Global Species Programme: Publication Office of the European Commission. doi: 10.2779/77003.

Norton, A. M., Remnant, E. J., Tom, J., Buchmann, G., Blacquiere, T. and Beekman, M. (2021). Adaptation to vector-based transmission in a honeybee virus, *Journal of Animal Ecology* 90(10), 2254–2267. doi: 10.1111/1365-2656.13493.

Oldroyd, B. P. (2007). What's killing American honey bees?, *PLoS Biology* 5(6), e168. doi: 10.1371/journal.pbio.0050168.

Ongus, J. R., Peters, D., Bonmatin, J. M., Bengsch, E., Vlak, J. M. and van Oers, M. M. (2004). Complete sequence of a picorna-like virus of the genus Iflavirus replicating in the mite *Varroa destructor*, *Journal of General Virology* 85(12), 3747–3755. doi: 10.1099/vir.0.80470-0.

Ono, M., Igarashi, T., Ohno, E. and Sasaki, M. (1995). Unusual thermal defence by a honeybee against mass attack by hornets, *Nature* 377(6547), 334–336. doi: 10.1038/377334a0.

Osterman, J., Aizen, M. A., Biesmeijer, J. C., Bosch, J., Howlett, B. G., Inouye, D. W., Jung, C., Martins, D. J., Medel, R., Pauw, A., Seymour, C. L. and Paxton, R. J. (2021). Global trends in the number and diversity of managed pollinator species, *Agriculture, Ecosystems and Environment* 322, 107653. doi: 10.1016/j.agee.2021.107653.

Otti, O. and Schmid-Hempel, P. (2007). *Nosema bombi*: a pollinator parasite with detrimental fitness effects, *Journal of Invertebrate Pathology* 96(2), 118–124. doi: 10.1016/j.jip.2007.03.016.

Palmer-Young, E. C., Raffel, T. R. and Evans, J. D. (2021). Hot and sour: parasite adaptations to honeybee body temperature and pH, *Proceedings of the Royal Society B: Biological Sciences* 288(1964). doi: 10.1101/2021.07.03.447385.

Pascall, D. J., Tinsley, M. C., Obbard, D. J. and Wilfert, L. (2019). Host evolutionary history predicts virus prevalence across bumblebee species, *bioRxiv*, 498717. doi: 10.1101/498717.

Paxton, R. J., Fries, I., Pieniazek, N. J. and Tengö, J. (1997). High incidence of infection of an undescribed microsporidium (microspora) in the communal bee *Andrena scotica* (Hymenoptera, Andrenidae), *Apidologie* 28(3–4), 129–141. doi: 10.1051/apido:19970304.

Paxton, R. J. (2020). A microbiome silver bullet for honey bees, *Science* 367(6477), 504–506. doi: 10.1126/science.aba6135.

Paxton, R. J., Tengö, J. and Hedstrom, L. (1996). Dipteran parasites and other associates of a communal bee, *Andrena scotica* (Hymenoptera: Apoidea), on Öland, SE Sweden, *Entomologisk Tidskrift* 117, 165–178.

Piot, N., Meeus, I., Kleijn, D., Scheper, J., Linders, T. and Smagghe, G. (2019). Establishment of wildflower fields in poor quality landscapes enhances micro-parasite prevalence in wild bumble bees, *Oecologia* 189(1), 149–158. doi: 10.1007/s00442-018-4296-y.

Pitts-Singer, T. L. and Cane, J. H. (2011). The alfalfa leafcutting bee, Megachile rotundata: the world's most intensively managed solitary bee. *Annual Review of Entomology* 56, 221–237. doi: 10.1146/annurev-ento-120709-144836.

Plischuk, S., Meeus, I., Smagghe, G. and Lange, C. E. (2011). *Apicystis bombi* (Apicomplexa: Neogregarinorida) parasitizing *Apis mellifera* and *Bombus terrestris* (Hymenoptera: Apidae) in Argentina, *Environmental Microbiology Reports* 3(5), 565–568. doi: 10.1111/j.1758-2229.2011.00261.x.

Poinar, G. O. and van der Laan, P. A. (1972). Morphology and life history of *Sphaerularia bombi*, *Nematologica* 18(2), 239–252. doi: 10.1163/187529272X00476.

Potts, S. G., Imperatriz-Fonseca, V., Ngo, H. T., Aizen, M. A., Biesmeijer, J. C., Breeze, T. D., Dicks, L. V., Garibaldi, L. A., Hill, R., Settele, J. and Vanbergen, A. J. (2016). Safeguarding pollinators and their values to human well-being, *Nature* 540(7632), 220–229. doi: 10.1038/nature20588.

Pritchard, D. J. (2016). Grooming by honey bees as a component of varroa resistant behavior, *Journal of Apicultural Research* 55(1), 38–48. doi: 10.1080/00218839.2016.1196016.

Pritchard, Z. A., Hendriksma, H. P., St Clair, A. L., Stein, D. S., Dolezal, A. G., O'Neal, M. E. and Toth, A. L. (2021). Do viruses from managed honey bees (Hymenoptera: Apidae) endanger wild bees in native prairies?, *Environmental Entomology* 50(2), 455–466. doi: 10.1093/ee/nvaa181.

Proesmans, W., Albrecht, M., Gajda, A., Neumann, P., Paxton, R. J., Pioz, M., Polzin, C., Schweiger, O., Settele, J., Szentgyörgyi, H., Thulke, H. H. and Vanbergen, A. J. (2021). Pathways for novel epidemiology: plant-pollinator-pathogen networks and global change, *Trends in Ecology and Evolution* 36(7), 623–636. doi: 10.1016/j.tree.2021.03.006.

Rafael Braga, A., G. Gomes, D., Rogers, R., E. Hassler, E. M., Freitas, B. and Cazier, J. A. (2020). A method for mining combined data from in-hive sensors, weather and

apiary inspections to forecast the health status of honey bee colonies, *Computers and Electronics in Agriculture* 169, 105161. doi: 10.1016/j.compag.2019.105161.
Ramsey, S. D., Ochoa, R., Bauchan, G., Gulbronson, C., Mowery, J. D., Cohen, A., Lim, D., Joklik, J., Cicero, J. M., Ellis, J. D., Hawthorne, D. and van Engelsdorp, D. (2019). Varroa destructor feeds primarily on honey bee fat body tissue and not hemolymph. *Proceedings of the National Academy of Sciences* 116, 1792–1801. doi: 10.1073/pnas.1818371116.
Raven, P. H. and Wagner, D. L. (2021). Agricultural intensification and climate change are rapidly decreasing insect biodiversity, *Proceedings of the National Academy of Sciences of the United States of America* 118(2), e2002548117. doi: 10.1073/PNAS.2002548117.
Ravoet, J., De Smet, L., Meeus, I., Smagghe, G., Wenseleers, T. and de Graaf, D. C. (2014). Widespread occurrence of honey bee pathogens in solitary bees, *Journal of Invertebrate Pathology* 122, 55–58. doi: 10.1016/j.jip.2014.08.007.
Ravoet, J., Maharramov, J., Meeus, I., De Smet, L., Wenseleers, T., Smagghe, G. and de Graaf, D. C. (2013). Comprehensive bee pathogen screening in Belgium reveals *Crithidia mellificae* as a new contributory factor to winter mortality, *PLoS ONE* 8(8), e72443. doi: 10.1371/journal.pone.0072443.
Remnant, E. J., Mather, N., Gillard, T. L., Yagound, B. and Beekman, M. (2019). Direct transmission by injection affects competition among RNA viruses in honeybees, *Proceedings. Biological Sciences* 286(1895), 20182452. doi: 10.1098/rspb.2018.2452.
Richards, K. W. (1984). *Alfalfa leafcutter bee management in Canada*, Ottowa, Canada: Research Station Alberta, Agriculture Canada. doi: 10.1080/0005772X.1987.11098930.
Rinderer, T. E., Harris, J. W., Hunt, G. J. and de Guzman, L. I. (2010). Breeding for resistance to *Varroa destructor* in North America, *Apidologie* 41(3), 409–424. doi: 10.1051/apido/2010015.
Rosenkranz, P., Aumeier, P. and Ziegelmann, B. (2010). Biology and control of *Varroa destructor*, *Journal of Invertebrate Pathology* 103 (Suppl. 1), S96–S119. doi: 10.1016/j.jip.2009.07.016.
Rothenbuhler, W. C. (1964). Behavior genetics of nest gleaning in honey bees. IV. responses of F1 and backcross generations to disease-killed brood, *American Zoologist* 4(2), 111–123. doi: 10.1093/icb/4.2.111.
Royan, M. (2019). Mechanisms of probiotic action in the honeybee, *Critical Reviews in Eukaryotic Gene Expression* 29(2), 95–103. doi: 10.1615/CritRevEukaryotGeneExpr.2019025358.
Ruiz-González, M. X., Bryden, J., Moret, Y., Reber-Funk, C., Schmid-Hempel, P. and Brown, M. J. (2012). Dynamic transmission, host quality, and population structure in a multihost parasite of bumblebees, *Evolution; International Journal of Organic Evolution* 66(10), 3053–3066. doi: 10.1111/j.1558-5646.2012.01655.x.
Runckel, C., Flenniken, M. L., Engel, J. C., Ruby, J. G., Ganem, D., Andino, R. and DeRisi, J. L. (2011). Temporal analysis of the honey bee microbiome reveals four novel viruses and seasonal prevalence of known viruses, *Nosema*, and *Crithidia*, *PLoS ONE* 6(6), e20656. doi: 10.1371/journal.pone.0020656.
Rutrecht, S. T. and Brown, M. J. F. (2008). The life-history impact and implications of multiple parasites for bumble bee queens, *International Journal for Parasitology* 38(7), 799–808. doi: 10.1016/j.ijpara.2007.11.004.

Ryabov, E. V., Wood, G. R., Fannon, J. M., Moore, J. D., Bull, J. C., Chandler, D., Mead, A., Burroughs, N. and Evans, D. J. (2014). A virulent strain of deformed wing virus (DWV) of honeybees (*Apis mellifera*) prevails after *Varroa destructor*-mediated, or *in vitro*, transmission, *PLoS Pathogens* 10(6), e1004230. doi: 10.1371/journal.ppat.1004230.

Schmickl, T. and Crailsheim, K. (2001). Cannibalism and early capping: strategy of honeybee colonies in times of experimental pollen shortages, *Journal of Comparative Physiology. A, Sensory, Neural, and Behavioral Physiology* 187(7), 541-547. doi: 10.1007/s003590100226.

Schmid-Hempel, P. (1998). *Parasites in Social Insects*, Princeton, NJ: Princeton University Press. doi: 10.2307/j.ctvs32rn5.

Schmid-Hempel, R., Eckhardt, M., Goulson, D., Heinzmann, D., Lange, C., Plischuk, S., Escudero, L. R., Salathé, R., Scriven, J. J. and Schmid-Hempel, P. (2014). The invasion of southern South America by imported bumblebees and associated parasites, *Journal of Animal Ecology* 83(4), 823–837. doi: 10.1111/1365-2656.12185.

Schmid-Hempel, P., Müller, C., Schmid-Hempel, R. and Shykoff, J. A. (1990). Frequency and ecological correlates of parasitism by conopid flies (Conopidae, Diptera) in populations of bumblebees, *Insectes Sociaux* 37(1), 14–30. doi: 10.1007/BF02223812.

Schoonvaere, K., Brunain, M., Baeke, F., De Bruyne, M., De Rycke, R. and de Graaf, D.C. (2020). Comparison between Apicystis cryptica sp. n. and Apicystis bombi (Arthrogregarida, Apicomplexa): Gregarine parasites that cause fat body hypertrophism in bees. *European Journal of Protistology* 73, 125688. doi: 10.1016/j.ejop.2020.125688.

Schoonvaere, K., Smagghe, G., Francis, F. and de Graaf, D. C. (2018). Study of the metatranscriptome of eight social and solitary wild bee species reveals novel viruses and bee parasites, *Frontiers in Microbiology* 9, 177. doi: 10.3389/fmicb.2018.00177.

Schulz, M., Ścibior, R., Grzybek, M., Łoś, A., Paleolog, J. and Strachecka, A. (2019). A new case of honeybee *Apis mellifera* infection with bumblebee parasite *Apicystis bombi* (Apicomplexa: Neogregarinorida), *Comparative Parasitology* 86(1), 65-67. doi: 10.1654/1525-2647-86.1.65.

Shen, M., Cui, L., Ostiguy, N. and Cox-Foster, D. (2005). Intricate transmission routes and interactions between picorna-like viruses (Kashmir bee virus and Sacbrood virus) with the honeybee host and the parasitic varroa mite, *Journal of General Virology* 86(8), 2281-2289. doi: 10.1099/vir.0.80824-0.

Singh, R., Levitt, A. L., Rajotte, E. G., Holmes, E. C., Ostiguy, N., Vanengelsdorp, D., Lipkin, W. I., Depamphilis, C. W., Toth, A. L. and Cox-Foster, D. L. (2010). RNA viruses in hymenopteran pollinators: evidence of inter-taxa virus transmission via pollen and potential impact on non-*Apis* hymenopteran species, *PLoS ONE* 5(12), e14357. doi: 10.1371/journal.pone.0014357.

Siviter, H., Bailes, E. J., Martin, C. D., Oliver, T. R., Koricheva, J., Leadbeater, E. and Brown, M. J. F. (2021). Agrochemicals interact synergistically to increase bee mortality, *Nature* 596(7872), 389–392. doi: 10.1038/s41586-021-03787-7.

Smith, D. R. (Ed) (1991). *Diversity in the Genus Apis*. Boulder, Colorado: Westview Press.

Smith, V. (2007). Host resource supplies influence the dynamics and outcome of infectious disease, *Integrative and Comparative Biology* 47(2), 310–316. doi: 10.1093/icb/icm006.

Starks, P. T., Blackie, C. A. and Seeley, T. D., P. T. (2000). Fever in honeybee colonies, *Naturwissenschaften* 87(5), 229–231. doi: 10.1007/s001140050709.

Staveley, J. P., Law, S. A., Fairbrother, A. and Menzie, C. A. (2014). A causal analysis of observed declines in managed honey bees (*Apis mellifera*), *Human and Ecological Risk Assessment: HERA* 20(2), 566–591. doi: 10.1080/10807039.2013.831263.

Steffan, S. A., Dharampal, P. S., Danforth, B. N., Gaines-Day, H. R., Takizawa, Y. and Chikaraishi, Y. (2019). Omnivory in bees: elevated trophic positions among all major bee families, *American Naturalist* 194(3), 414–421. doi: 10.1086/704281.

Sumpter, D. J. T. and Martin, S. J. (2004). The dynamics of virus epidemics in varroa-infested honey bee colonies, *Journal of Animal Ecology* 73(1), 51–63. doi: 10.1111/j.1365-2656.2004.00776.x.

Tang, C.-K., Tsai, C.-H., Lin, Y. C., Wu, C.-P. and Wu, Y.-L. (2021). Real-time monitoring of DWV-infected bee foraging behavior following histone deacetylase inhibitor treatment, *SSRN Electronic Journal* 24(10), 103056. doi: 10.2139/ssrn.3830018.

Tehel, A., Streicher, T., Tragust, S. and Paxton, R. J. (2020). Experimental infection of bumblebees with honeybee-associated viruses: no direct fitness costs but potential future threats to novel wild bee hosts, *Royal Society Open Science* 7(7), 200480. doi: 10.1098/rsos.200480.

Tehel, A., Vu, Q., Bigot, D., Gogol-Döring, A., Koch, P., Jenkins, C., Doublet, V., Theodorou, P. and Paxton, R. (2019). The two prevalent genotypes of an emerging infectious disease, deformed wing virus, cause equally low pupal mortality and equally high wing deformities in host honey bees, *Viruses* 11(2). doi: 10.3390/v11020114.

Tepedino, V. J. and Portman, Z. M. (2021). Intensive monitoring for bees in North America: indispensable or improvident? *Insect Conservation and Diversity* 14(5), 535–542. doi: 10.1111/icad.12509.

Tian, T., Piot, N., Meeus, I. and Smagghe, G. (2018). Infection with the multi-host micro-parasite *Apicystis bombi* (Apicomplexa: Neogregarinorida) decreases survival of the solitary bee *Osmia bicornis*, *Journal of Invertebrate Pathology* 158, 43–45. doi: 10.1016/j.jip.2018.09.005.

Tokarev, Y. S., Huang, W. F., Solter, L. F., Malysh, J. M., Becnel, J. J. and Vossbrinck, C. R. (2020). A formal redefinition of the genera *Nosema* and *Vairimorpha* (Microsporidia: Nosematidae) and reassignment of species based on molecular phylogenetics, *Journal of Invertebrate Pathology* 169, 107279. doi: 10.1016/j.jip.2019.107279.

Traynor, K. S., Rennich, K., Forsgren, E., Rose, R., Pettis, J., Kunkel, G., Madella, S., Evans, J., Lopez, D. and vanEngelsdorp, D. (2016). Multiyear survey targeting disease incidence in US honey bees, *Apidologie* 47(3), 325–347. doi: 10.1007/s13592-016-0431-0.

Traynor, K. S., Mondet, F., de Miranda, J. R., Techer, M., Kowallik, V., Oddie, M. A. Y., Chantawannakul, P. and McAfee, A. (2020). *Varroa destructor*: a complex parasite, crippling honey bees worldwide, *Trends in Parasitology* 36(7), 592–606. doi: 10.1016/j.pt.2020.04.004.

Tsvetkov, N., MacPhail, V. J., Colla, S. R. and Zayed, A. (2021). Conservation genomics reveals pesticide and pathogen exposure in the declining bumble bee *Bombus* terricola, *Molecular Ecology* 30(17), 4220–4230. doi: 10.1111/mec.16049.

Tyl, J., Kamler, M. and Titera, D. (2014). Pictorial Atlas for Honeybee Disease Diagnostics 2014. Výzkumný ústav včelařský, s.r.o., Dol, Czech Republic. www.beedol.cz.

Urbieta-Magro, A., Higes, M., Meana, A., Gómez-Moracho, T., Rodríguez-García, C., Barrios, L. and Martín-Hernández, R. (2019). The levels of natural *Nosema* spp. infection in *Apis mellifera iberiensis* brood stages, *International Journal for Parasitology* 49(8), 657–667. doi: 10.1016/j.ijpara.2019.04.002.

Vanbergen, A. J. and the Insect Pollinators Initiative (2013). Threats to an ecosystem service: pressures on pollinators, *Frontiers in Ecology and the Environment* 11(5), 251–259. doi: 10.1890/120126.

vanEngelsdorp, D., Evans, J. D., Saegerman, C., Mullin, C., Haubruge, E., Nguyen, B. K., Frazier, M., Frazier, J., Cox-Foster, D., Chen, Y., Underwood, R., Tarpy, D. R. and Pettis, J. S. (2009). Colony collapse disorder: a descriptive study, *PLoS ONE* 4(8), e6481. doi: 10.1371/journal.pone.0006481.

van Klink, R., Bowler, D. E., Gongalsky, K. B., Swengel, A. B., Gentile, A. and Chase, J. M. (2020). Meta-analysis reveals declines in terrestrial but increases in freshwater insect abundances, *Science* 368(6489), 417–420. doi: 10.1126/science.aax9931.

Vaudo, A. D., Fritz, M. L. and López-Uribe, M. M. (2018). Opening the door to the past: accessing phylogenetic, pathogen, and population data from museum curated bees, *Insect Systematics and Diversity* 2(5), 1–14. doi: 10.1093/isd/ixy014.

Wagner, D. L., Grames, E. M., Forister, M. L., Berenbaum, M. R. and Stopak, D. (2021). Insect decline in the Anthropocene: death by a thousand cuts, *Proceedings of the National Academy of Sciences of the United States of America* 118(2), e2023989118. doi: 10.1073/PNAS.2023989118.

Walter, J. (2020). Dryness, wetness and temporary flooding reduce floral resources of plant communities with adverse consequences for pollinator attraction, *Journal of Ecology* 108(4), 1453–1464. doi: 10.1111/1365-2745.13364.

Wcislo, W. T. (1987). The roles of seasonality, host synchrony, and behaviour in the evolutions and distributions of nest parasites in hymenoptera (Insecta), with special reference to bees (Apoidea), *Biological Reviews* 62(4), 515–542. doi: 10.1111/j.1469-185X.1987.tb01640.x.

Westphal, C., Bommarco, R., Carré, G., Lamborn, E., Morison, N., Petanidou, T., Potts, S. G., Roberts, S. P. M., Szentgyörgyi, H., Tscheulin, T., Vaissière, B. E., Woyciechowski, M., Biesmeijer, J. C., Kunin, W. E., Settele, J. and Steffan-Dewenter, I. (2008). Measuring bee diversity in different European habitats and biogeographical regions, *Ecological Monographs* 78(4), 653–671. doi: 10.1890/07-1292.1.

White, G. F. (1912). *The Cause of European Foul Brood*, Washington, DC: United States Department of Agriculture, Bureau of Entomology. doi: 10.5962/bhl.title.63494.

Wilfert, L., Long, G., Leggett, H. C., Schmid-Hempel, P., Butlin, R., Martin, S. J. and Boots, M. (2016). Deformed wing virus is a recent global epidemic in honeybees driven by varroa mites, *Science* 351(6273), 594–597. doi: 10.1126/science.aac9976.

Williams, P. H. and Osborne, J. L. (2009). Bumblebee vulnerability and conservation world-wide, *Apidologie* 40(3), 367–387. doi: 10.1051/apido/2009025.

Wilson-Rich, N., Spivak, M., Fefferman, N. H. and Starks, P. T. (2009). Genetic, individual, and group facilitation of disease resistance in insect societies, *Annual Review of Entomology* 54(1), 405–423. doi: 10.1146/annurev.ento.53.103106.093301.

Yañez, O., Jaffé, R., Jarosch, A., Fries, I., Moritz, R. F. A., Paxton, R. J. and de Miranda, J. R. (2012). Deformed wing virus and drone mating flights in the honey bee (*Apis mellifera*): implications for sexual transmission of a major honey bee virus, *Apidologie* 43(1), 17–30. doi: 10.1007/s13592-011-0088-7.

Yañez, O., Piot, N., Dalmon, A., de Miranda, J. R., Chantawannakul, P., Panziera, D., Amiri, E., Smagghe, G., Schroeder, D. and Chejanovsky, N. (2020). Bee viruses: routes of infection in Hymenoptera, *Frontiers in Microbiology* 11, 943. doi: 10.3389/fmicb.2020.00943.

Yang, X. and Cox-Foster, D. L. (2005). Impact of an ectoparasite on the immunity and pathology of an invertebrate: evidence for host immunosuppression and viral amplification, *Proceedings of the National Academy of Sciences of the United States of America* 102(21), 7470–7475. doi: 10.1073/pnas.0501860102.

Zaragoza-Trello, C., Vilà, M., Botías, C. and Bartomeus, I. (2021). Interactions among global change pressures act in a non-additive way on bumblebee individuals and colonies, *Functional Ecology* 35(2), 420–434. doi: 10.1111/1365-2435.13703.

Zattara, E. E. and Aizen, M. A. (2021). Worldwide occurrence records suggest a global decline in bee species richness, *One Earth* 4(1), 114–123. doi: 10.1016/j.oneear.2020.12.005.

Chapter 6

How neonicotinoid insecticides affect bees and other pollinators

D. Susan Willis Chan, University of Guelph, Canada

1 Introduction

Agricultural production faces a dilemma that pits the practice of using pesticides to protect crops against the need to protect the insect pollinators that provide crop pollination services from exposure to those pesticides (Knapp et al., 2022). Among pesticides (insecticides, fungicides, herbicides), insecticides pose the greatest, though not the only, risk to insect pollinators simply because they are engineered to kill insects (Sánchez-Bayo and Goka, 2014).

The pesticide regulatory system that exists in many countries is designed to balance protection of crops from pests with protection of pollinators from pesticide exposure and resulting negative impacts (see Chapter 7 of this book). In a farming context, the question is not if neonicotinoids are toxic to individual pollinators but rather if the approved agricultural uses lead to negative consequences for desired crop pollination services, pollinator populations, or, more broadly, biodiversity, which are not fully captured by the regulatory risk assessment process.

http://dx.doi.org/10.19103/AS.2022.0111.28

2 What are neonicotinoids?

The neonicotinoids are a group of broad-spectrum insecticides related to botanical nicotines that are used extensively across the world to control insect pests (Jeschke et al., 2011). They are applied on plant foliage, by injection into tree trunks, as soil drenches, in irrigation water (chemigation), and as coatings on seeds, with seed and soil applications dominating (Jeschke et al., 2011).

There are currently seven neonicotinoids commercially available as active ingredients in crop protection products: acetamiprid, clothianidin, dinotefuran, imidacloprid, nitenpyram, thiacloprid, and thiamethoxam, starting with the introduction of imidacloprid in 1991, followed by clothianidin and thiamethoxam in the early 2000s (Jeschke and Nauen, 2008; Thompson et al., 2020). Thiacloprid, the least toxic of the neonicotinoids, was voluntarily removed from the US market by registrants in 2016 (Minnesota Department of Agriculture, 2016). Neonicotinoid insecticide products are sold under a broad range of brand names by many companies, and brand names may vary from country to country (Elbert et al., 2008). A good summary of the properties, producers, and brand names of neonicotinoids sold in the United States is provided at: https://www.mda.state.mn.us/sites/default/files/inline-files/neonicreviewrpt2016.pdf.

All neonicotinoids are regulated where regulatory bodies exist. However, in most jurisdictions, including Canada, Europe, and the United States, seeds treated with neonicotinoids are not classified as pesticides but as 'treated articles' (like treated lumber), thereby avoiding some regulations. However, in one jurisdiction in Canada (i.e. the province of Ontario), this approach has been revisited, and neonicotinoid-coated corn and soybean seeds are considered a pesticide class that can only be used if the user demonstrates that pest pressure exceeds an established integrated pest management threshold on their land (Government of Ontario, 2012).

Although neonicotinoids exhibit relatively low toxicity to mammals, including humans, most are highly toxic to insects, including pollinators, causing death at concentrations less than 2 ug ai/bee (Minnesota Department of Agriculture, 2016). They act as insect neurotoxins by binding strongly to the nicotinic acetylcholine receptors in the nervous system of insects, causing overfiring of neurons followed by shaking and eventual paralysis and death at lethal doses (Jeschke et al., 2011; https://ncceh.ca/environmental-health-in-canada/health-agency-projects/neonicotinoid-pesticides).

Although they are not the focus of this chapter, mention should be made of other insecticides that share the same mode of action as neonicotinoids, such as flonicamid, flupyradifurone, and sulfoxaflor, and these substances have been referred to as neonicotinoid-like (Thompson et al., 2020). These insecticides are being used as alternatives to neonicotinoids but have been described as 'regrettable substitutions' (Maertens et al., 2021).

3 Why are neonicotinoids such a concern for pollinators?

Neonicotinoids are particularly important in any discussion of pollinators in an agricultural context because of their use patterns, physical properties, environmental persistence, and their high toxicity to bees, making them stand apart from other pesticides in relation to pollinators (Simon-Delso et al., 2015). These characteristics, including their versatility and ease of application, make neonicotinoids the most extensively used group of insecticides across the globe on a wide variety of crop plants (Jeschke et al., 2011). Their water solubility makes them easy for farmers to apply but also increases the risk of leakage into the environment through water (Schaafsma et al., 2015). Their systemic uptake in both wild and crop plants often results in contamination of pollen and nectar in those plants (Dively and Kamal, 2012; Botías et al., 2015; Traynor et al., 2021). Their relative stability makes them persistent in the environment, especially in soil where 70% of solitary bees nest (Bonmatin, et al., 2015; Willis Chan et al., 2019). Their broad toxicity to insects may cause unintended negative impacts on non-target beneficial insects including bees and most other pollinators such as hoverflies, butterflies, and moths (reviewed in, e.g. Godfray et al., 2014; Pisa et al., 2014; Hladik et al., 2018).

The bee-related risks of neonicotinoids were first brought to broader public attention when German (Pistorius et al., 2009) and Italian (Mutinelli et al., 2009) beekeepers observed weakening and loss of colonies during planting of neonicotinoid-treated maize seeds in 2008. Dust from the neonicotinoid-coated seeds landed on and contaminated flowers in the surroundings of the planted maize fields (Greatti et al., 2006; Krupke et al., 2012). Pollinators that visit crops or live in the vicinity of crops are especially likely to be exposed to neonicotinoids used in agriculture, especially those pollinators that nest within cropping areas (Willis Chan et al., 2019; Rondeau et al., 2022). In North America, 739 of the estimated 3500 species of bees have been found in association with crops, including crops such as corn or soybean that are not insect pollinated, 59% of which visit crop flowers and 54% of which nest in the ground (Rondeau et al., 2022).

Most knowledge about both acute exposure and toxicity of neonicotinoids to insect pollinators comes from bioassays on western honey bees (*Apis mellifera)* to determine mortality endpoints (i.e. median lethal dose-LD50) and other effects on individual bees and from studies on honey bee colonies to determine colony-level impacts. However, sensitivity to neonicotinoids may vary greatly among pollinators, especially for small bees (Arena and Sgolastra,

2014; Hayward et al., 2019). This discrepancy can be mitigated to some degree by applying a 10x safety factor applied to the honey bee LD50 (Arena and Sgolastra, 2014).

Extra caution is needed in interpreting the results of honey bee colony studies because the western honey bee is a highly unusual bee species that forms large, perennial colonies that enjoy a functional redundancy of labor that can buffer short-term impacts (Wu-Smart and Spivak, 2016). This natural buffer against colony or population-level impacts of neonicotinoid exposure on honey bees is not enjoyed by bumble bees who form smaller, annual colonies with a solitary queen stage or by solitary bees that do not form colonies (Franklin and Raine, 2019).

As such, the western honey bee is likely a useful model for evaluating direct acute contact or oral exposure for individual bees in laboratory conditions (Thompson and Pamminger, 2019), but it is probably less useful as a model for evaluating field exposure and population-level effects among other species of bees or pollinator taxa, depending on the diversity and variation in life histories (Boyle et al., 2018; Sgolastra et al., 2019). The variation in exposure routes and levels as well as toxicity and resulting risks for other bee species or insect pollinators is still not well understood and deserves serious attention (Pisa et al., 2017; Lundin et al., 2015; Boyle et al., 2018).

Although acute toxicity is important, cumulative chronic exposure to low doses of pesticides over time can also result in mortality and is poorly understood and poorly represented in existing regulatory protocols (Sánchez-Bayo and Tennekes, 2020; Pisa et al., 2017). Because of their physical-chemical properties, neonicotinoids pose a special risk to pollinators of chronic low-level exposure with an associated potential for serious sublethal impacts on individuals and populations over time (Sánchez-Bayo and Goka, 2014; Willis Chan and Raine, 2021a) and carryover effects on reproduction across generations (Stuligross and Williams, 2020). Long-lived pollinators with annual social colonies such as bumble bees or bees such as squash bees that collect pollen only from crop plants may suffer more severe consequences (Gill and Raine, 2014; Willis Chan and Raine, 2021a), but such consequences have also been demonstrated for generalist solitary bee species (Stuligross and Williams, 2020).

4 Environmental contamination and pollinator exposure

The water solubility and systemic uptake and dispersal of neonicotinoids in plants results in residues in plant tissue, nectar, and pollen – all major routes of pesticide exposure to pollinators (Kopit and Pitts-Singer, 2018; Main et al., 2018). The ubiquity of neonicotinoid residues in the environment is well illustrated by the detection of one or more neonicotinoids in 75% of 198 honey samples from around the world (Mitchell et al., 2017). Even if neonicotinoids account for only a small fraction of the detected pesticides in pollen collected by honey bees across the United States, those residues are often more hazardous than other detected residues (Traynor et al., 2021). In a study of pollen, nectar, and soil samples taken from cucurbit growing systems in Ontario, Canada, 99% of the calculated hazard from insecticides in soil to adult ground-nesting bees was from neonicotinoids, with 100% of hazard in cucurbit pollen and 40% of hazard in cucurbit nectar (Willis Chan et al., 2019). In other cropping systems, neonicotinoid residues are commonly found in soil at concentrations that can cause mortality in small ground-nesting bees (Willis Chan et al., 2019).

Estimates suggest that when seeds are coated with neonicotinoids, only 2–20% of the neonicotinoids are taken up by the crop (Sur and Stork, 2003; Alford and Krupke, 2017), leaving the remaining residues in the soil where they can be further distributed in the environment. The water solubility of neonicotinoids means they may leak from soil via runoff or leaching into water systems and potentially expose the aqueous life stages of some pollinators (Hallman et al., 2014; Hladik et al., 2018). Within cropping systems, neonicotinoid residues can persist in the soil from season to season at levels that represent a high acute and chronic risk to bees that nest in the ground in some scenarios (Goulson, 2013; Willis Chan et al., 2019). Soils in orchards and other perennial crops can be particularly polluted because pesticide use is intensive and no tillage or crop rotation occurs in those contexts (Tang and Maggi, 2021).

Honey bees and their colonies encounter a mixture of pesticides in agricultural landscapes For example, 120 pesticide residues including herbicides, insecticides, and fungicides were detected in thousands of pollen samples collected by honey bees over seven years in the United States with only 18% of pollen samples being pesticide free (Traynor et al., 2021). Similar evidence is emerging for other pollinators such as wild bees (David et al., 2016; Hladik et al., 2016; Willis Chan et al., 2019) or butterflies (Olaya-Arenas and Kaplan, 2019). This raises concerns about the potential for synergism between neonicotinoids and their metabolites and other pesticides and chemicals. For example, neonicotinoids are often applied alongside fungicides in seed treatments with possible potentiation of neonicotinoids with some fungicides (e.g. Tsvetkov et al., 2017; Carnesecchi et al., 2019; Iverson et al., 2019).

Occurrence of neonicotinoids in the environment is the outcome of grower use practices and patterns, with pollinator exposure to neonicotinoids and other pesticides the result of the intersection of grower use and pollinator activity in space and time (Sponsler et al., 2019). This intersection is determined by life-history traits of the pollinators with expected variation among and within pollinators groups (reviewed for solitary bees: Kopit and Pitts-Singer, 2018; Sgolastra et al., 2019; bumble bees: Gradish et al., 2018; butterflies: Braak et al., 2018; non-target terrestrial arthropods: Main et al., 2018). For example, for pollinator species that nest or overwinter in soil, that intersection includes not only crop nectar and pollen available during the growing season but also residues present in soil during all seasons (Willis Chan et al., 2019; Rondeau et al., 2022).

In summary, to accurately understand the risk and potential impact of neonicotinoids, the acute toxicity of each neonicotinoid as well as the exposure time (acute or chronic), exposure route (oral or contact), and the synergistic toxicity of co-exposures to neonicotinoids and other pesticides must be considered (Sánchez-Bayo and Goka, 2014).

5 Toxicity and effects across pollinator taxa and contexts

Pollinators can be exposed to neonicotinoids topically through contact with residues on their cuticle or through their breathing spiracles, and/or they can be exposed orally through ingestion of residues in the food they consume. Toxicity of neonicotinoids is generally greater for oral exposure versus topical (contact) exposure (Alkassab and Kirchner, 2017).

5.1 Topical exposure

Topical neonicotinoid exposure routes for pollinators include overspray, dust, contaminated air breathed in through spiracles, soil and plant material, and also water for aquatic life stages (e.g. some fly larvae: Morrissey et al., 2015; Hladik et al., 2018). Since pollinators are engaged in the transfer of pollen from anthers to stigmas (see Chapter 1 of this book), all pollinators can be topically exposed to neonicotinoid residues in pollen (Willis Chan et al., 2019). Pollinators can also be topically exposed to pesticide residues as they collect other resources (nectar, resins), find mates, collect nesting materials, excavate soil to make ground nests, and care for their offspring – with variability in importance among routes of exposure related to pollinator life history traits (Sgolastra et al., 2019). The relative importance of these different routes of exposure is not well understood among species and groups of pollinators despite attempts to quantify this among bees for their adult and larval life stages (e.g. Boyle et al., 2018; Willis Chan et al., 2019).

5.2 Oral exposure

Pollinators can be orally exposed to neonicotinoids primarily through consumption of pollen and nectar at both adult and larval stages, although exposure can also occur for some pollinators through water, honey dew, plant materials, resins, or prey. For example, plant material, including resins, is an especially relevant oral exposure route for butterfly larvae (Braak et al., 2018) and leafcutter and resin bees (Kopit and Pitts-Singer, 2018). Furthermore, predatory pollinator species such as phidophagous hover fly larvae (Dunn et al., 2020) may be exposed through their prey, which are routinely controlled in crops by neonicotinoids.

6 Sublethal effects on bees and interactions among multiple stressors

Neonicotinoids can also have sublethal effects below the lethal dose, which impair the health of pollinators as individuals or populations but may not cause mortality outright (Alkassab and Kirchner, 2017). The sublethal effects of neonicotinoids are increasingly well understood for bees, with most studies conducted on honey bees and to a lesser extent on bumble bees, stingless bees, and solitary bees (Table 1).

Effects of exposure to neonicotinoids on bees occur at the cellular or subcellular level with knock-on effects at individual, colony, and population levels (Godfray et al., 2014, 2015; Lundin et al., 2015; Alkassab and Kirchner, 2017), and these effects can be detected across generations (Stuligross and Williams, 2020). These include negative impacts on gene expression, learning, locomotion, memory, foraging, hygienic behavior, queen viability, reproduction, growth, nest initiation, colony establishment, population stability, and crop pollination services (Decourtye and Devillers, 2010; Godfray et al., 2014, 2015; Stanley et al., 2015a,b; Rundlöf et al., 2015; Williams et al., 2015; Alkassab and Kirchner, 2017; Baron et al., 2017; Tesovnik et al., 2017; Tison et al., 2017; Jacob et al., 2019a,b; Willis Chan and Raine, 2021a; Stuligross and Williams, 2020). A meta-analysis of sublethal effects on non-*Apis* bees from 'field-realistic' neonic exposure is provided by Siviter and colleagues (2021).

6.1 Interactions among multiple stressors

Clearly, the pollinator species, the exposure route (topical or oral), the exposure concentration (lethal or sublethal), the length of exposure (acute or chronic), and whether synergism is occurring with other pesticides all contribute to the severity of effects of neonicotinoids to pollinators (Decourtye and Devillers, 2010; Arena and Sgolastra, 2014; Sánchez-Bayo and Goka, 2014; Tsvetkov et al., 2017; Sgolastra et al., 2019). However, effects may be further complicated by the additive or synergistic effects of neonicotinoids on other stressors such as nutritional stress or parasite and disease pressure that also impair pollinator health (e.g. meta-analysis by Siviter et al., 2021 and Tosi et al., 2022).

Table 1 Examples of effects on bees of exposure to neonicotinoids

Pollinator taxa	Species	Neonicotinoid active ingredient	Sublethal effect on bees	Reference
Honey bees	*Apis mellifera*	imidacroprid thiacloprid,	Locomotion	Jacob et al. (2019b)
	Apis mellifera	thiacloprid	Learning and memory	Tison et al. (2017)
	Apis mellifera	clothianidin, thiamethoxam	Queen viability	Williams et al. (2015)
		imidacloprid		Wu-Smart and Spivak (2016)
	Apis mellifera	clothianidin	Interannual colony viability	Woodcock et al. (2017)
	Apis mellifera	clothianidin	Hygienic behavior, reproduction	Tsvetkov et al. (2017)
	Apis mellifera	clothianidin, thiacloprid	Nurse bee behavior, larval development	Siefert et al. (2020)
Stingless bees	*Tetragonisca angustula*	imidacloprid thiacloprid	Locomotion	Jacob et al. (2019a)
	Scaptotrigona postica	acetomiprid, imidacloprid thiacloprid, thiamethoxam	Hyperactivity	Jacob et al. (2019a)
	Scaptotrigona aff. *depilis*	thiamethoxam	Survival and development of larvae	Rosa et al. (2016)
Bumble bees	*Bombus terrestris*	clothianidin, thiamethoxam	Reproduction	Woodcock et al. (2017)
	Bombus terrestris	imidacloprid	Colony growth and queen production	Whitehorn et al. (2012)
	Bombus terrestris	thiamethoxam	Foraging, colony growth	Stanley et al. (2015a)
	Bombus terrestris	thiamethoxam	Pollination services	Stanley et al. (2015b)
	Bombus impatiens	imidacloprid	Queen survival, nest initiation	Wu-Smart and Spivak (2017)
	Bombus terrestris	imidacloprid, thiamethoxam	Thermal physiology	Potts et al. (2018)
	Bombus impatiens	imidacloprid	Foraging	Muth and Leonard (2019)
Solitary bees	*Eucera pruinosa*	ilmidacloprid	Foraging, nest initiation, population decline	Willis Chan and Raine (2021a)
	Osmia lignaria	imidacloprid	Nesting behavior	Fortuin et al. (2021)
	Osmia bicornis	clothianidin	Nesting behavior	Rundlöf et al. (2015)
	Megachile rotundata	imidacloprid	Foraging, reproduction	Cecala and Wilson Rankin (2021)
	Osmia lignaria	imidacloprid	Reproduction, intergenerational effects	Stuligross and Williams (2020)

7 Case studies illustrating consequences for pollinator populations and pollination

7.1 Consequences for bumble bees, solitary Osmia sp. and honey bees in field conditions

The effects on bees of direct exposure to neonicotinoids do not necessarily reflect exposure in the field where exposure may be via interaction with a treated crop. To test the effect of indirect exposure via a crop, a study that examined the effect of neonicotinoid exposure on bumble bees (*Bombus terrestris*), solitary bees (*Osmia bicornis*), and honey bee colonies (*A. mellifera*) under field-realistic conditions was carried out (Rundlöf et al., 2015). The study used eight pairs of replicated study sites in matched landscapes in which bees were either exposed to oilseed rape crops that were treated or untreated with clothianidin as seed coatings. In landscapes where oilseed rape was grown from neonicotinoid-treated seed, the overall density of wild bees visiting the crop was lower even when crop flowering density was considered. Results also showed a clear reduction in colony growth and reproduction for bumble bees and reduced nesting activity by solitary bees. However, no effects were seen on the colony strength of honey bees. This suggests that crops grown from clothianidin-treated seed have a negative impact on wild bees and that honey bees are not necessarily a reliable model for testing population-level effects on wild bees in field conditions.

7.2 Effect of exposure to soil-applied imidacloprid on solitary ground-nesting squash bees in semi-field conditions

The ground-nesting solitary hoary squash bee (*Eucera pruinosa*) is an important pollinator of pumpkins and squash (*Cucurbita* species) in North America. The hoary squash bee is unusual because it is a specialist on wild and cultivated *Cucurbita*, and across most of its range, its sole source of pollen is from pumpkin and squash crops (López-Uribe et al., 2016). As such, it is particularly vulnerable to exposure to pesticides used on those crops. In Canada, over a 3-year period, the effects of exposure to a soil-applied imidacloprid or a seed-applied thiamethoxam on hoary squash bee populations were evaluated (Willis Chan and Raine, 2021b). Measures of nest initiation, pollen harvesting, and offspring production by the hoary squash bee were evaluated in field-realistic conditions in large enclosures. In the enclosures, the bees were exposed to squash plants that had been treated with neonicotinoids 8 weeks prior to bee activity on the flowers. Over 3 generations, bees exposed to plants treated with imidacloprid harvested much less pollen, initiated 86% fewer nests, and produced 89% fewer offspring than did unexposed bees. Bees exposed to plants grown from seed treated with thiamethoxam showed no difference from unexposed bees in pollen harvesting, nest initiation, and offspring production.

7.3 Effect of oral exposure to thiamethoxam on establishment of bumble bee nests

Founding queens of bumble bee colonies represent a vulnerable yet critical stage of the bumble bee lifecycle. In the spring, founder queens emerge from overwintering sites, locate a suitable nesting site, and begin to initiate a colony by creating brood cells within the nest, foraging for nectar and pollen, and laying eggs. The success of the subsequent colony and the bumble bee population in a location depends upon the founding queen's ability to initiate a nest and lay eggs. A study done in the UK with the bumble bee *Bombus terrestris* exposed founding queens to a field-relevant dose of the neonicotinoid thiamethoxam while exposing the queens to two natural stressors: the parasite *Crithidia bombi* and varying lengths of hibernation (Baron et al., 2017). For queens treated with thiamethoxam, the proportion of founding queens that laid eggs was reduced by 26%. When this reduction in egg laying was modeled mathematically, it revealed a dramatic increase in the likelihood of population extinction for founding queens exposed to thiamethoxam.

7.4 Intergenerational reproduction effects of soil-applied imidacloprid on the above-ground cavity-nesting solitary bee Osmia lignaria

Although the direct impacts of exposure to neonicotinoids on pollinators are of great importance, the carryover effects from one generation to another, especially for solitary bees, can have implications for long-term population persistence.

In a study over 2 years on the solitary orchard bee, *Osmia lignaria*, researchers evaluated the effects of exposure to the neonicotinoid imidacloprid on individual vital rates and the carryover effects on population growth in subsequent generations (Stuligross and Williams, 2020). In the study, populations of orchard bees were either exposed by foraging on imidacloprid-treated wildflowers in a flight cage or not exposed in year one. In year 2, bees exposed in year 1 were either exposed to imidacloprid again or not exposed, and unexposed bees were either unexposed again or exposed to imidacloprid. Bees directly exposed produced 30% fewer offspring and the following unexposed generation produced 20% fewer offspring than bees that had never been exposed directly or in a previous generation. Bees exposed in both years produced 44% fewer offspring than unexposed bees, and the proportion of female offspring they produced was reduced.

7.5 Consequences for bumble bees on apple pollination

Providing pollination services to crops is often cited as the most important role that pollinators play in agriculture. In a study in the UK, researchers demonstrated that bumble bees (*Bombus terrestris*) exposed to the neonicotinoid thiamethoxam visited apple flowers at a lower rate and collected pollen less often than unexposed bees (Stanley et al., 2015b). The resulting apples contained fewer seeds indicating that pollination services were impaired by exposure to thiamethoxam and that exposure of individuals led to a colony-level effect on pollination services.

8 Conclusion

Neonicotinoids clearly have the potential to pose health risks to agricultural pollinators, including both managed and wild populations, based on a range of laboratory, semi-field, and field studies. As such, a re-evaluation of their use in agriculture is warranted, especially because they are so widely used on a broad range of crops. Although neonicotinoids are highly regulated, because of differences in exposure, sensitivity, and life history between the western honey bee and other important bee and non-bee pollinators, the use of honey bees as the sole insect model to represent other pollinators in risk assessments of neonicotinoids may be a flawed approach that needs improvement. Although neonicotinoids are the focus of this chapter, other closely related systemic insecticides such as flonicamid, flupyradifurone, and sulfoxaflor may also pose risks, and their use should be guided by the lessons learned from neonicotinoids.

9 Where to look for further information

Future research is needed to more fully understand the consequences of long-term exposure to neonicotinoids and other pesticides on a broader range of pollinator taxa. The importance of exposure via soil needs to be more fully documented for ground-nesting bees, with particular emphasis on the mechanism of translocation of residues from soil to the bees. Also needed is evidence supporting specific ways of mitigating or reducing exposure to all systemic pesticides for all pollinator groups. For more information about this topic, the European Academies Science Advisory Council's (EASAC) 2015 Report on ecosystem services, agriculture and neonicotinoids (https://easac.eu/publications/details/ecosystem-services-agriculture-and-neonicotinoids/) is an excellent resource. For more information about how to protect pollinators in specific crops see Pollinator Partnership Canada's Agricultural Guides (https://pollinatorpartnership.ca/en/ag-guides). The main academic reviews of the topic are by Godfray et al. (2014, 2015), Lundin et al. (2015), and Pisa et al. (2014, 2017).

10 Acknowledgements

The author wishes to acknowledge the contribution to the chapter made by Dr. Maj Rundlöf (Lund University, Sweden) in suggesting an original structure and contributing initial text and references.

11 References

Alford, A. and Krupke, C. H. (2017). Translocation of the neonicotinoid seed treatment clothianidin in maize. *PLoS ONE* 12(3), e0173836.

Alkassab, A. T. and Kirchner, W. H. (2017). Sublethal exposure to neonicotinoids and related side effects on insect pollinators: Honeybees, bumblebees, and solitary bees. *Journal of Plant Diseases and Protection* 124(1), 1–30.

Arena, M. and Sgolastra, F. (2014). A meta-analysis comparing the sensitivity of bees to pesticides. *Ecotoxicology* 23(3), 324–334. https://doi.org/10.1007/s10646-014-1190-1.

Baron, G. L., Jansen, V. A. A., Brown, M. J. F. and Raine, N. E. (2017). Pesticide reduces bumblebee colony initiation and increases probability of population extinction. *Nature Ecology and Evolution* 1(9), 1308–1316. https://doi.org/10.1038/s41559-017-0260-1.

Bonmatin, J. M., Giorio, C., Girolami, V., Goulson, D., Kreutzweiser, D. P., Krupke, C., Liess, M., Long, E., Marzaro, M., Mitchell, E. A. D., Noome, D. A., Simon-Delso, N. and Tapparo, A. (2015). Environmental fate and exposure: Neonicotinoids and fipronil. *Environmental Science and Pollution Research International* 22(1), 35–67. https://doi.org/10.1007/s11356-014-3332-7.

Botías, C., David, A., Horwood, J., Abdul-Sada, A., Nicholls, E., Hill, E. and Goulson, D. (2015). Neonicotinoid residues in wildflowers, a potential route of chronic exposure for bees. *Environmental Science and Technology* 49(21), 12731–12740. https://doi.org/10.1021/acs.est.5b03459.

Boyle, N. K., Pitts-Singer, T. L., Abbott, J., Alix, A., Cox-Foster, D. L., Hinarejos, S., Lehmann, D. M., Morandin, L., O'Neill, B., Raine, N. E., Singh, R., Thompson, H. M., Williams, N. M. and Steeger, T. (2018). Workshop on pesticide exposure assessment paradigm for non-apis bees: Foundation and summaries. *Environmental Entomology* 48(1), 4–11. https://doi.org/10.1093/ee/nvy103.

Braak, N., Neve, R., Jones, A. K., Gibbs, M. and Breuker, C. J. (2018). The effects of insecticides on butterflies - A review. *Environmental Pollution* 242(A), 507–518. https://doi.org/10.1016/j.envpol.2018.06.100.

Carnesecchi, E., Svendsen, C., Lasagni, S., Grech, A., Quignot, N., Amzal, B., Toma, C., Tosi, S., Rortais, A., Cortinas-abrahantes, J., Capri, E., Kramer, N., Benfenati, E., Spurgeon, D., Guillot, G. and Dorne, J. L. C. M. (2019). Investigating combined toxicity of binary mixtures in bees: Meta-analysis of laboratory tests, modelling, mechanistic basis and implications for risk assessment. *Environment International* 133(B), 105256. https://doi.org/10.1016/j.envint.2019.105256.

Cecala, J. M. and Wilson Rankin, E. E. (2021). Pollinators and plant nurseries: How irrigation and pesticide treatment of native ornamental plants impact solitary bees. *Proceedings Biological Sciences* 288(1955), 20211287. https://doi.org/10.1098/rspb.2021.1287.

David, A., Botías, C., Abdul-Sada, A., Nicholls, E., Rotheray, E. L., Hill, E. M. and Goulson, D. (2016). Widespread contamination of wildflower and bee-collected pollen with complex mixtures of neonicotinoids and fungicides commonly applied to crops. *Environment International* 88, 169–178. https://doi.org/10.1016/j.envint.2015.12.011.

Decourtye, A. and Devillers, J. (2010). Ecotoxicity of neonicotinoid insecticides to bees. In: Thany, S. (Ed.), *Advances in Experimental Medicine and Biology: Insect Nicotinic Acetylcholine Receptors, LLC Dual Imprint/Springer Series* (Landes Bioscience/Springer Science & Business Media) 683, 96–106.

Dively, G. P. and Kamel, A. (2012). Insecticide residues in pollen and nectar of a cucurbit crop and their potential exposure to pollinators. *Journal of Agricultural and Food Chemistry* 60(18), 4449–4456. https://www.doi.org/10.1021/jf205393x.

Dunn, L., Lequerica, M., Reid, C. R. and Latty, T. (2020). Dual ecosystem services of syrphid flies (Diptera: Syrphidae): Pollinators and biological control agents. *Pest Management Science* 76(6), 1973–1979. https://doi.org/10.1002/ps.5807.

Elbert, A., Haas, M., Springer, B., Thielert, W. and Nauen, R. (2008). Applied aspects of neonicotinoid uses in crop protection. *Pest Management Science* 64(11), 1099-1105. https://www.doi.org/10.1002/ps.1616.

Fortuin, C. C., McCarty, E. and Gandhi, K. J. (2021). Acute contact with Imidacloprid in soil affects the nesting and survival success of a solitary wild bee, *Osmia lignaria* (Hymenoptera: Megachilidae). *Chemosphere* 264(2), 128572–128572. https://doi.org/10.1016/j.chemosphere.2020.128572.

Franklin, E. L. and Raine, N. E. (2019). Moving beyond honeybee-centric pesticide risk assessments to protect all pollinators. *Nature Ecology and Evolution* 3(10), 1373–1375. https://doi.org/10.1038/s41559-019-0987-y.

Gill, R. J. and Raine, N. E. (2014). Chronic impairment of bumblebee natural foraging behaviour induced by sublethal pesticide exposure. *Functional Ecology* 28(6), 1459-1471. https://www.doi.org/10.1111/1365-2435.12292.

Godfray, H. C. J., Blacquière, T., Field, L. M., Hails, R. S., Petrokofsky, G., Potts, S. G., Raine, N. E., Vanbergen, A. J. and McLean, A. R. (2014). A restatement of the natural science evidence base concerning neonictoinoid insecticides and insect pollinators. *Proceedings. Biological Sciences* 281(1786), 20140558. https://www.doi.org/10.1098/rspb.2014.0558.

Godfray, H. C. J., Blacquière, T., Field, L. M., Hails, R. S., Petrokofsky, G., Potts, S. G., Raine, N. E., Vanbergen, A. J. and McLean, A. R. (2015). A restatement of recent advances in the natural science evidence base concerning neonicotinoid insecticides and insect pollinators. *Proceedings of the Royal Society Series B: Biological Sciences* 282, 20151821. https://www.doi.org/10.1098/rspb.2015.1821.

Goulson, D. (2013). REVIEW: An overview of the environmental risks posed by neonicotinoid insecticides. *Journal of Applied Ecology* 50(4), 977–987. https://www.doi.org/10.1111/1365-2664.12111.

Government of Ontario. (2012). Neonicotinoid rules for seed vendors. Available at: https://www.ontario.ca/page/neonicotinoid-rules-seed-vendors.

Gradish, A., Steen, J. van der, Scott-Dupree, C. D., Cabrera, A. R., Cutler, G. C., Goulson, D., Klein, O., Lehmann, D. M., Lückmann, J., O'Neill, B., Raine, N. E., Sharma, B. and

Thompson, H. (2018). Comparison of pesticide exposure in honey bees (Hymenoptera: Apidae) and bumble bees (Hymenoptera: Apidae): Implications for risk assessments. *Environmental Entomology* 48(1), 12–21. https://doi.org/10.1093/ee/nvy168.

Greatti, M., Barbattini, R., Stravisi, A., Sabatini, A. G. and Rossi, S. (2006). Presence of the ai Imidacloprid on vegetation near corn fields sown with Gaucho® dressed seeds. *Bulletin of Insectology* 59, 99.

Hallmann, C. A., Foppen, R. P. B., van Turnhout, C. A. M., de Kroon, H. and Jongejans, E. (2014). Declines in insectivorous birds are associated with high neonicotinoid concentrations. *Nature* 511(7509), 341–343.

Hayward, A., Beadle, K., Singh, K. S., Exeler, N., Zaworra, M., Almanza, M. T., Nikolakis, A., Garside, C., Glaubitz, J., Bass, C. and Nauen, R. (2019). The leafcutter bee, Megachile rotundata, is more sensitive to N-cyanoamidine neonicotinoid and butenolide insecticides than other managed bees. *Nature Ecology and Evolution* 3(11), 1521–1524. https://doi.org/10.1038/s41559-019-1011-2.

Hladik, M. L., Vandever, M. and Smalling, K. L. (2016). Exposure of native bees foraging in an agricultural landscape to current-use pesticides. *Science of the Total Environment* 542(A), 469–477.

Hladik, M. L., Main, A. R. and Goulson, D. (2018). Environmental risks and challenges associated with neonicotinoid insecticides. *Environmental Science and Technology* 52(6), 3329–3335.

Iverson, A., Hale, C., Richardson, L., Miller, O. and McArt, S. (2019). Synergistic effects of three sterol biosynthesis inhibiting fungicides on the toxicity of a pyrethroid and neonicotinoid insecticide to bumble bees. *Apidologie* 50(5), 733–744.

Jacob, C. R. O., Zanardi, O. Z., Malaquias, J. B., Souza Silva, C. A. and Yamamoto, P. T. (2019a). The impact of four widely used neonicotinoid insecticides on Tetragonisca angustula (Latreille) (Hymenoptera: Apidae). *Chemosphere* 224, 65–70.

Jacob, C. R. O., Malaquias, J. B., Zanardi, O. Z., Silva, C. A. S., Jacob, J. F. O. and Yamamoto, P. T. (2019b). Oral acute toxicity and impact of neonicotinoids on Apis mellifera L. and Scaptotrigona postica Latreille (Hymenoptera: Apidae). *Ecotoxicology* 28(7), 744–753. https://doi-org.subzero.lib.uoguelph.ca/10.1007/s10646-019-02070-w.

Jeschke, P. and Nauen, R. (2008). Neonicotinoids – From zero to hero in insecticide chemistry. *Pest Management Science* 64(11), 1084–1098. https://doi.org/10.1002/ps.1631.

Jeschke, P., Nauen, R., Schindler, M. and Elbert, A. (2011). Overview of the status and global strategy for neonicotinoids. *Journal of Agricultural and Food Chemistry* 59(7), 2897–2908. https://doi.org/10.1021/jf101303g.

Knapp, J. L., Bates, A., Jonsson, O., Klatt, B., Krausl, T., Sahlin, U., Svensson, G. P. and Rundlöf, M. (2022). Pollinators, pests and yield—Multiple trade-offs from insecticide use in a mass-flowering crop. *Journal of Applied Ecology* 59(9), 2419–2429. https://doi.org/10.1111/1365-2664.14244.

Kopit, A. M. and Pitts-Singer, T. L. (2018). Routes of pesticide exposure in solitary, cavity-nesting bees. *Environmental Entomology* 47(3), 499–510. https://doi.org/10.1093/ee/nvy034Krupke.

Krupke, C. H., Hunt, G. J., Eitzer, B. D., Andino, G. and Given, K. (2012). Multiple routes of pesticide exposure for honey bees living near agricultural fields. *PLoS ONE* 7(1), e29268. https://doi.org/10.1371/journal.pone.0029268.

López-Uribe, M. M., Cane, J. H., Minckley, R. L. and Danforth, B. N. (2016). Crop domestication facilitated rapid geographical expansion of a specialist pollinator, the squash bee *Peponapis pruinosa*. *Proceedings Biological Sciences* 283(1833), 20160443. https://doi.org/10.1098/rspb.2016.0443.

Lundin, O., Rundlöf, M., Smith, H. G., Fries, I. and Bommarco, R. (2015). Neonicotinoid insecticides and their impacts on bees: A systematic review of research approaches and identification of knowledge gaps. *PLoS ONE* 10(8), e0136928. https://doi.org/10.1371/journal.pone.0136928.

Maertens, A., Golden, E. and Hartung, T. (2021). Avoiding regrettable substitutions: Green toxicology for sustainable chemistry. *ACS Sustainable Chemistry and Engineering* 9(23), 7749–7758. https://doi.org/10.1021/acssuschemeng.0c09435.

Main, A. R., Webb, E. B., Goyne, K. W. and Mengel, D. (2018). Neonicotinoid insecticides negatively affect performance measures of non-target terrestrial arthropods: A meta-analysis. *Ecological Applications: A Publication of the Ecological Society of America* 28(5), 1232–1244. https://doi.org/10.1002/eap.1723.

Minnesota Department of Agriculture. (2016). Review of neonicotinoid use, registration and insect pollinator impacts in Minnesota. Available at: https://www.mda.state.mn.us/sites/default/files/inline-files/neonicreviewrpt2016.pdf.

Mitchell, E. A. D., Mulhauser, B., Mulot, M., Mutabazi, A., Glauser, G. and Aebi, A. (2017). A worldwide survey of neonicotinoids in honey. *Science* 358(6359), 109–111. https://doi.org/10.1126/science.aan3684.

Morrissey, C. A., Mineau, P., Devries, J. H., Sánchez-Bayo, F., Liess, M., Cavallaro, M. C. and Liber, K. (2015). Neonicotinoid contamination of global surface waters and associated risk to aquatic invertebrates: A review. *Environment International* 74, 291–303.

Muth, F. and Leonard, A. S. (2019). A neonicotinoid pesticide impairs foraging, but not learning, in free-flying bumblebees. *Scientific Reports* 9(1), 4764. https://doi.org/10.1038/s41598-019-39701-5.

Mutinelli, F., Costa, C., Lodesani, M., Baggio, A., Medzycki, P., Formato, G. and Porrini, C. (2009). Honeybee colony losses in Italy. In: *Proceedings of the 4th COLOSS conferences* (vols. 26–27), Faculty of Agriculture, University of Zagreb, Croatia, pp. 3–4.

Olaya-Arenas, P. and Kaplan, I. (2019). Quantifying pesticide exposure risk for monarch caterpillars on milkweeds bordering agricultural land. *Frontiers in Ecology and Evolution Sec. Conservation and Restoration Ecology* 7. https://doi.org/10.3389/fevo.2019.00223.

Pisa, L. W., Amaral-Rogers, V., Belzunces, L. P., Bonmatin, J. M., Downs, C. A., Goulson, D., Kreutzweiser, C. and Krupke, M. (2014). Effects of neonicotinoids and fipronil on non-target invertebrates. *Environmental Science and Pollution Research* 22, 68–102.

Pisa, L., Goulson, D., Yang, E. C., Gibbons, D., Sánchez-Bayo, F., Mitchell, E., Aebi, A., van der Sluijs, J., MacQuarrie, C. J. K., Giorio, C., Long, E. Y., McField, M., Bijleveld van Lexmond, M. and Bonmatin, J. M. (2017). An update of the Worldwide Integrated Assessment (WIA) on systemic insecticides. Part 2: Impacts on organisms and ecosystems. *Environmental Science and Pollution Research International* 28(10), 11749–11797. https://doi.org/10.1007/s11356-017-0341-3.

Pistorius, J., Bischoff, G., Heimbach, U. and Stähler, M. (2009). Bee poisoning incidents in Germany in spring 2008 caused by abrasion of active substance from treated seeds during sowing of maize. *Julius-Kühn-Archiv* 423, 118.

Potts, R., Clarke, R. M., Oldfield, S. E., Wood, L. K., Hempel de Ibarra, N. and Cresswell, J. E. (2018). The effect of dietary neonicotinoid pesticides on nonflight thermogenesis in worker bumble bees (Bombus terrestris). *Journal of Insect Physiology* 104, 33–39.

Rondeau, S., Baert, N., McArt, S. and Raine, N. E. (2022). Quantifying exposure of bumblebee (*Bombus* spp.) queens to pesticide residues when hibernating in agricultural soils. *Environmental Pollution* 309, 119722. https://doi.org/10.1016/j.envpol.2022.119722.

Rondeau S., Willis Chan D. S. and Pindar, A. (2022). Identifying wild bee visitors of major crops in North America with notes on potential threats from agricultural practices. *Frontiers in Sustainable Food Systems* 6, 943237. https://doi: 10.3389/fsufs.2022.943237.

Rosa, A. d. S., Teixeira, J. S. G., Vollet-Neto, A., Queiroz, E. P., Blochtein, B., Pires, C. S. S. and Imperatriz-Fonseca, V. L. (2016). Consumption of the neonicotinoid thiamethoxam during the larval stage affects the survival and development of the stingless bee, *Scaptotrigona* aff. depilis. *Apidologie* 47(6), 729–738. https://doi.org/10.1007/s13592-015-0424-4.

Rundlöf, M., Andersson, G. K. S., Bommarco, R., Fries, I., Hederström, V., Herbertsson, L., Jonsson, O., Klatt, B. K., Pedersen, T. R., Yourstone, J. and Smith, H. G. (2015). Seed coating with a neonicotinoid insecticide negatively affects wild bees. *Nature* 521(7550), 77–80. https://doi.org/10.1038/nature14420.

Sánchez-Bayo, F. and Goka, K. (2014). Pesticide residues and bees – A risk assessment. *PLoS ONE* 9(4), e94482. https://doi.org/10.1371/journal.pone.0094482.

Sánchez-Bayo, F. and Tennekes, H. A. (2020). Time-cumulative toxicity of neonicotinoids: Experimental evidence and implications for environmental risk assessments. *International Journal of Environmental Research and Public Health* 17(5), 1629. https://doi.org/10.3390/ijerph17051629.

Schaafsma, A., Limay-Rios, V., Baute, T., Smith, J. and Xue, Y. (2015). Neonicotinoid insecticide residues in surface water and soil associated with commercial maize (corn) fields in southwestern Ontario. *PLoS ONE* 10(2), e0118139. https://doi.org/10.1371/journal.pone.0118139.

Sgolastra, F., Hinarejos, S., Pitts-Singer, T. L., Boyle, N. K., Joseph, T., Lūckmann, J., Raine, N. E., Singh, R., Williams, N. M. and Bosch, J. (2019). Pesticide exposure assessment paradigm for solitary bees. *Environmental Entomology* 48(1), 22–35. https://doi.org/10.1093/ee/nvy105.

Siefert, P., Hota, R., Ramesh, V. and Grünewald, B. (2020). Chronic within-hive video recordings detect altered nursing behaviour and retarded larval development of neonicotinoid treated honey bees. *Scientific Reports* 10(1), 8727. https://doi.org/10.1038/s41598-020-65425-y.

Simon-Delso, N., Amaral-Rogers, V., Belzunces, L. P., Bonmatin, J. M., Chagnon, M., Downs, C., Furlan, L., Gibbons, D. W., Giorio, C., Girolami, V., Goulson, D., Kreutzweiser, D. P., Krupke, C. H., Liess, M., Long, E., McField, M., Mineau, P., Mitchell, E. A. D., Morrissey, C. A., Noome, D. A., Pisa, L., Settele, J., Stark, J. D., Tapparo, A., Van Dyck, H., Van Praagh, J., Van der Sluijs, J. P., Whitehorn, P. R. and Wiemers, M. (2015). Systemic insecticides (neonicotinoids and fipronil): Trends, uses, mode of action and metabolites. *Environmental Science and Pollution Research International* 22(1), 5–34. https://doi.org/10.1007/s11356-014-3470-y.

Siviter, H., Richman, S. K., Muth, F. and Gomez, J. M. (2021). Field-realistic neonicotinoid exposure has sub-lethal effects on non-Apis bees: A meta-analysis. *Ecology Letters* 24(12), 2586–2597. https://doi.org/10.1111/ele.13873.

Sponsler, D. B., Grozinger, C. M., Hitaj, C., Rundlöf, M., Botías, C., Code, A., Lonsdorf, E. V., Melathopoulos, A. P., Smith, D. J., Suryanarayanan, S., Thogmartin, W. E.,

Williams, N. M., Zhang, M. and Douglas, M. R. (2019). Pesticides and pollinators: A socioecological synthesis. *The Science of the Total Environment* 662, 1012–1027. https://doi.org/10.1016/j.scitotenv.2019.01.016.

Stanley, D. A., Smith, K. E. and Raine, N. E. (2015a). Bumblebee learning and memory is impaired by chronic exposure to a neonicotinoid pesticide. *Scientific Reports* 5, 16508. https://doi.org/10.1038/srep16508.

Stanley, D. A., Garratt, M. P. D., Wickens, J. B., Wickens, V. J., Potts, S. G. and Raine, N. E. (2015b). Neonicotinoid pesticide exposure impairs crop pollination services provided by bumblebees. *Nature* 528(7583), 548–550. https://doi.org/10.1038/nature16167.

Stuligross, C. and Williams, N. M. (2020). Pesticide and resource stressors additively impair wild bee reproduction. *Proceedings Biological Sciences* 287(1935). http://doi.org/10.1098/rspb.2020.1390.

Sur, R. and Stork, A. (2003). Uptake, translocation and metabolism of Imidacloprid in plants. *Bulletin of Insectology* 56, 35–40.

Tang, F. H. M. and Maggi, F. (2021). Pesticide mixtures in soil: A global outlook. *Environmental Research Letters* 16(4), 044051. https://doi.org/10.1088/1748-9326/abe5d6.

Tesovnik, T., Cizelj, I., Zorc, M., Čitar, M., Božič, J., Glavan, G. and Narat, M. (2017). Immune related gene expression in worker honey bee (Apis mellifera carnica) pupae exposed to neonicotinoid thiamethoxam and Varroa mites (Varroa destructor). *PLoS ONE* 12(10), e0187079. https://doi.org/10.1371/journal.pone.0187079.

Thompson, D. A., Lehmler, H. J., Kolpin, D. W., Hladik, M. L., Vargo, J. D., Schilling, K. E., LeFevre, G. H., Peeples, T. L., Poch, M. C., LaDuca, L. E., Cwiertny, D. M. and Field, R. W. (2020). A critical review on the potential impacts of neonicotinoid insecticide use: Current knowledge of environmental fate, toxicity, and implications for human health. *Environmental Science. Processes and Impacts* 22(6), 1315–1346. https://doi.org/10.1039/c9em00586b.

Thompson, H. M. and Pamminger, T. (2019). Are honeybees suitable surrogates for use in pesticide risk assessment for non-Apis bees? *Pest Management Science* 75(10), 2549–2557. https://doi.org/10.1002/ps.5494.

Tison, L., Holtz, S., Adeoye, A., Kalkan, Ö., Irmisch, N. S., Lehmann, N. and Menzel, R. (2017). Effects of sublethal doses of thiacloprid and its formulation Calypso® on the learning and memory performance of honey bees. *Journal of Experimental Biology* 220(20), 3695–3705.

Tosi, S., Sfeir, C., Carnesecchi, E., vanEngelsdorp, D. and Chauzat, M. P. (2022). Lethal, sublethal, and combined effects of pesticides on bees: A meta-analysis and new risk assessment tools. *Science of the Total Environment* 844, 156857. https://doi.org/10.1016/j.scitotenv.2022.156857.

Traynor, K. S., Tosi, S., Rennich, K., Steinhauer, N., Forsgren, E., Rose, R., Kunkel, G., Madella, S., Lopez, D., Eversole, H., Fahey, R., Pettis, J., Evans, J. D. and vanEngelsdorp, D. (2021). Pesticides in honey bee colonies: Establishing a baseline for real world exposure over seven years in the USA. *Environmental Pollution* 279, 116566–116566. https://doi.org/10.1016/j.envpol.2021.116566.

Tsvetkov, N., Samson-Robert, O., Sood, K., Patel, H. S., Malena, D. A., Gajiwala, P. H., Maciukiewicz, P., Fournier, V. and Zayed, A. (2017). Chronic exposure to neonicotinoids reduces honey bee health near corn crops. *Science* 356(6345), 1395–1397. https://doi.org/10.1126/science.aam7470.

Whitehorn, P. R., O'Connor, S., Wackers, F. L. and Goulson, D. (2012). Neonicotinoid pesticide reduces bumble bee colony growth and queen production. *Science* 336(6079), 351-352. https://doi.org/10.1126/science.1215025.

Williams, G. R., Troxler, A., Retschnig, G., Roth, K., Yañez, O., Shutler, D., Neumann, P. and Gauthier, L. (2015). Neonicotinoid pesticides severely affect honey bee queens. *Scientific Reports* 5(1), 14621-14621. https://doi.org/10.1038/srep14621.

Willis Chan, D. S., Prosser, R. S., Rodríguez-Gil, J. L. and Raine, N. E. (2019). Assessment of risk to hoary squash bees (Peponapis pruinosa) and other ground-nesting bees from systemic insecticides in agricultural soil. *Scientific Reports* 9(1), 11870. https://doi.org/10.1038/s41598-019-47805-1.

Willis Chan, D. S. and Raine, N. E. (2021a). Population decline in a ground-nesting solitary squash bee (*Eucera pruinosa*) following exposure to a neonicotinoid insecticide treated crop (*Cucurbita pepo*). *Scientific Reports* 11(1), 4241. https://doi.org/10.1038/s41598-021-83341-7.

Willis Chan, D. S. and Raine, N. E. (2021b). Hoary squash bees (*Eucera pruinosa*: Hymenoptera: Apidae) provide abundant and reliable pollination services to Cucurbita crops in Ontario (Canada). *Environmental Entomology* 50(4), 968-981. https://doi.org/10.1093/ee/nvab045.

Woodcock, B. A., Bullock, J. M., Shore, R. F., Heard, M. S., Pereira, M. G., Redhead, J., Ridding, L., Dean, H., Sleep, D., Henrys, P., Peyton, J., Hulmes, S., Hulmes, L., Sárospataki, M., Saure, C., Edwards, M., Genersch, E., Knäbe, S. and Pywell, R. F. (2017)Country-specific effects of neonicotinoid pesticides on honey bees and wild bees. *Science* 356(6345), 1393-1395. https://doi.org/10.1126/science.aaa1190.

Wu-Smart, J. and Spivak, M. (2016). Sub-lethal effects of dietary neonicotinoid insecticide exposure on honey bee queen fecundity and colony development. *Scientific Reports* 6, 32108. https://doi.org/10.1038/srep32108.

Wu-Smart, J. and Spivak, M. (2017). Effects of neonicotinoid Imidacloprid exposure on bumble bee (Hymenoptera: Apidae) queen survival and nest initiation. *Environmental Entomology* 47(1), 55-62. https://doi.org/10.1093/ee/nvx175.

Chapter 7

Assessing the impact of pesticides on pollinators

Christian Maus, Bayer AG, Germany; Anne Alix, Corteva Agriscience, UK; and Daniel R. Schmehl, Bayer CropScience LP, USA

1 Introduction

In agriculture, pesticides are widely used to control hazardous pests, diseases, and weeds which would otherwise affect the health of crops and agricultural productivity (Popp et al., 2013). The need to control these factors varies across crops and regions, depending on pest pressure, crop variety and its tolerance to stress, and local agronomic and climatic conditions. However, where no or insufficient control is provided, yield losses of up to more than 80% may be the consequence (Oerke, 2006). On the other hand, pesticides are designed to control target populations of organisms which are damaging crops (e.g. insect pests, fungi, weeds) by killing individuals of these organisms or acting on their development or presence in a field. Due to these inherent properties, pesticides may have undesired side effects to the environment if they are not applied with the necessary caution.

The selectivity of pesticides to a group of target organisms is variable and depends on a number of factors such as mode of action, mechanisms of

http://dx.doi.org/10.19103/AS.2022.0111.14

uptake, metabolization, and administration. The inclusion of a large number of screening tests in the product development has overall improved the selectivity of pesticides over time. However, only very few pesticides are so selective that they inherently would only affect pest species and would not have the potential to harm non-target species taxonomically related to pests. Therefore, environmental safety requirements, along with regulatory systems to enforce them, have been established for the registration of pesticides to ensure that the use of all pesticide products in the field will not cause unacceptable effects to the environment. For this purpose, the use and commercialization of pesticides is subject to the most stringent regulatory systems which are globally in place for chemicals, which are defining principles and requirements of environmental risk assessment and management, and directions for the implementation of environmental risk assessment and management. One of the most debated and most intensively researched examples of non-target organisms and their protection from unwanted pesticide side effects in this regulatory context are bees and other pollinators.

The reasons for this are diverse: first, bees are important pollinators and, as an organism group exhibiting mutually beneficial trophic interactions with crops, an essential production factor in agriculture (e.g. IPBES, 2016). Therefore, growers have a genuine interest in having them protected from possible pesticide side effects. Second, honey bees are managed as livestock, especially in regions like Europe and North America (e.g. Pirk et al., 2017), and for managed colonies, there is a beekeeper attending them who is concerned about their health and safety, which fostered the specific attention of pesticide safety to bees earlier than for many other organisms of the environment. Finally, honey bees can be readily bred and managed, and there is long-standing experience with handling their colonies (van Engelsdorp and Meixner, 2010). As such, they could be used for tests under semi-natural conditions from the beginning of ecotoxicological research, without having to upfront develop techniques for breeding and handling.

These specific features ensure a special position of the honey bee in risk assessment systems. Whereas most other standard testing organisms in risk assessment have from the beginning been established as surrogate species, representing a more or less broad spectrum of other species that could not be directly tested, the honey bee was, at least initially, considered as a standalone species, representing in the first place itself rather than other taxa. Although in the meantime honey bees are debated as surrogate organisms for other pollinators, especially solitary bees and bumblebees (e.g. European Food Safety Authority, 2013a, United States Environmental Protection Agency et al., 2014), this unique feature still involves specificities in the risk assessment (e.g. the Hazard Quotient), which will be discussed in more detail in subsequent sections.

2 History of bee testing and risk assessment

The first steps to investigate the side effects of pesticides on bees started early in the history of ecotoxicology. The historic development of bee testing and risk assessment in Germany and in the European Union, which is well-documented, serves as a case study, yet the prerequisites and processes which governed these developments were basically comparable in other countries.

When the use of pesticides became prevalent in agriculture in the early twentieth century, many of the products used to control insect pests contained highly toxic substances like arsenic or lead. Their application in flowering crops led in some cases to bee kills, and subsequently to complaints from beekeepers, which prompted authorities to develop solutions for this issue. As an example, in Germany, a Testing Center for Pesticides ('Prüfstelle für Pflanzenschutzmittel') at the Imperial Institution for Biology (Biologische Reichsanstalt) was established in 1920, which dealt with the assessment of pesticide side effects to bees, initially on the basis of simple oral, and later also contact acute tests. At this time, a specific registration process for pesticides with integral environmental safety assessment was not in place and thus bee testing was not mandatory. The Testing Center issued, based on their findings, recommendations for risk mitigation measures to protect bees (Brasse, 2007).

Later, testing approaches became more sophisticated, and tent tests with bee colonies were used in safety assessments (e.g. Finkenbrink, 1940). The 1950s saw an increasing need to further optimize the approaches to protect bees from potential pesticide side effects, and, in this context, to generate accurate figures describing intrinsic bee toxicity. In 1956, the International Commission on Bee Botany organized a meeting in Bern (Switzerland) on the theme of the protection of bees from pesticide effects (Pettinga, 1980), which triggered further work and follow-up consultations and congresses by various organizations to further elaborate approaches for testing solutions. At the same time, approaches to systematically compile methods to analyze pesticide residues in dead bees were initiated (Stute, 1956).

In the 1970s, the progressive development of testing methods triggered the need for method harmonization between European countries and research organizations. The Symposium on the Harmonization of Methods for Testing the Toxicity of Pesticides to Bees took place in 1980 in Wageningen (The Netherlands) (Pettinga, 1980, Stevenson, 1980). Further international consultations on the topic followed, and a few years later, Oomen (1986) published a sequential scheme for evaluating the hazard of pesticides to bees. On the basis of the fundamentals elaborated during these exchanges, the European and Mediterranean Plant Protection Organization issued harmonized testing guidelines (EPPO, 1992) and a risk evaluation scheme (EPPO, 1993) for honey bees in the early 1990s. The EPPO Guidelines and Schemes have been subject to revisions since that time, reflecting scientific knowledge and experience gained from their use. The latest version was issued in 2010 (EPPO, 2010a,b). The EPPO Risk Assessment Scheme was officially withdrawn in 2018 (European and Mediterranean Plant Protection Organization, 2021) but is effectively still in use in the European Union.

In 2012, the European Food Safety Authority took over the review of risk assessment guidance for bees and proposed a new Guidance Document (EFSA, 2013a). The document expands the risk assessment to bumble bees and solitary bees but so far could not be fully implemented, among other reasons because of the lack of complete testing schemes for these groups and the lack of methods to experimentally measure the protection goals proposed in the document. A revised version has been prepared by EFSA (EFSA, 2019) and is up for public commenting (EFSA, 2019).

3 Testing pesticide effects to pollinators

In order to evaluate the risk of a pesticide to pollinators, one of the essential steps is determining the intrinsic toxicity of the pesticide to the pollinator. In regulatory ecotoxicology, this is done by means of an elaborate system of testing methods and approaches, which are described in the following.

3.1 Test species

Assessing the effects of chemicals on non-target species populations and communities is not a simple task. Evaluating effects on the environment, biodiversity and ecosystems must concede that testing every species in any type of ecosystem is impossible. Thus, risk assessments, and particularly regulatory risk assessments that support regulatory decision making, typically rely on a set of standardized and reliable data aimed at providing a representative indication of a chemical's potential effects on the environment, with a high level of certainty.

In this context, standardized ecotoxicity testing has seen an increased development in the last decades, offering a range of testing methods on standard species that provide a sensitive assessment of the intrinsic toxicity of chemicals, in a reproducible and robust way (OECD, 2021a).

For pollinators, most regulatory studies are performed on the honey bee, *Apis mellifera*. The choice of the honey bee as a standard or surrogate species in the risk assessment of chemicals has been extensively discussed (e.g. Boyle et al., 2018, Hinarejos et al., 2018) and is based on several rationales as follows:

- Historically, honeybees appear to be the first non-target insect species farmers and beekeepers were adamant to protect in a crop, because of its role in pollination and the provision of hive products such as honey.

- Non-intentional effects of pesticides on honeybees have hence been studied since the early twentieth century (e.g. Brasse, 2007, Palmer-Jones, 1958, Everts, 1990) even before testing methods on non-target arthropods used in biocontrol were developed (Barrett et al., 1994).
- Testing methods could be rapidly developed, as honeybees are easy to raise and maintain under human management, and well suitable for the development of accurate, reproducible, and measurable exposure dosing, such as controlled oral administration and topical application of a test item in the laboratory.
- Honey bee behavior enabled testing under confined laboratory conditions, in tunnels in the field, and in the open field, allowing testing of effects in a broad range of scenarios and testing tiers from the laboratory at an individual level to the field scale at the colony level.
- Individual bees can learn to respond to a stimulus and thus testing can involve behavioral effects such as foraging behavior (OECD, 2021c, Decourtye et al., 2005) or learning capacity.
- The biology of the honey bee is well described which allows for an easy interpretation of testing results.
- Ecotoxicity data indicate that the honey bee is usually an inherently sensitive species to pesticides, and thus a good indicator of pesticide intrinsic toxicity. Exceptions exist that usually relate to differences in body weight and/or differences in the metabolic pathways that degrade pesticides (Arena and Sgolastra, 2014, Pamminger, 2021).

Differences in biology and behavior between bee species may greatly influence their factual sensitivities to pesticides under field conditions, which has stimulated the development of new testing methods on bee species other than the honey bee. To date, acute oral and contact toxicity testing guidelines have been developed for bumblebees *Bombus terrestris* (OECD, 2017b,c). Method development is ongoing for a chronic bumblebee laboratory test (Exeler et al., 2020), an acute contact (Roessink et al., 2018) and oral (Hodapp and Kimmel, 2018, Roessink et al., 2018, 2020) laboratory test with the mason bee (*Osmia* spp.), a larval laboratory test with *Osmia cornuta* (Exeler and Quambusch, 2020), and an acute laboratory contact test (Nocelli et al., 2020) as well as a larval laboratory test (Rosa-Fontana et al., 2020) with stingless bees (Meliponini). Moreover, higher-tier testing methods are under development for bumble bees and solitary bees (Cabrera et al., 2015, Knaebe et al., 2018, Franke et al., 2020).

Overall, the robustness of a risk assessment to pollinators relies on the possibility to assess whether the effects that were observed in laboratory studies for a tested species will express under higher-tier testing conditions reproducing the conditions of exposure expected after the intended use, addressing the uncertainties identified in lower-tier studies, and enabling extrapolation to other (untested) species. While ideally complete testing schemes for each species could reduce uncertainties in the risk assessment and adapt the testing requirements to the most representative pollinator of a cultivated crop, alternative ways to perform the risk assessment are necessary in real life, since testing each possibly exposed species is impossible in practice. A possible way forward is to design risk assessment scenarios according to the concept of 'focal species', based on the model developed for example in the risk assessment performed for birds and mammals (European Food Safety Authority, 2009). In such systems, effect testing is performed through methods developed on two standard species of different bird groups. The risks to the bird species expected to use a specific crop as a habitat and thus being the most likely to be exposed are evaluated through exposure scenario and biological traits defined for those species, to account for differences in traits, diet, and behavior. The possibilities to develop such a system for pollinators were discussed in a dedicated workshop, and recommendations were published by Boyle et al. (2018).

3.2 Testing methodologies and designs

Ecotoxicological tests with bees expose bees or bee colonies to a pesticide in a standardized and defined way and include regular observations to evaluate whether effects are induced and which effect the tested substance may have on the test organisms. All testing types include an untreated control group, one or several treatment groups which mostly evaluate different test concentrations, and a toxic standard group with a substance that causes known, defined, and reproducible effects to the test organisms. Only in open field tests, a toxic standard is usually not applied to avoid the deliberate release of toxic concentrations of pesticides into the environment (see e.g. Lewis et al., 2001, 2009). Mostly, tests are conducted in a replicated design (i.e. each treatment group is run in a number of replicates) to allow for statistical evaluation.

For all tests, there are defined parameters to be evaluated, which may be affected by the test substance. Typical parameters measured in laboratory tests are mortality, food uptake, and behavior. In more complex tests involving entire honey bee colonies, other parameters such as flight and foraging activity, colony strength, hive weight, nectar and pollen storage, breeding activity and hatching success, or even overwintering success can additionally be tested. The more complex a test design is, the more parameters can generally be assessed. The values of these parameters measured for the test substance are subsequently evaluated by comparison with the data from the control and the toxic standard groups. From those measured parameters and their comparative evaluation, a study endpoint is derived (for a more detailed definition, see e.g. United States

Environmental Protection Agency, 2003). Each test type has a defined endpoint, e.g. an LD50 (concentration that is lethal for 50% of the test organisms), a Non-Observable Effect Concentration (NOEC) (concentration that causes no observable effects), an NOED (individual dose that causes no observable effects), or an NOER (application rate that causes no observable effects).

Most risk evaluation systems are based on studies of different complexity. Studies of lower complexity are characterized by a simplified design, whereas the more complex studies reflect more realistic exposure scenarios. These different levels of complexity are referred to as study 'tiers'. Lower-tier studies are mostly conducted in the laboratory under artificial exposure conditions, and, in the case of honeybees, usually with individual worker bees. By their simplistic design, they exclude as far as possible factors that may influence exposure and therefore generate variability in the results. These studies focus on the test organism and its intrinsic biological traits, and on the test substance, and are typically well standardized. Lower-tier studies have the advantage that they can usually be conducted relatively easily and frequently without great efforts and that their interpretation is accordingly straightforward.

Higher-tier studies, in contrast, are typically exposing bees under more realistic conditions in the semi-field or field, in the case of honeybees with entire colonies. They assess the effects of a substance in a field-relevant scenario, considering additional relevant factors other than the test organism and the substance. The bees are typically exposed to the test substance through matrices that are exposed to the treatment, such as flowers, nectar, or pollen. However, in certain study types, e.g. colony feeding studies, artificial standard exposure designs can be employed in order to achieve controlled exposure conditions (e.g. Oomen et al. 1992).

In honey bee risk assessment, the following study tiers are distinguished (United States Environmental Protection Agency, 2021a):

1 Tier 1: laboratory testing (individual bees tested, simplistic design, artificial exposure).
2 Tier 2: cage studies, semi-field and colony feeding studies (bee colonies tested in a compartment (tunnel/tent) and exposed to a realistically treated crop, or free-flying colonies exposed to an artificially spiked diet).
3 Tier 3: field testing (free-flying bee colonies exposed to realistically treated crops in a controlled design).
4 Tier 4: post-registration field monitoring studies, involving observations under practical use conditions without a controlled design.

The progression across the tiers from 1 to 3 in a risk assessment is illustrated in Fig. 1.

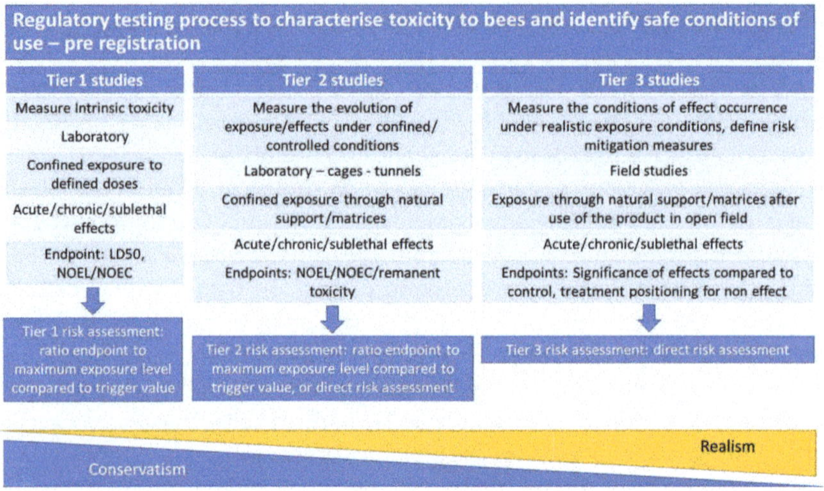

Figure 1 Illustration of the organization of studies in a tiered risk assessment.

A broad spectrum of different study types has been developed, which are testing different endpoints at different testing tiers. Some important study types for honey bee and bumblebee tests are listed in Table 1.

3.3 Testing guidelines and risk assessment guidance documents

In order to generate test results that are comparable to each other and with benchmarks relevant to the safety assessment, ecotoxicological tests have to be conducted in a standardized way, according to defined testing guidelines. A testing guideline is a detailed, officially approved method description for a test procedure. The degree of detail with which a guideline stipulates test design and procedures depends on type and complexity of the test; the simpler the test setup (e.g. laboratory tests), the more detailed and prescriptive the guideline can be, whereas in more complex setups (e.g. field testing), more flexibility is needed to adjust the design for specificities of the crop, the tested product, and prevailing environmental conditions.

The studies to be used for the pesticide safety assessment usually have to follow an officially recognized guideline which sets out the test design and procedure, but also the prerequisites under which a test and its results can be considered valid (so-called Validity Criteria). Guidelines are issued or approved by international organizations, for example, the Organisation for Economic Co-operation and Development (OECD) or the European and Mediterranean Plant Protection Organization (EPPO), or authorities (e.g. United States Environmental Protection Agency). Some important testing guidelines

Table 1 Some important study types and the guidelines according to which they are conducted

Study Type	Guideline	Source
Tier 1		
Acute oral toxicity (honey bee adult) (Fig. 2)	OECD 213	OECD (1998a)
Acute contact toxicity (honey bee adult) (Fig. 2)	OECD 214	OECD (1998b)
	OCSPP 850.3020 (USA)	United States Environmental Protection Agency (2012a)
Acute toxicity/single exposure (honey bee larvae) (Fig. 3)	OECD 237	OECD (2013)
Chronic oral toxicity (honey bee adult)	OECD 245	OECD (2017a)
Chronic toxicity/repeated exposure (honey bee larvae)	OECD GD 239	OECD (2021b)
Acute oral toxicity (bumblebee adult)	OECD 247	OECD (2017c)
Acute contact toxicity (bumblebee adult)	OECD 246	OECD (2017b)
Toxicity of foliar residues (honey bees)	OCSPP 850.30309 (USA)	United States Environmental Protection Agency (2012b)
Tier 2		
Semi-field (tunnel or tent) testing (Figs. 4–6)	EPPO 170	European and Mediterranean Plant Protection Organization (2010a)
	CEB 230 (France)	Végéphyl (2013)
	EFSA Guidance Document (European Union)[a]	European Food Safety Authority (2013a)
Semi-field (tunnel) brood testing	OECD GD 75	OECD (2014)
Brood feeding test	No officially validated guideline	Oomen et al. (1992); Lückmann and Schmitzer (2019)
Colony feeding test	No officially validated guideline	Overmyer et al. (2018), Thompson et al. (2019)
Tier 3		
Field testing (Fig. 7)	EPPO 170	European and Mediterranean Plant Protection Organization (2010a)
	OCSPP 850.3040	United States Environmental Protection Agency (2012c)
	EFSA Guidance Document (European Union)[a]	European Food Safety Authority (2013a)
Tier 4		
Field monitoring	No officially validated guideline	

[a] Draft guidelines for semi-field and field testing.

Figure 2 Acute laboratory test with adult honey bees according to OECD 213/214.

are compiled in Table 1. Besides internationally accepted testing guidelines, there are also some specific guidelines on a national level in certain countries (e.g. the Foliar Residues Test according to OCSPP 850.3030 (USA, Brazil, South Korea) or the semi-field test according to CEB 230 (France)).

3.4 Test method development and validation

For a test design to be useful in risk assessment, it must be capable of yielding reproducible and consistent results. This means that wherever and by whomever a study of a certain type with given parameters like test substance and test concentrations is conducted, it must be assured that the results are always the same or at least do not exceed a defined range of variation (e.g. a certain maximum control mortality). This sounds trivial but is not. Experience has shown that it is relatively straightforward for a testing laboratory to implement and successfully apply a new testing design; however, using a new method in a way that ensures results consistent with those of other testing facilities is much more difficult, and in most cases, it has taken years for experienced laboratories to implement new methods according to consistent standards. This is because even small details in the test design (e.g. air humidity, light regimen, food composition, etc.), the relevance of which may not be initially evident, can have a substantial influence on the results of a study. In order to address this, new

Figure 3 Honey bee larval laboratory test according to OECD 237.

Figure 4 Semi-field (tunnel) test with honey bees in oilseed rape according to EPPO 170.

Figure 5 Semi-field (tunnel) test with honey bees in phacelia according to EPPO 170.

Figure 6 Application in a semi-field (tunnel) test with honey bees in phacelia according to EPPO 170.

Published by Burleigh Dodds Science Publishing Limited, 2023.

Figure 7 Field test with honey bees in oilseed rape according to EPPO 170.

testing guidelines normally go through a validation process called ring testing (OECD, 2005) before they can be finalized and officially endorsed. Ring testing is a stepwise process where different testing facilities apply a new test design in a coordinated way in parallel and compare the results. When the results between the different laboratories are not consistent, parameters are identified which may be causing this variability, and eliminated to harmonize the design. This is done iteratively until consistently reproducible results can be produced by all participating testing units. Considering the importance of validation for the generation of significant results, risk assessment should always be based on validated test methods.

3.5 Good Laboratory Practice compliance ensuring quality standards for regulatory studies

To ensure that safety studies are conducted according to globally comparable quality standards and to ascertain that study procedures and study results are clearly traceable and reproducible, the principles of Good Laboratory Practice (GLP) have been developed and established as a framework for conducting and reporting studies. The principles of GLP set out clear regulations for responsibilities and documentation in the course of study conduction and evaluation. Adherence to the GLP regulation prevents unreproducible study results and precludes data falsification and fraud. The principles of GLP were established in the 1970s and adopted in many countries around the world in the 1980s and 1990s under the auspices of OECD (OECD, 1998c, 2021d). Pesticide studies for regulatory purposes must be conducted under GLP in most countries.

4 Ecotoxicological risk assessment for pollinators

Once the intrinsic toxicity of a pesticide to a pollinator has been determined, the risk posed to the pollinator by the use of the product can be determined by respective risk assessment approaches. One of the basic features of these risk assessment approaches is that it compares effect data and exposure data (see e.g. European Food Safety Authority, 2012). This step is essential, since not only the intrinsic toxicity of a substance but also the exposure to the substance drive the risk. In the risk assessment, both parameters are brought together and integrated.

4.1 Fundamental principles of risk assessment

In the registration process for pesticides, studies that have been conducted to assess the toxicity of a product are followed by procedures that assess the product for its risk to pollinators. Risk assessment is based on some fundamental principles and follows respective rules and procedures which have been stipulated by the legislator. Principles, details, and backgrounds of the most important risk assessment approaches are outlined in the following sections.

4.1.1 Protection goals

Protection goals are the benchmarks used to interpret the outcome of a risk assessment for regulatory decision-making purposes.

In the pesticide regulatory process of the European Union (European Parliament and Council of the European Union 2009, Article 4 (3)(e)), uniform principles for the placing of pesticides on the market are defined. They stipulate that 'a pesticide must not have unacceptable effects on the environment, and no impact on biodiversity and ecosystems'. The compliance of a pesticide to these protection goals is verified by the risk assessment, according to dedicated guidance documents. The data needed to perform a risk assessment are also listed as regulatory data requirements (European Commission, 2013a,b) in the regulation itself. Adherence to this list is part of a completeness check process for the active substance, its metabolites or degradation products and the

Published by Burleigh Dodds Science Publishing Limited, 2023.

formulated product.[1] A risk assessment is performed for each anticipated use of a pesticide and thus for each use, the outcome of the risk assessment is compared to 'Specific Protection Goals'.

Specific Protection Goals for pollinators are defined to ensure a high level of protection for each intended use. More specifically, they describe what to protect, where to protect it, and the timescale over which transient effects might be tolerated without affecting the protection goal, where relevant.

In the European Union, the Specific Protection Goals developed for pollinators are based on ecosystem services that need to be preserved in cultivated areas, such as pollination, the production of hive products (for honeybees only), and biodiversity (specifically addressed under genetic resources and cultural services). This approach was also proposed by experts of the Pellston Workshop on the Risk Assessment to Pollinators in North America (Fischer and Moriarty, 2014), and in most regions globally (e.g. APVMA, 2017, Cham et al., 2017, United States Environmental Protection Agency et al., 2014).

Protection goals valid for pollinators (and other insects) always relate to entities like the colony (in the case of honeybees), or the population (in the case of solitary bees), or to services like pollination, but never to the individual. This means that the individual bee is not the target of the protection goals, and effects on individual bees can be considered acceptable, as long as the population, the colony, or relevant functional endpoints are not affected. This is because organisms like insects are characterized by an intrinsically high loss rate (and colonies by an intrinsically high turnover of individuals), therefore effects on individuals will not necessarily result in biologically relevant effects at the population or colony level.

In the European Union, the European Food Safety Authority (EFSA) defines the Specific Protection Goals attributes, such as survival and development of honey bee colonies and effects on larvae and bee behavior to preserve those ecosystem services (European Food Safety Authority Scientific Committee, 2016).

For those attributes, experts have derived the magnitude of effects that must be met for the ecosystem services to be protected and thus for the risks to be acceptable. Effects can e.g. be defined as a percentage of colony size affected (e.g. 10%, 15%, or 20%). Exposure also enters in the Specific Protection Goals definition, through a given percentage of colonies located at the edge of a field whose exposure is to remain below the levels of exposure calculated in the risk assessment.[2] Specific Protection Goals can be defined for other groups such as solitary bees and can be similar to that for the honey bee or adapted to their particular conditions.

1 A dossier typically counts ca 450–500 studies, 150 of which are ecotoxicological studies.
2 Specific Protection Goals were, for example, set to 7% colony size, and 90th of colonies at the edge of a field would be exposed to lower levels than the calculated exposure levels for a particular use.

An important criterion in defining Specific Protection Goals and in particular the magnitude of effects to be used is the distinction of effects induced by the exposure to a pesticide from other effects and thus to be able to measure the compliance of a pesticide use to the Specific Protection Goal in situ/in the field. While Specific Protection Goals can be directly assessed in a field study, it may be necessary to define benchmark, or trigger values, to be used when the risk assessment relies on laboratory studies and on a comparison of ecotoxicological reference values derived in laboratory studies to calculated or expected exposure levels. Those benchmark or trigger values are then calibrated based on the magnitude of effect for acceptable risks. These benchmark or trigger values become an indicator that the acceptability criteria of the regulation are met on the basis of the outcome of a screening risk assessment, relying on screening data, and a high level of conservatism in the exposure assessment.

4.1.2 Hazard vs. risk

One of the fundamentals of each ecotoxicological risk assessment is the interrelation between hazard, exposure, and risk. Hazard is the potential of a substance to cause an effect in certain organisms; it is determined by the intrinsic toxicity of the substance. Risk is the product of hazard times exposure. Consequently, the risk may be low when exposure is lower than the thresholds for effects, even in case of inherently toxic substances, whereas, when exposure is higher than these thresholds, there may be a risk even in case of low intrinsic toxicity. When exposure is zero or negligible, the risk is negligible too. As for any other chemistry including pharmaceuticals as well as for foodstuffs, this also means that toxicity does not equal risk. This is commonly and intuitively understood for things we consume (e.g. sodium chloride, fluoride, sugar, caffeine, and ethanol). Paracelsus captured the concept well, 'All things are poison, and nothing is without poison, the dosage alone makes it so a thing is not a poison', but in the public discussion of pesticide risks, the difference between hazard and risk is frequently misconceived.

4.1.3 Effects and exposure

As in all risk assessment approaches, the ultimate goal of a pollinator risk assessment is to compare the exposure of the pollinator species under consideration to a substance under defined use conditions with the respective toxicity thresholds of the substance to the species under evaluation (e.g. honeybees) and to derive from this comparison the risk that the use of products containing the substance poses to the evaluated species.

In the case of laboratory data, measured toxicity data are compared to practically measured or mathematically modeled exposure data. Exposure can be expressed in different dimensions, e.g. concentration (quantity of substance per unit of a medium (e.g. diet, water, or soil)), dose (quantity of substance taken up per individuum of the assessed species (or per body mass unit of the individuum)), or rate (quantity of substance applied per area unit). Which dimension is relevant depends on the test design. For instance, in studies where animals can be individually dosed, a specific dose can be tested, whereas where the exact exposure per individual cannot be accurately determined (e.g. in water for aquatic organisms, or in honey bee colony feeding studies), exposure is expressed as a concentration. Where the test substance is applied by overspray, a relevant application rate can be established (Fig. 1). Likewise, effects can be measured by a variety of parameters and endpoints, the most important of which have been described above (Section 3.2). In higher-tier studies, where the exposure of the test organism according to realistic scenarios is already a part of the study design, a quantitative comparison between exposure and effects can be omitted since both elements are already integral parts of the study design (Fig. 1).

4.1.4 Assessment factors and risk quotients

The result of a study is expressed as an endpoint, i.e. an exposure concentration, dose, or rate which is causing a defined effect (or no effect) to the test organism. In risk assessment, there are frequently additional assessment factors (or safety factors) applied against an endpoint. Assessment factors serve to introduce an additional margin of safety, to account for natural variability of the test system, and to compensate for inter-species variabilities where a test organism is a representative surrogate organism, covering other species too. Typically, the lower the tier of the study used to generate an endpoint, the higher the assessment (safety) factor. Lower-tier study endpoints may not be directly defined as a safe concentration in the assessment unless an assessment factor (or safety factor) is applied against them. In toxicology and ecotoxicology in general, assessment factors span from one up to 1000, depending on the study type, the number of studies that have been performed of the respective type, and the test organism.[3] LD50 endpoints are almost always used in the connection with a substantial assessment factor, accounting for the fact that the underlying studies are lower-tier, and to account for the relevant protection goals in the risk assessment. Endpoints derived from higher-tier studies can typically be directly used in the risk assessment and mostly do not require an assessment factor since their design is close to realistic field conditions.

3 For example, the derivation of a PNEC or Predicted No Effect Concentration for aquatic organisms uses a factor of 1,000 when a limited number of acute toxicity studies is available, 100 and 10 for acute and chronic studies where both are available and the PNEC will be the lower of the two values (European Chemicals Agency 2008).

Especially in lower-tier risk assessment, the application of an assessment factor against a toxicity endpoint is frequently done in the calculation of a risk quotient. Risk quotients include the Hazard Quotient (HQ), the Exposure-Toxicity Ratio (ETR), and the Toxicity-Exposure Ratio (TER). All these quotients are based on a division of an exposure value by a toxicity value (HQ, ETR) or reversely (TER). These are compared to a trigger value reflecting the protection goal or can directly include the assessment factor.[4]

Whether the outcome of this calculation is above or below a pre-defined trigger value indicates whether the evaluated use can, at the respective tier of the risk assessment, be considered safe, or whether a risk cannot be excluded at this tier. For example, in the assessment of chronic risks to adult bees, with a TER trigger of 10, a dietary concentration of the evaluated substance of 2 mg/kg diet, and an NOEC of 30 mg/kg diet as the relevant toxicological endpoint, the TER would be calculated as 30 divided by 2, i.e. 15, meaning that the exposure is 15 times lower than the threshold for toxicity. In such a case the use could therefore be considered safe.

Where the dimensions of the toxicity endpoint and the exposure endpoint are not the same (e.g. effect data expressed as mg substance ingredient/bee and exposure data expressed as mg active substance/kg diet), conversion factors have to be included in the equation in order to make them directly comparable. A specific case is the HQ as used in the EPPO Risk Assessment Scheme (European and Mediterranean Plant Protection Organization, 2010b). In this approach, exposure (expressed as an application rate of the product or substance under evaluation in g/ha) is divided by an effect endpoint (an LD50 expressed in µg/bee). Here, the different dimensions of numerator and denominator are not adapted by an adjustment factor. In contrast, the approach is based on a systematic evaluation of historic incident data undertaken by Aldridge and Hart (1993) that established a relationship between the HQ and the likelihood of a respective pesticide use to be involved in bee intoxications in the field. Pesticide uses with a HQ greater than the trigger value of 2500 were frequently implicated in incidents. Uses with an HQ smaller than 50, however, can be considered intrinsically safe according to their analysis. Thus far, the HQ in this approach is one of the few evaluation factors in ecotoxicology which have been validated against realistic field data.

4.2 Tiered risk assessment systems

A risk assessment system based on studies of different tiers is called a tiered risk assessment system. Therein, the different study tiers as described in the previous

4 Note that additional refinement or correction factors may be included in the expression of exposure in the risk equation.

section are mirrored, and their specific strengths and limitations accordingly inform the respective risk assessment steps or tiers (see e.g. Alix and Lewis, 2010) (Fig. 1). Lower-tier (especially laboratory) studies are most suitable to determine the intrinsic toxicity of a substance to a test organism; on the other hand, they are normally not appropriate to directly deduce information about the existence of a risk that a substance may pose under realistic conditions. Since they exclude all naturally mitigating factors, their results, if directly extrapolated to field conditions, would frequently lead to an overestimation of risks. Higher-tier studies, in contrast, are suitable for directly deducing information about potential risks of a test substance; the more realistic the scenario is, the better. Especially field studies are complex in their setup, their conduction requires significant efforts and resources, and the interpretation of their results needs to be based on expert judgment. Whereas lower-tier studies, since focusing on intrinsic features of the test substance and the test organism, are universally valid, the extrapolatability of higher-tier studies may be more limited, especially when they do not represent worst-case conditions in all parameters, or when specific scenarios (e.g. in terms of crop, environmental conditions, application of the test substance) are tested, which do not reflect standard conditions.

Risk assessment systems in which lower-tier and higher-tier studies are related to each other in a hierarchical configuration are called tiered, hierarchical systems (Fig. 1). They aim at assessing the compliance of pesticide products with defined protection goals, based on a stepwise approach to assessing effects and risks, from highly standardized laboratory-derived parameters to realistic field studies. All existing major risk assessment schemes for bees belong to this category. The rationale of such a system is that different studies and their results have a different weight and a different significance for the understanding of a potential risk. The more realistic a study design and an exposure regime is, the higher weight is assigned to a study tier, and the results of a higher-tier study can override the results of a lower-tier study.

In terms of study conduction in the framework of a risk assessment scheme, most systems are operating as follows: the first studies to be conducted are lower-tier studies. In the case of substances that are intrinsically non-toxic to bees (as many herbicides and fungicides), there are no adverse effects seen even under the extreme conditions of lower-tier studies. In such cases, testing can stop, if the findings of the toxicity test in connection with the expected exposure under field conditions clearly suggest the absence of a risk. If the results of the tested tier do not attest to the absence of a risk, the next testing tier is triggered. This process continues up to the tier where the safety of the product can be demonstrated, including risk mitigation measures where needed. In other words, if the outcome of the risk assessment does not meet the benchmark or trigger value indicating a safe use, this suggests that a refined risk assessment is needed, which assesses whether or not the pesticide meets the Specific Protection Goals under the proposed conditions of use, based on high-tier studies in tunnels or open field (European Commission, 2011a).

To some degree, studies of a lower-tier may be waived, if studies of the next higher-tier are available (e.g. Alix and Lewis, 2010). For instance, if a field study is provided, a corresponded semi-field study may not be necessary. If there are adverse effects even in the highest tiers, use restrictions for the product to minimize exposure need to be designed, or the product cannot be registered for the respective use.

The advantage of the tiered system is that intrinsically low-toxic substances can be filtered out early and with a limited effort, so that the focus can be directed to substances that have a higher intrinsic toxicity, allowing unnecessary complex higher-tier tests on intrinsically safe substances to be avoided.

An example of the application of the tiered system is provided in Box 1.

Box 1 Case study: Honey bee safety of the neonicotinoids

One of the most controversially discussed topics related to pesticide environmental compatibility is the honey bee safety of the neonicotinoids (e.g. Eisenstein, 2015). This is a class of insecticides which selectively act on insect nicotinic acetylcholine receptors and which are on the market since the 1990s (Jeschke and Nauen, 2010). Due to their efficacy against pest insects, their systemicity in plant tissues, which allows their use as seed and soil treatment, and their high level of operator safety, they are widely used in a broad variety of crops (Jeschke and Nauen, 2010). A sub-class of the neonicotinoids, the so-called nitro-substituted neonicotinoids exhibit a high intrinsic toxicity to bees, yet of a comparable order of magnitude as other insecticide classes; imidacloprid, thiamethoxam, and clothianidin for instance belong to this group (Iwasa et al., 2003).

In the mid-1990s, seed treatment applications of imidacloprid in sunflower in France were suspected as a cause of honey bee colony losses due to systemic residues in nectar and pollen (Maus et al., 2003). In 2008, a severe incident took place in Germany where a significant number of bee colonies were intoxicated by dust particles from clothianidin-coated corn seeds of deficient seed treatment quality (Forster, 2009, Nikolakis et al., 2009). In the early 2010s, some studies about sublethal effects of neonicotinoids to bees attracted the attention of the public and the scientific and regulatory communities (Henry et al., 2012, Whitehorn et al., 2012), which triggered political and regulatory processes which eventually led to far-reaching regulatory restrictions for imidacloprid, thiamethoxam, and clothianidin in the European Union (European Commission, 2013c, 2018a,b,c), whereas they are continued to be used in other regions of the world, yet in some countries with use adaptations in order to optimize their bee safety.

The regulatory risk assessment for the neonicotinoids under discussion is driven by higher-tier data, which exemplifies the principles of risk assessment as outlined in this chapter. The substances usually do not pass the lower-tier risk assessment due to their intrinsic toxicity to bees. Therefore, higher-tier studies had to be performed, or risk mitigation measures minimizing exposure were stipulated. For seed treatment uses which can lead to trace level residues in nectar and pollen of treated crops (and, to some degree also of succeeding crops), the regulatory risk assessment is largely based on colony feeding study data in combination with residue data from bee-relevant matrices in treated and succeeding crops (second tier), and field data (third tier). These data largely revealed no or little concern about exposed honey bee colonies (e.g. Maus et al., 2003, Schmuck and Keppler, 2003, Schmuck et al., 2005, Pilling et al., 2013, Thompson et al., 2019).[5] In the case of foliar uses, the risk to bees can in practice be easily mitigated by the avoidance of application during flowering of the crop and the restriction of pre-flowering applications.

Public concerns about the neonicotinoids were predominantly triggered by studies on sublethal effects which were conducted off the regulatory study system and mostly off the regulatory standards for safety studies. Hundreds of studies on neonicotinoid sublethal effects have been published (e.g. by Godfray et al., 2014, 2015, Siviter et al., 2021). The existence of sublethal effects as such is neither new nor unexpected, since any intrinsically toxic substance which is administered at a non-lethal level will inherently cause sublethal effects at certain exposure concentrations. This principle is well known to us from substances of our everyday life (caffeine, sodium chloride, ethanol, glucose). However, it is important to consider that most of the studies which have been conducted on sublethal effects were done in the laboratory (or under otherwise unrealistic exposure conditions) and with individual honeybees rather than with honey bee colonies. Moreover, exposure concentrations were not reflecting realistic field exposure scenarios in the vast majority of cases (see e.g. Carreck and Ratnieks, 2014). Accordingly, most of these studies would in a regulatory system be categorized as first-tier studies (see Fig. 1), which can inform about the intrinsic toxicity of the test substance, but which do not allow conclusions about the safety of a product under realistic conditions. Interestingly, a majority of studies which were conducted off the regulatory system but would correspond to higher-tier studies largely confirmed the conclusions of the regulatory risk assessment: most field studies either showed no or only insignificant, or only inconclusive effects to exposed honey bee colonies (Stadler et al., 2003, Cutler and Scott-Dupree, 2007, Nguyen et al., 2009, Cutler et al. 2014, Rundlöf et al., 2015, Rolke et al., 2016, Woodcock et al., 2017). This conclusion has in the meantime been confirmed by a comprehensive meta-analysis of available data

5 The EFSA risk assessment came to a different conclusion since, among other reasons, large part of the available higher-tier studies was not taken into consideration (European Food Safety Authority 2013b, c, d).

on the honey bee safety of the neonicotinoids (Stephenson and Solomon, 2017, Solomon and Stephenson, 2017a,b). This case study illustrates the critical importance of interpreting individual studies in their context and according to their purpose and context, which can be fundamentally different between the regulatory framework and academic research.

4.3 Risk assessment schemes and guidance documents

Risk assessment schemes or guidance documents provide guidance on the types of studies to conduct depending on the uses or product type, which procedures are to be used, how the study results are to be interpreted, and how the different study types and their results relate to each other in the risk assessment.

Like guidelines, risk assessment schemes and guidance documents are issued by international organizations or authorities. Examples are the EPPO Risk Assessment Scheme (European and Mediterranean Plant Protection Organization, 2010b) (Europe), the EFSA Guidance Document on the Risk Assessment of Plant Protection Products on Bees (European Food Safety Authority, 2013a) (European Union), the Guidance for Assessing Pesticide Risks to Bees (United States Environmental Protection Agency et al., 2014) (USA, Canada), the Guidance on Exposure and Effects Testing for Assessing Risks to Bees (United States Environmental Protection Agency, 2016) (USA, Canada), the Manual de Avaliação de Risco Ambiental de Agrotóxicos para Abelhas (Cham et al., 2017) (Brazil), and the Roadmap for Insect Pollinator Risk Assessment in Australia (Australian Pesticide and Veterinary Medicines Authority, 2017) (Australia).

Since the publication of the first risk assessment scheme for pollinators by EPPO in 1992 (European and Mediterranean Plant Protection Organization, 1992), various countries and geographical regions have developed their own guidance documents for pollinator risk assessment. A summary of their main contents including protection goals, testing requirements, and risk assessment principles are outlined in Table 2 for the European Union, North America, Brazil, and Japan, which taken together represent the diversity that may be observed in the existing risk assessment schemes. The risk assessment schemes in other countries such as Australia, New Zealand, South Korea, and Mexico have adapted and updated the North American standards to meet their own specificities.

Risk assessment schemes in all regions are based on comparable protection goals and thus share similar perspectives on the role and importance of pollinators to agriculture, beekeeping, and biodiversity. For

Table 2 Exemplary overview of risk assessment schemes in different countries and regions

Country/geographical region	Protection goals	Fundamental lower-tier testing requirements	Risk assessment principles	High-tier studies
Europe: EPPO Risk Assessment Scheme[a] (European and Mediterranean Plant Protection Organization, 2010b)	Colony development and survival	OECD 213, 214	Sub-schemes for foliar and soil-systemic treatments HQ as decision criterion for chronic and higher-tier testing	Residue refinement, semi-field and field studies
European Union: Guidance Document on Terrestrial Ecotoxicology Under Council Directive 91/414/EEC (European Commission 2002)	Overall survival and productivity of the hive	OECD 213, 214	Adaptation of the EPPO Scheme. HQ as decision criterion for higher-tier testing	Residue refinement, semi-field and field studies
European Union: EFSA Guidance Document (European Food Safety Authority, 2013a)	Colony development and survival; brood development and bee behavior. Provision of pollination services Production of hive products Non-*Apis* bees' populations (biodiversity)	OECD 213, 214, 239, and 245	Risk assessment for major and minor exposure routes Trigger values developed for each bee stage and assessment scale	Residue refinement, semi-field and field studies
USA and Canada: Guidance for Assessing Pesticide Risks to Bees (United States Environmental Protection Agency et al., 2014)	Colony and individual level assessments Honey bees as proxy for biodiversity Provision of pollination services Production of hive products	OECD 213, 214, 237, 239, and 245	Oral and contact exposure via treated crop Trigger values developed for acute and chronic risks	Residue refinement, semi-field and field studies
Brazil: Manual de Avaliação de Risco Ambiental de Agrotóxicos para Abelhas (Cham et al., 2017)	Colony development and survival; brood development and bee behavior. Provision of pollination services Production of hive products Non-Apis bees' populations (biodiversity)	OECD 213, 214, 237, 239, and 245	Oral and contact exposure via treated crop Trigger values developed for acute and chronic risks	Residue refinement, semi-field and field studies
Japan: Data Requirements for Registration of Agricultural ChemicalsFood and Agricultural Materials Inspection Center (2019)	Maintenance of honey bee colony for beekeeping Colonies to be protected Focus on Apis mellifera (plan to introduce guidance for native pollinators e.g. *Apis cerena*)	OECD 213, 214, 237, 239, and 245	Oral and contact via treated crop Trigger values developed for acute and chronic risks	Residue refinement, semi-field and field studies

[a] The EPPO Risk Assessment Scheme has officially been withdrawn in 2018 (European and Mediterranean Plant Protection Organization 2021), but is effectively still in use in the European Union.

study requirements, OECD guidelines remain the standard for all regions to address intrinsic toxicity assessments. For risk assessment calculations, most schemes primarily focus on direct exposure to a treated crop through oral and contact exposure routes, though e.g. in the European Union other possible exposure routes have been added to the scheme (see Chapter 15). Trigger values, also called levels of concern in North America, usually distinguish acute and chronic risk assessments. The EFSA Scheme has introduced additional distinctions between bee life stages and species, based on the models and calculations used to derive those trigger values.

Despite the differences between the approaches, the systems are based on the same principles and fundamentals, and theoretically lead to largely similar results in most cases, yet the margin of discretion which is granted in all systems may in practice lead to diverging conclusions in individual cases. All systems consider higher-tier tests under realistic exposure conditions, which are mostly required for intrinsically toxic substances, as overriding other study types. In contrast to the other systems, the EFSA system does not discriminate and filter out intrinsically non-toxic products in the first tiers, while simultaneously complicating higher-tier testing options by extremely rigorous requirements for study validity. The EFSA Scheme has so far not been fully implemented in the European Union, among other reasons because methods to experimentally measure the protection goals proposed in the document are lacking. A revised version has been prepared by EFSA (EFSA, 2019) and is up for public commenting (EFSA 2022).

5 Indirect and sublethal effects

Indirect effects are an important source of uncertainty in the risk assessment. For pollinators, indirect effects may result from effects of pesticides on habitat and food resource or quality. Hence weed management (chemical or mechanical) could result in indirect effects that affect the survival of a colony to a similar extent as direct effects.

In the regulatory risk assessment of pesticides, indirect effects on pollinators are addressed through the risk assessment performed for non-target plants that are present at the edge of a field and may be exposed to spray drift, for example. This risk assessment is performed for all types of products including herbicides and plant-growth regulators, and guidance documents are available that provide detailed recommendations on how to perform laboratory and field studies (European Commission, 2002, Arts et al., 2017).

Besides effects on mortality or reproduction, pesticides may affect other parameters such as behavior, longevity, immune defense, sensory perception, cognitive skills, and many others. Such effects are referred to as sublethal effects. Relatively little is known about how far sublethal effects may affect bee colonies or populations in a way that would be measurable on the basis of population parameters. In fact, there are very few documented cases where an adverse impact at the honey bee colony level has been seen to be caused by sublethal effects in a field-realistic scenario. Therefore, sublethal effects that are measurable at the level of individual honeybees may not necessarily be biologically relevant effects, or do at least not necessarily affect the protection goal (see e.g. Henry et al., 2015, Siede et al., 2017). In existing risk assessment systems, sublethal effects are mostly accounted for in the same study types that are also assessing mortality and other standard endpoints (e.g. Thompson and Maus, 2007). Some study designs have been developed to test for specific sublethal effects (e.g. OECD GD 332, effects on homing capability (OECD, 2021c)), but they are not mandatory in most risk assessment schemes, as our understanding of the biological significance of sublethal effects is limited. However, sublethal effects can generally be considered to be essentially covered by the existing testing methods, as they would either manifest in effects on other endpoints (e.g. mortality, foraging, reproduction, and colony performance) or alternatively would not be relevant in relation to the protection goal.

6 Risk mitigation

Where the outcome of the risk assessment does not exclude the presence of risks to pollinators under practical use conditions of a product, risk mitigations have to be defined to ensure a safe use by minimizing exposure. The registration process for pesticides in the European Union defines specific risk mitigation measures to accompany the registration of pesticides where they are triggered by the outcome of the risk assessment (European Parliament and Council of the European Union, 2009). Risk mitigation measures for pesticides may be implemented at various levels. The regulations stipulate a range of precautionary or safety phrases describing appropriate conditions of use on the product's labeling (European Commission, 2011b). Labeling recommendations can hence reduce exposure of both managed and wild bees (for examples see **Box 2** Safety phrases on product labels). For managed species, beekeepers can also efficiently manage colonies in space and time thanks to adapted recommendations. Mitigation measures may include a variety of crop or hive management approaches that reduce the exposure of bees to the treatment, and which are defined on a case-by-case basis, considering the intrinsic properties and characteristics of the product, the crop, and pollinator species of concern in the mitigation measure. Examples are avoiding applications during flowering and during pre-flowering safety intervals depending on the residual toxicity of the product, applications in the evening after daily bee flight, avoiding spray drift to bee-attractive crops or weeds, removal of flowering weeds in the crop's understory before application, or covering or removal of bee hives in or nearby the crop prior to the treatment (see e.g. Australian Government – Rural Industries Research and Development Corporation, 2012, Hooven et al., 2013, CropLife International, 2017, Johansen and Wu-Smart, 2021). Label information is mandatory and is implemented in all countries, adapted to national situations and farming practices, and designed specifically for each product. These mitigation methods have been developed with a focus on honeybees and may not in all cases applicable to wild pollinators.

Box 2 Safety phrases on product labels

Examples of safety phrases on product labels to protect pollinators

European Union (European Commission, 2003):

SPe 8: Dangerous to bees./To protect bees and other pollinating insects do not apply to crop plants when in flower./Do not use where bees are actively foraging./Remove or cover beehives during application and for (state time) after treatment./Do not apply when flowering weeds are present./Remove weeds before flowering./Do not apply before (state time).

Australia (Australian Pesticide and Veterinary Medicines Authority, 2018):

Mandatory label for 'Protection of Livestock' with the precaution to not spray onto bees or beehives and the mention that 'once the spray deposit has dried, foraging bees will not be affected'.

New Zealand (New Zealand Ministry for Primary Industries, 2020):

'POLLINATORS: When applied during non-foraging periods the Product will not interfere with the activity of honeybees once the spray has dried. At least 3 hours drying time should occur before bee foraging is expected. At times when bees aggregate in large numbers outside the hive, ensure they are not directly contacted by the spray.'

United States (United States Environmental Protection Agency, 2021b):

'This product is toxic to bees exposed to treatment during the 3 hours following treatment. Do not apply this pesticide to blooming, pollen-shedding or nectar-producing parts of plants if bees may forage on the plants during this time period.'

Besides labelling of pesticides, crop management practices adopted by farmers at the farm scale may greatly reduce exposure of pollinating species visiting crops (Alix and Garrido, 2015, Alix et al., 2017). A range of farm management tools beneficial to pollinators has been identified, ranging from establishing natural and semi-natural field margins to managed field margins, dedicated pollen and nectar seed mixes, wildflower sown margins, grass strips, or conservation headlands. Each provides advantages to pollinating insects, either as a refuge area or in providing a dedicated source of food or nesting habitat. A ranking of the benefits represented by each type of farm management was provided by the MAgPIE Project.

An inventory of mitigation measures for pollinators has been compiled by the OECD-PEIP working group (OECD, 2021e). The inventory includes regulatory risk mitigation recommendations as communicated through the information on pesticide labels, and education and training programs. Education and training programs for farmers and beekeepers are also key components of risk management as they drive the accuracy with which risk mitigation measures are implemented. These programs may be organized by any stakeholder and are most often voluntary initiatives, thus indicating a real commitment of countries.

7 Pesticide incident monitoring

Monitoring approaches aim to assess the potential effects of pesticides used in crop protection according to label recommendations under practical conditions on specific populations and communities. These studies complement the risk assessment performed under relevant regulations (European Parliament and Council of the European Union, 2009) and describe the conditions of exposure of organisms in their environment, the conditions of occurrence of risks, and provide in situ assessment of the efficacy of risk mitigation measures that were recommended thus enabling further adaptation based on the collected information.

Despite an increasing interest in field observations of pesticide impacts on pollinators, there is to date no harmonized regulatory guidance or monitoring methodology for honeybees or other pollinating species, nor is there regulatory guidance on the use of generated data in support of risk assessment or decision making. An inventory of published monitoring studies was undertaken by the International Commission on Plant-Pollinator Relationship (ICPPR) a few years ago (Alix and Garrido, 2015), which analyzed 24 monitoring studies and was further expanded to 56 studies in 2016, as a basis for a possible guidance on good monitoring practices. The analysis of the data revealed distinct approaches depending on the species monitored and on the purpose of the study. For managed species, the baseline can be defined at the introduction or release of pollinators in the cultivated system whereas wild bees require setting study designs reflecting ecological approaches to first define the expected occurrence and diversity from the collected environmental features of the ecosystem. In both cases, a dedicated description of the environment is critical to the interpretation of the data. There is a need to gather existing knowledge on the specific landscape management initiatives and their benefits to pollinators in cropping systems in order to build a decision tree to assess the potential of a crop to induce pollinator exposure and to identify the risk mitigation measures and crop management actions with the highest potential to limit exposure. However, limited knowledge is available on species traits and relationship to specific flora to accurately predict the expected community in a particular crop system, which would define which species need to be protected and where.

Projects that are initiated should allow a better implementation of monitoring data collection, as for example in the European Union (European Commission, 2021) and propose protocols. A first financial analysis shows that despite the relatively high financial costs of systematic monitoring data generation, the benefits to pollinated crops and systems from the resulting learning can largely compensate these expenses (Breeze et al., 2020). Besides a harmonized method to collect data that can be compared and used to feedback to regulatory processes, basic ecological data are needed to better understand plant-bee relationships and depict the species and assemblage of species that can be expected in particular systems, to inform protection goals and design risk assessments and monitoring.

In some countries, there are passive monitoring systems in place, where reported incidents of bee intoxications are investigated by official institutions to analyze whether an incidence of mortality was caused by pesticide intoxication and, if so, which products and application practices were involved. Examples for such incident monitoring approaches are the British Wildlife Incident Investigation Scheme, the Incident Monitoring of the Julius Kühn Institute in Germany, and the Pesticide Incident Reporting Program in Canada (Seefeld, 2006, Barnett et al., 2007, Thompson and Thorbahn, 2009, Mineau et al., 2008, Julius-Kühn-Institut, 2021a,b, Government of Canada, 2018, 2021).

8 Conclusion

Overall, the development of the described approaches and efforts, which have been continuously extended and refined over the last decades, underscores the importance of bees and other pollinators for agriculture and crop protection. Incident monitoring systems confirm that in countries where systematic long-term figures are available, numbers of bee intoxications with pesticides have shown a decreasing trend over the last decades and are currently on low absolute numbers (e.g. Thompson and Thorbahn, 2009, Carreck and Ratnieks, 2014). This reflects the improvement in the level of protection provided by regulatory systems developed over the years as well as the increasing awareness of the importance of pollinator protection among the involved stakeholders. Nonetheless, environmental safety testing and risk assessment schemes in regulatory ecotoxicology are not static constructs, but dynamic systems, which can be constantly optimized to always reflect the latest state of science and address emerging research issues; in so far, we can expect further new developments and optimizations in the years to come.

9 Where to look for further information

Key resources for further information on the topics outlined in this chapter are in the first place the guidelines and guidance documents cited in the main text. Beyond this, a broad overview of methodologies for scientific studies on honeybees (not restricted to pesticide assessment and evaluation) is provided in the COLOSS BEEBOOK (https://coloss.org/beebook/). General information can also be found in Whitford et al. (2017).

Further information about the regulation of pesticides with regard to pollinator safety can be found on the websites of a variety of authorities and agencies in charge of environment, agriculture, or pesticide registration, such as the United States Environmental Protection Agency (US EPA) (https://www.epa.gov/pollinator-protection/epa-actions-protect-pollinators#:~:text=EPA%20has%20taken%20the%20following,contract%20to%20provide%20pollination%20services.), the United States Department of Agriculture (e.g. https://www.usda.gov/sites/default/files/documents/pollinator-priorities-2021R4-508-version.pdf), the European Food Safety Authority (EFSA) (https://www.efsa.europa.eu/en/topics/topic/bee-health), the French Agency for Food, Environmental and Occupational Health & Safety (ANSES) (https://www.anses.fr/en/content/bee-health), the German Julius Kühn-Institut (https://www.julius-kuehn.de/en/bs/), and others.

Standardized testing methods for pollinators are issued by the Organisation for Economic Co-operation and Development (OECD) (https://www.oecd.org/chemicalsafety/testing/work-related-beespollinators.htm).

Major scientific programs and networks which are looking at a broad spectrum of aspects related to pollinator health (including, but going far beyond pesticide topics) comprise COLOSS (https://coloss.org/), PoshBee (https://www.poshbee.eu/), Safeguard (https://www.safeguard.biozentrum.uni-wuerzburg.de/), STEP (http://www.step-project.net/), Salud Apicola (https://en.saludapicola.com/), and others.

Important scientific organizations working on pollinator safety topics in agriculture include the International Commission of Plant-Pollinator Relationships (ICPPR (https://www.icppr.com/), the Society of Environmental Toxicology and Chemistry (SETAC) (https://www.setac.org/), Project Apis m. (https://www.projectapism.org/honey-bee-research.html), the Pollinator Partnership (https://www.pollinator.org/), the Pollinator Research Task Force (https://pollinatorresearchtaskforce.com/), the Honey Bee Health Coalition (https://honeybeehealthcoalition.org/), the Bee Informed Partnership (https://beeinformed.org/), and ABELHA (https://abelha.org.br/en/).

Information about the engagement of the crop protection industry in pollinator protection can be found on the websites of CropLife International (e.g. https://croplife.org/case-study/pollinator-protection-vital-to-crop-production/, https://croplife.org/crop-protection/pollinators-2/), CropLife Europe (e.g. https://croplifeeurope.eu/report/pollinators-and-agriculture/), CropLife America (e.g. https://www.croplifeamerica.org/resources), CropLife Latin America (https://www.croplifela.org/en/whats-new/pollinators), and CropLife Asia (https://www.croplifeasia.org/video_tag/pollinators/#/). CropLife International also provides an overview of stewardship approaches to protect pollinators (https://croplife.org/wp-content/uploads/2017/04/Protecting-Pollinators-Through-Good-Stewardship-Practices-v7.pdf).

10 Acknowledgements

The authors would like to thank Thomas Steeger (United States Environmental Protection Agency) for providing information about the development of the development of pollinator risk assessment in North America, and Maria Teresa Almanza (Bayer AG) for proof-reading parts of the chapter and input about latest developments regarding testing and risk assessment procedures in the EU.

11 References

Aldridge, C. A. and Hart, A. D. M. (1993). Validation of the EPPO/CoE risk assessment scheme for Honeybees. *Proceedings of the Fifth International Symposium on the Hazards of Pesticides to Bees*. October 26–28, 1993. Plant Protection Services, Wageningen, The Netherlands. Appendix 5: 37–42.

Alix, A., Brown, Ch., Capri, E., Goerlitz, G., Golla, B., Knauer, K., Laabs, V., Mackay, N., Marchis, M., Poulsen, V., Prados, E. A., Reinert, W. and Streloke, M. (2017). *Mitigating the Risks of Plant Protection Products in the Environment: MAgPIE. SETAC Editions*. ISBN: 978-1-880611-99-9.

Alix, A. and Garrido, C. (2015). Monitoring effects of pesticides on pollinators - a review of methods and outcomes by the ICPPR working group. *Julius-Kühn-Archiv* 450: 284–295.

Alix, A. and Lewis, G. (2010). Guidance for the assessment of risks to bees from the use of plant protection products under the framework of Council Directive 91/414 and Regulation 1107/2009. *Bulletin OEPP/EPPO Bulletin* 40: 196–203.

APVMA (2017). Roadmap for insect pollinator risk assessment in Australia. Available at: apvma.gov.au/node/27551.

Arena, M. and Sgolastra, F. (2014). A meta-analysis comparing the sensitivity of bees to pesticides. *Ecotoxicology* 23(3): 324–334.

Arts, G. H. P., Kohlschmid, E., Maltby, L., Mayer, Ch., Meregalli, G., Poulsen, V. and Streissl, F. (2017). An ecosystem services approach to pesticide risk assessment and risk management of non-target terrestrial plants: recommendations from a 2nd SETAC Europe Workshop. Society of Environmental Toxicology and Chemistry (SETAC). Available at: https://cdn.ymaws.com/www.setac.org/resource/resmgr/docs/Workshop_report_2nd_NTTP_wor.pdf.

Australian Government - Rural Industries Research and Development Corporation (2012). Honeybee pesticide poisoning: a risk management tool for Australian farmers and beekeepers. Publication No. 12/043. Available at: https://www.agrifutures.com.au/wp-content/uploads/publications/12-043.pdf.

Australian Pesticide and Veterinary Medicines Authority (2017). Roadmap for insect pollinator risk assessment in Australia. Available at: https://apvma.gov.au/sites/default/files/publication/27551-27551-1_apvma_ra_road_map_-_final-october_2017_a656939_2.pdf.

Australian Pesticide and Veterinary Medicines Authority (2018). Label content. Available at: https://apvma.gov.au/node/934.

Barnett, E. A., Charlton, A. J. and Fletcher, M. R. (2007). Incidents of bee poisoning with pesticides in the United Kingdom, 1994–2003. *Pest Management Science* 63(11): 1051–1057.

Barrett, K. L., Grandy, N., Harrison, E. G., Hassan, S. and Oomen, P. A. (1994). Guidance document on regulatory testing procedures for pesticides with non-target arthropods. Report of the SETAC/ESCORT Workshop., 51. Wageningen, The Netherlands. SETAC-Europe, Brussels, Belgium.

Boyle, N. K., Pitts-Singer, Th. L., Abbott, J., Alix, A., Cox-Foster, D. L., Hinarejos, S., Lehmann, D. M., Morandin, L., O'Neill, B., Raine, N. E., Singh, R., Thompson, H. M., Williams, N. M. and Steeger, Th. (2018). Workshop on pesticide exposure assessment paradigm for non-apis bees: foundation and summaries. *Environmental Entomology* 48(1): 4–11. DOI: 10.1093/ee/nvy103.

Brasse, D. (2007). Der Arbeitsbereich Bienenschutz in der Geschichte der BBA. *Mitteilungen der Biologischen Bundesanstalt für Land- Forstwirtschaft* 410: 14–25.

Breeze, T. D., Bailey, A. P., Balcombe, K. G., Brereton, T., Comont, R., Edwards, M., Garratt, M. P., Harvey, M., Hawes, C., Isaac, N., Jitlal, M., Jones, C. M., Kunin, W. E., Lee, P., Morris, R. K. A., Musgrove, A., O'Connor, R. S., Peyton, J., Potts, S. G., Roberts, S. P. M., Roy, D. B., Roy, H. E., Tang, C. Q., Vanbergen, A. J. and Carvell, C. (2020) Pollinator monitoring more than pays for itself. *Journal of Applied Ecology* 58(1): 44–57. DOI: 10.1111/1365-2664.13755.

Cabrera, A. R., Almanza, M. T., Cutler, G. Ch., Fischer, D. L., Hinarejos, S., Lewis, G., Nigro, D., Olmstead, A., Overmyer, J., Potter, D. A., Raine, N. E., Stanley-Stahr, C., Thompson, H. and van der Steen, J. (2015). Initial recommendations for higher-tier risk assessment protocols for bumble bees, Bombus spp. (Hymenoptera: Apidae). *Integrated Environmental Assessment and Management* 12(2): 222–229.

Carreck, N. L. and Ratnieks, F. L. W. (2014). The dose makes the poison: have "field realistic" rates of exposure of bees to neonicotinoid insecticides been overestimated in laboratory studies?. *Journal of Apicultural Research* 53(5): 607–614. DOI: 10.3896/IBRA.1.53.5.08.

Cham, K. de O., Rebelo, R. M., Oliveira, R. de P., Ferro, A. A., Vianasilva, F. E. de C., Borges, L. de O., Saretto, C. O. S. D., Tonelli, C. A. M. and Macedo, T. C. (2017). Manual de Avaliação de Risco ambiental de agrotóxicos para abelhas. Instituto brasilero do Medio ambiente e dos recursos naturais Renoáveis. IBAMA/Diqua, Brasília. Available at: http://www.ibama.gov.br/phocadownload/agrotoxicos/reavaliacao-ambiental/2017/2017-07-25-Manual-IBAMA-ARA-Abelhas-IN0217-WEB.pdf.

CropLife International (2017). *Protecting Pollinators Through Good Stewardship Practices. A Compendium of Guidelines and Other Documents Supporting Pollinator Stewardship Version 7*. Available at: https://croplife.org/wp-content/uploads/2017/04/Protecting-Pollinators-Through-Good-Stewardship-Practices-v7.pdf.

Cutler, G. Ch. and Scott-Dupree, C. D. (2007). Exposure to clothianidin seed-treated canola has no long-term impact on honey bees. *Journal of Economic Entomology* 100(3): 765–772.

Cutler, G. Ch., Scott-Dupree, C. D., Sultan, M., McFarlane, A. D. and Brewer, L. (2014). A large-scale field study examining effects of exposure to clothianidin seed-treated canola on honey bee colony health, development, and overwintering success. *PeerJ* 2: e652; DOI: 10.7717/peerj.652.

Decourtye, A., Devillers, J., Genecque, E., Le Menach, K., Budzinski, H., Cluzeau, S. and Pham-Delègue, M. H. (2005). Comparative sublethal toxicity of nine pesticides on olfactory learning performances of the honeybee Apis mellifera. *Archives of Environmental Contamination and Toxicology* 48(2): 242–250. DOI: 10.1007/s00244-003-0262-7.

Eisenstein, M. (2015). Pesticides: seeking answers amid a toxic debate. *Nature* 521(7552): S52–S55.

European and Mediterranean Plant Protection Organization (1992). Guideline on test methods for evaluating the side effects of plant protection products on Honeybees. *Bulletin OEPP/EPPO Bulletin* 22: 203–215.

European and Mediterranean Plant Protection Organization (1993). Decision-making scheme for the environmental risk assessment of plant protection products. Chapter 10: Honeybees. *Bulletin OEPP/EPPO Bulletin* 23: 151–165.

European and Mediterranean Plant Protection Organization (2010a). Efficacy evaluation of plant protection products: side-effects on Honeybees. *PP 1/170 (4). - Bulletin OEPP/EPPO Bulletin* 40: 313–319.

European and Mediterranean Plant Protection Organization (2010b). Environmental risk assessment scheme for plant protection products. Chapter 10: Honeybees. *PP 3/10 (3). - Bulletin OEPP/EPPO Bulletin* 40: 323–331.

European and Mediterranean Plant Protection Organization (2021). EPPO standards. Available at: www.eppo.int/RESOURCES/eppo_standards.

European Chemicals Agency (2008). Guidance on information requirements and chemical safety assessment. Chapter R.10: characterization of Dose [Concentration]-Response for Environment. Available at: https://echa.europa.eu/documents/10162/17224/information_requirements_r10_en.pdf/bb902be7-a503-4ab7-9036-d866b8ddce69?t=1322594768638.

European Commission (2002). Guidance document on terrestrial ecotoxicology in the context of the directive 91/414/EEC. SANCO/10329/2002, rev.2 final, 17 October 2002: 1-39. Available at: https://ec.europa.eu/food/system/files/2016-10/pesticides_ppp_app-proc_guide_ecotox_terrestrial.pdf.

European Commission (2003). Commission directive 2003/82/EC of 11 September 2003 Amending Council Directive 91/414/EEC as regards standard phrases for special risks and safety precautions for plant-protection products. *Official Journal of the*

European Union L 228/11-28. Available at: https://eur-lex.europa.eu/LexUriServ/LexUriServ.do?uri=OJ:L:2003:228:0011:0028:EN:PDF.

European Commission (2011a). Commission Regulation (EU) No 546/2011 of 10 June 2011 Implementing Regulation (EC) No 1107/2009 of the European Parliament and of the Council as Regards Uniform Principles for Evaluation and Authorisation of Plant Protection Products. *Official Journal of the European Union* L155: 127–175. Available at: https://eur-lex.europa.eu/legal-content/EN/TXT/PDF/?uri=CELEX:32011R0546&from=DE.

European Commission (2011b). Commission Regulation (EU) No 547/2011 of 8 June 2011 Implementing Regulation (EC) No 1107/2009 of the European Parliament and of the Council as Regards Labelling Requirements for Plant Protection Products. *Official Journal of the European Union* L155: 176–205. Available at: https://eur-lex.europa.eu/legal-content/EN/TXT/PDF/?uri=CELEX:32011R0547&from=EN.

European Commission (2013a). Commission Regulation (EU) No 283/2013 of 1 March 2013 setting out the data requirements for active substances, in accordance with Regulation (EC) No 1107/2009 of the European Parliament and of the Council Concerning the Placing of Plant Protection Products on the Market. *Official Journal of the European Union* L93: 1-84. Available at: https://eur-lex.europa.eu/legal-content/EN/TXT/PDF/?uri=CELEX:32013R0283&from=de.

European Commission (2013b). Commission Regulation (EU) No 284/2013 of 1 March 2013 Setting out the Data Requirements for Plant Protection Products, in Accordance with Regulation (EC) No 1107/2009 of the European Parliament and of the Council Concerning the Placing of Plant Protection Products on the Market. *Official Journal of the European Union* L93: 95–152. Available at: https://eur-lex.europa.eu/legal-content/EN/TXT/PDF/?uri=CELEX:32013R0284&from=DE.

European Commission (2013c). Commission Implementing Regulation (EU) No 485/2013 of 24 May 2013 Amending Implementing Regulation (EU) No 540/2011, as Regards the Conditions of Approval of the Active Substances Clothianidin, Thiamethoxam and Imidacloprid, and Prohibiting the Use and Sale of Seeds Treated with Plant Protection Products Containing Those Active Substances. *Official Journal of the European Union* 25.5.2013: L 139/12-26. Available at: https://eur-lex.europa.eu/LexUriServ/LexUriServ.do?uri=OJ:L:2013:139:0012:0026:EN:PDF.

European Commission (2018a). Commission Implementing Regulation (EU) 2018/783 of 29 May 2018 Amending Implementing Regulation (EU) No 540/2011 as Regards the Conditions of Approval of the Active Substance Imidacloprid. *Official Journal of the European Union* 30.5.2018: L 132/31-34. Available at: https://eur-lex.europa.eu/legal-content/EN/TXT/PDF/?uri=CELEX:32018R0783&from=DA.

European Commission (2018b). Commission Implementing Regulation (EU) 2018/784 of 29 May 2018 Amending Implementing Regulation (EU) No 540/2011 as Regards the Conditions of Approval of the Active Substance Clothianidin. *Official Journal of the European Union* 30.5.2018: L 132/35-39. Available at: https://eur-lex.europa.eu/legal-content/EN/TXT/PDF/?uri=CELEX:32018R0784.

European Commission (2018c). Commission Implementing Regulation (EU) 2018/785 of 29 May 2018 Amending Implementing Regulation (EU) No 540/2011 as Regards the Conditions of Approval of the Active Substance Thiamethoxam. *Official Journal of the European Union* 30.5.2018: L 132/40-44. Available at: https://eur-lex.europa.eu/legal-content/EN/TXT/PDF/?uri=CELEX:32018R0785&from=ET.

European Commission (2021). A proposal for an EU Pollinator Monitoring Scheme (EU-PoMS). Available at: https://ec.europa.eu/jrc/en/science-update/proposal-eu-pollinator-monitoring-scheme-eu-poms.

European Food Safety Authority (2009). Risk assessment for birds and mammals. On request from EFSA, Question No EFSA-Q-2009-00223 First published on 17 December 2009. *EFSA Journal* 7(14383): 1-139. Available at: https://efsa.onlinelibrary.wiley.com/doi/pdf/10.2903/j.efsa.2009.1438.

European Food Safety Authority (2012). Scientific opinion on the science behind the development of a risk assessment of plant protection products on bees (Apis mellifera, Bombus spp. and solitary bees). *EFSA Journal* 10(2668): 1-275.

European Food Safety Authority (2013a). EFSA guidance document on the risk assessment of plant protection products on bees (Apis mellifera, Bombus spp. and solitary bees). *EFSA Journal* 11(7): 3295: 266 pp., DOI: 10.2903/j.efsa.2013.3295.

European Food Safety Authority (2013b). Conclusion on the peer review of the pesticide risk assessment for bees for the active substance Imidacloprid. *EFSA Journal* 11(1): 1-55. Available at: https://efsa.onlinelibrary.wiley.com/doi/pdf/10.2903/j.efsa.2013.3068.

European Food Safety Authority (2013c). Conclusion on the peer review of the pesticide risk assessment for bees for the active substance clothianidin. *EFSA Journal* 11(1): 1-58. Available at: https://efsa.onlinelibrary.wiley.com/doi/pdf/10.2903/j.efsa.2013.3066.

European Food Safety Authority (2013d). Conclusion on the peer review of the pesticide risk assessment for bees for the active substance thiamethoxam. *EFSA Journal* 11(1): 1-68. Available at: https://efsa.onlinelibrary.wiley.com/doi/pdf/10.2903/j.efsa.2013.3067.

European Food Safety Authority (2022). Bees and pesticides: draft guidance update for public consultation. Available at: https://www.efsa.europa.eu/de/news/bees-and-pesticides-draft-guidance-update-public-consultation.

European Food Safety Authority Scientific Committee (2016). Guidance to develop specific protection goals options for environmental risk assessment at EFSA, in relation to biodiversity and ecosystem services. *EFSA Journal* 14(6): 1-50. DOI: 10.2903/j.efsa.2016.4499.

European Food Safety Authority (2019). Bees and pesticides: stakeholders to participate in guidance review. Available at: www.efsa.europa.eu/en/press/news/190508.

European Parliament and Council of the European Union (2009). Regulation (EC) No 1107/2009 of the European Parliament and of the Council of 21 October 2009 concerning the placing of plant protection products on the market and repealing council directives 79/117/EEC and 91/414/EEC. *Official Journal of the European Union* L 309: 1-50. Available at: https://eur-lex.europa.eu/LexUriServ/LexUriServ.do?uri=OJ:L:2009:309:0001:0050:EN:PDF.

Everts, J. W. (1990). Sensitive indicators of side-effects of pesticides on the epigeal fauna of arable land. PhD thesis. University of Wageningen, 114 pp.

Exeler, N. and Quambusch, A. (2020). Method development for a larval test design for the solitary bee *Osmia cornuta* - first experiences with different larval pollen provisions. *Julius-Kühn-Archiv* 465: 132.

Exeler, N., Quambusch, A., Hanewald, N., Zicot, A., Soler, E., Kling, A., Vinall, S., Dressler, K., Taenzler, V., Kimmel, S., Lehmann, D. M., Patnaude, M. and Cabrera, A. R. (2020).

Development and validation of a bumble bee adult chronic oral test. *Julius-Kühn-Archiv* 465: 132.

Finkenbrink, W. (1940). Neue Insektizide. *Anzeiger für Schädlingskunde* 16(6): 63–67.

Fischer, D. and Moriarty, T. (Eds.) (2014). Pesticide risk assessment for pollinators - Society of Environmental Toxicology and Chemistry (SETAC). Online ISBN:9781118852408. doi:10.1002/9781118852408.

Food and Agricultural Materials Inspection Center (2019). Data requirements for registration of agricultural chemicals. Available at: https://www.acis.famic.go.jp/eng/shinsei/6278_2nd_e.pdf.

Forster, R. (2009). Bee poisoning caused by insecticidal seed treatment of maize in Germany in 2008. *Julius-Kühn-Archiv* 423: 126–131.

Franke, L., Elston, Ch., Jütte, T., Klein, O., Knäbe, S., Lückmann, J., Roessink, I., Persigehl, M., Cornement, M., Exeler, N., Giffard, H., Hodapp, B., Kimmel, S., Kullmann, B., Schneider, Ch. and Schnurr, A. (2020). Results of 2-year ring testing of a semifield study design to investigate potential impacts of plant protection products on the solitary bees osmia bicornis and osmia Cornuta and a proposal for a suitable test design. *Environmental Toxicology and Chemistry* 40(1): 236–250. DOI: 10.1002/etc.4874.

Godfray, H. Ch. J., Blacquière, T., Field, L. M., Hails, R. S., Petrokofsky, G., Potts, S. G., Raine, N. E., Vanbergen, A. J. and McLean, A. R. (2014). A restatement of the natural science evidence base concerning neonicotinoid insecticides and insect pollinators. *Proceedings. Biological Sciences* 281(1786): 20140558. Available at: http://dx.doi.org/10.1098/rspb.2014.0558.

Godfray, H. Ch. J., Blacquière, T., Field, L. M., Hails, R. S., Potts, S. G., Raine, N. E., Vanbergen, A. J. and McLean, A. R. (2015). A restatement of recent advances in the natural science evidence base concerning neonicotinoid insecticides and insect pollinators. *Proceedings: Biological Sciences* 282(1818): 20151821. Available at: http://dx.doi.org/10.1098/rspb.2015.1821.

Government of Canada (2018). Report a pesticide incident. Available at: www.canada.ca/en/health-canada/services/consumer-product-safety/pesticides-pest-management/public/protecting-your-health-environment/report-pesticide-incident.html.

Government of Canada (2021). Pollinator protection. Available at: www.canada.ca/en/health-canada/services/consumer-product-safety/pesticides-pest-management/growers-commercial-users/pollinator-protection.html.

Henry, M., Beguin, M., Requier, F., Rollin, O., Odoux, J. F., Aupinel, P., Aptel, J., Tchamitchian, S. and Decourtye, A. (2012). A common pesticide decreases foraging success and survival in honey bees. *Science* 336(6079): 348–350.

Henry, M., Cerrutti, N., Aupinel, P., Decourtye, A., Gayrard, M., Odoux, J. F. and Pissard, A., Rüger, C. and Bretagnolle, V. Ch. (2015). Reconciling laboratory and field assessments of neonicotinoid toxicity to honeybees. *Proceedings. Biological Sciences* 282(1819): 20152110. DOI: 10.1098/rspb.2015.2110.

Hinarejos, S., Abbott, J., Alix, A., Bibek, S., Cabrera, A., Joseph, T., O'Neill, B., Singh, R. and Thompson, H. (2018). Non-*Apis* Bee exposure workshop: industry participants' view. *Environmental Entomology* 48(1): 49–52. DOI: 10.1093/ee/nvy138.

Hodapp, B. and Kimmel, S. (2018). Oral toxicity test with solitary bees: experiences on the acute feeding test. *Julius-Kühn-Archiv* 462: 158–160.

Hooven, L., Sagili, R. and Johansen, R. (2013). How to reduce bee poisoning from pesticides. Oregon State University, University of Idaho, Washington State University. A Pacific Northwest Extension Publication: PNW 591. Available at:

https://catalog.extension.oregonstate.edu/sites/catalog/files/project/pdf/pnw591.pdf.

IPBES (2016). The assessment report of the intergovernmental science-policy platform on biodiversity and ecosystem services on pollinators, pollination and food production. S. G. Potts, V. L. Imperatriz-Fonseca, and H. T. Ngo (Eds). *Secretariat of the Intergovernmental Science-Policy Platform on Biodiversity and Ecosystem Services*, Bonn, Germany, 552. DOI: 10.5281/zenodo.3402856.

Iwasa, T., Motoyama, N., Ambrose, J. T. and Roe, R. M. (2003). Mechanism for the differential toxicity of neonicotinoid insecticides in the honey bee, Apis mellifera. *Crop Protection* 23(5): 371–378.

Jeschke, P. and Nauen, R. (2010). Neonicotinoid insecticides. In L. I. Gilbert, K. Iatrou and S. S. Gill (Eds.) *(2005/2010): Comprehensive Molecular Insect Science*. Elsevier BV Oxford (Update 2010): 61–119.

Johansen, E. and Wu-Smart, J. (2021). Bee protection - how to reduce bee poisoning from pesticides. PNW insect management handbook: C1-C10. Available at: https://pnwhandbooks.org/sites/pnwhandbooks/files/insect/chapterpdf/bee-protection.pdf.

Julius-Kühn-Institut (2021a). Untersuchungsstelle für Bienenvergiftungen. Ergebnisse der Untersuchungen. Available at: https://bienenuntersuchung.julius-kuehn.de/index.php?menuid=74.

Julius-Kühn-Institut (2021b). Untersuchungsstelle für Bienenvergiftungen. Jahresberichte. Available at: https://bienenuntersuchung.julius-kuehn.de/index.php?menuid=89.

Knaebe, S., Alscher, A., Classen, Ch., Exeler, N., Franke, L., Fricke, J., Frommberger, M., Sorlí Guerola, J., Hecht-Rost, S., Kimmel, S., Klein, O., Lückmann, J., Molitor, C., Peters, B. and Schneider, Ch. (2018). Summary of an ICPPR non-*Apis* workshop – subgroup higher-tier (bumble bees and solitary bees) with recommendations for a semi-field experimental design. *Julius-Kühn-Archiv* 462: 142–145.

Lewis, G., Coulson, M., Vergnet, Ch., Maus, Ch., Thompson, H. M., Becker, R., Rogers, D. and Tornier, I. (2009). Proposed revision of the higher-tier testing requirements for EPPO standard PP1/170: test methods for evaluating the side-effects of plant protection products on honeybees. *Julius-Kühn-Archiv* 423: 34–42.

Lewis, G. B., Gough, H. J., Künast, Ch., Thompson, H. M. and Stevenson, J. H. (2001). The use of toxic standards in the honey bee acute toxicity test. In *Hazards of Pesticides to Bees*, Avignon (France), September 07–09, 1999: INRA, Paris: 149–156.

Lückmann, J. and Schmitzer, S. (2019). The Oomen bee brood feeding test – revision of the method to current needs and developments. *Bulletin OEPP/EPPO Bulletin* 49(1): 137–146.

Maus, Ch., Curë, G. and Schmuck, R. (2003). Safety of imidacloprid seed dressings to honey bees: a comprehensive overview and compilation of the current state of knowledge. *Bulletin of Insectology* 56: 51–57.

Mineau, P., Harding, K. M., Whiteside, M., Fletcher, M. R., Garthwaite, D. and Knopper, L. D. (2008). Using reports of bee mortality in the field to calibrate laboratory-derived pesticide risk indices. *Environmental Entomology* 37(2): 546–554.

Ministry for Primary Industries (2020). Labelling agricultural chemicals. Available at: https://www.mpi.govt.nz/dmsdocument/19481/direct.

Nguyen, B. K., Saegerman, C., Pirard, C., Mignon, J., Widart, J., Thirionet, B., Verheggen, F. J., Berkvens, D., De Pauw, E. and Haubruge, E. (2009). Does imidacloprid seed-treated maize have an impact on honey bee mortality?. *Journal of Economic Entomology* 102(2): 616–623.

Nikolakis, A., Chapple, A., Friessleben, R., Neumann, P., Schad, Th., Schmuck, R., Schier, H.-F., Schnorbach, H.-J., Schöning, R. and Maus, Ch. (2009). An effective risk management approach to prevent bee damage due to the emission of abraded seed treatment particles during sowing of seeds treated with bee toxic insecticides. *Julius-Kühn-Archiv* 423: 132–148.

Nocelli, R. C. F., Roat, Th. C., Miotelo, L., Lima, T. A., Rodrigues, A. G., Silva, G. M. and Malaspina, O. (2020). Stingless bee ring test: acute contact toxicity test. *Julius-Kühn-Archiv* 465: 46–47.

OECD (1998a). OECD guidelines for the testing of chemicals. 213: honeybees, acute oral toxicity test. DOI: 10.1787/20745761. Available at: https://www.oecd-ilibrary.org/docserver/9789264070165-en.pdf?expires=1629789263&id=id&accname=guest&checksum=A6E373EE705FE1FDA2C6E74F03907302.

OECD (1998b). OECD guidelines for the testing of chemicals. 213: honeybees, acute contact toxicity test. DOI: 10.1787/20745761. Available at: https://www.oecd-ilibrary.org/docserver/9789264070189-en.pdf?expires=1629789549&id=id&accname=guest&checksum=F9006AF2BC65E665C28946B91E8DA081.

OECD (1998c). OECD series on principles of good laboratory practice and compliance monitoring. Number 1: OECD Principles on Good Laboratory Practice (as revised in 1997). OECD Environmental Health and Safety Publications ENV/MC/CHEM (98) 17. https://www.oecd.org/officialdocuments/publicdisplaydocumentpdf/?cote=env/mc/chem(98)17&doclanguage=en.

OECD (2005). OECD guidance document on the validation and international acceptance of new or updated test methods for hazard assessment. OECD series on testing and assessment, number 34. Microsoft Word - GD34-FINAL-VERSION-18August05.doc (oecd.org). Available at: oecd.org.

OECD (2013). OECD guidelines for the testing of chemicals. 237: Honey bee (Apis mellifera) larval toxicity test, single exposure. DOI: 10.1787/20745761. Available at: https://www.oecd-ilibrary.org/docserver/9789264203723-en.pdf?expires=1629788686&id=id&accname=guest&checksum=7DD9D8CEB83B4607B558B1323A257264.

OECD (2014). Guidance document on the honey bee (Apis mellifera L.) Brood test under semi-field conditions, OECD Series on Testing and Assessment, No. 75. ENV/JM/MONO (2007)22. DOI: 10.1787/9789264085510-en.

OECD (2017a). OECD guidelines for the testing of chemicals. 245: Honey Bee (Apis mellifera L.), Chronic Oral Toxicity Test (10-Day Feeding). DOI: 10.1787/20745761. Available at: https://www.oecd-ilibrary.org/docserver/9789264284081-en.pdf?expires=1629788926&id=id&accname=guest&checksum=A8AA53A491D5AD80591EFBBDA885EDD5.

OECD (2017b). OECD guidelines for the testing of chemicals. 246: Bumblebee, Acute Contact Toxicity Test. DOI: 10.1787/9789264284104-en.

OECD (2017c). OECD guidelines for the testing of chemicals. 247: acute oral toxicity test. DOI: 10.1787/9789264284128-en.

OECD (2021a). OECD test guidelines for chemicals. Available at: https://www.oecd.org/chemicalsafety/testing/oecdguidelinesforthetestingofchemicals.htm www.oecd.org/chemicalsafety/testing/oecdguidelinesforthetestingofchemicals.htm.

OECD (2021b). Guidance document on honey bee (Apis mellifera) larval toxicity test, repeated exposure. *Series on Testing and Assessment. ENV/JM. MONO* 239: 34. Available at: www.oecd.org/officialdocuments/publicdisplaydocumentpdf/?cote=env/jm/mono(2016)34&doclanguage=en.

OECD (2021c). Guidance document on honey bee (Apis mellifera L.) Homing Flight Test, Using Single Oral Exposure to Sublethal Doses of Test Chemical. Series on Testing and Assessment, No. 332. ENV/CBC/MONO (2021)7. Available at: www.oecd.org/officialdocuments/publicdisplaydocumentpdf/?cote=ENV-CBC-MONO(2021)7%20&doclanguage=en

OECD (2021d). OECD Principles of Good Laboratory Practice (GLP) and GLP Compliance Monitoring. Available at: https:// www.oecd.org/chemicalsafety/testing/overview-of-good-laboratory-practice.htm.

OECD (2021e). Managing pesticide risk to insect pollinators. Available at: https://www.oecd.org/chemicalsafety/risk-mitigation-pollinators/.

Oerke, E.-C. (2006). Crop losses to pests. *Journal of Agricultural Science* 144(1): 31–43.

Oomen, P. A. (1986). A sequential scheme for evaluating the hazard of pesticides to bees, Apis mellifera. *Mededelingen van de Faculteit Landbouwwetenschappen Rijksuniversiteit Gent* 51: 1205–1213.

Oomen, P. A., Ruijter, A. and Steen, J. (1992). Method for honeybee brood feeding tests with insect growth-regulating insecticides. *OEPP/EPPO Bulletin* 22(4): 613–616.

Overmyer, J., Feken, M., Ruddle, N., Bocksch, S., Hill M. and Thompson H. (2018): Thiamethoxam Honey Bee Colony Feeding Study: Linking Effects at the Level of the Individual to Those at the Colony Level. *Environmental Toxicology* 37: 816–828. DOI: 10.1002/etc.4018

Palmer-Jones, P. (1958). Laboratory methods for measuring the toxicity of pesticides to honey bees. *New Zealand Journal of Agricultural Research* 1(3): 290–300. DOI: 10.1080/00288233.1958.10431411.

Pamminger, T. (2021). Extrapolating acute bee sensitivity to insecticides using a phylogenetically informed interspecies scaling framework. *Environmental Toxicology and Chemistry* 40: 2042–2050.

Pettinga, J. (1980). Introduction. In: Wageningen, The Netherlands: Report of the Meeting. International commission for bee botany (ed.): Symposium on the Harmonisation of Methods for Testing the Toxicity of Pesticides to Bees, 23–25 September 1980: 3–4.

Pilling, E., Campbell, P., Coulson, M., Ruddle, N. and Tornier, I. (2013). A four-year field program investigating long-term effects of repeated exposure of honey bee colonies to flowering crops treated with thiamethoxam. *PLoS ONE* 8(10): e77193. DOI: 10.1371/journal.pone.0077193.

Pirk, Ch. W. W., Crewe, R. M. and Moritz, R. F. A. (2017). Risks and benefits of the biological interface between managed and wild bee pollinators. *Functional Ecology* 31(1): 47–55.

Popp, J., Pető, K. and Nagy, J. (2013). Pesticide productivity and food security - a review. *Agronomy for Sustainable Development* 33(1): 243–255.

Roessink, I., Hanewald, N., Schneider, Ch., Exeler, N., Schnurr, A., Molitor, A. M., Soler, E., Kimmel, S., Molitor, C., Smagghe, G. and van der Steen, S. (2018). A method for a solitary bee (*Osmia* sp.) first tier acute contact and oral laboratory test: an update. *Julius-Kuhn-Archiv* 462: 158.

Roessink, I., Hanewald, N., Schneider, Ch., Quambusch, A., Exeler, N., Cabrera, A. R., Molitor, A. M., Taenzler, V., Hodapp, B., Albrecht, M., Brandt, A., Vinall, S., Rathke,

A.-K., Giffard, H., Soler, E., Schnurr, A., Patnaude, M., Couture, E. and Lehmann, D. (2020). Progress on the *osmia* acute oral test - findings of the ICPPR non-*Apis* subgroup solitary bee laboratory testing. *Julius-Kühn-Archiv* 465: 45.

Rolke, D., Fuchs, S., Grünewald, B., Gao, Z. and Blenau, W. (2016). Large-scale monitoring of effects of clothianidin-dressed oilseed rape seeds on pollinating insects in northern Germany: effects on Honey Bees (Apis mellifera). *Ecotoxicology* 25(9): 1648–1665. DOI: 10.1007/s10646-016-1725-8.

Rosa-Fontana, A., Dorigo, A., Galaschi-Teixeira, J. S., Nocelli, R. C. F. and Malaspina, O. (2020). Standardization of an in vitro rearing method for the stingless bee species *Scaptotrigona postica* larvae and its application for determining the toxicity of dimethoate on the larval phase. *Julius-Kühn-Archiv* 465: 47–48.

Rundlöf, M., Andersson, G. K. S., Bommarco, R., Fries, I., Hederström, V., Herbertsson, L., Jonsson, O. and Klatt, B. K., Pedersen, T. R., Yourstone, J., Smith, H. G., Yourstone, Th. R. and Smith, H. G. (2015). Seed coating with a neonicotinoid insecticide negatively affects wild bees. *Nature* 521(7550): 77–80. DOI: 10.1038/nature14420.

Schmuck, R. and Keppler, J. (2003). Clothianidin – Ecotoxicological Profile And Risk Assessment. *Pflanzenschutz-Nachrichten Bayer* 56: 26–58.

Schmuck, R., Schöning, R. and Sur, R. (2005). Studies on the effects of plant protection products containing imidacloprid on the honeybee, Apis mellifera L. In Forster, R., Bode, E. and Brasse, D. (Eds). *Das "Bienensterben" im Winter 2002/2003 in Deutschland – Zum Stand der wissenschaftlichen Erkenntnisse: Bundesamt für Verbraucherschutz und Lebensmittelsicherheit (BVL)*, Braunschweig: 68–92.

Seefeld, F. (2006). Chemical detection of damage to honey bees caused by pesticides. *Nachrichtenblatt des Deutschen Pflanzenschutzdienstes* 58: 59–66.

Siede, R., Faust, L., Meixner, M. D., Maus, Ch., Grünewald, B. and Büchler, R. (2017). Performance of honey bee colonies Under a long-lasting dietary exposure to sublethal concentrations of the neonicotinoid insecticide thiacloprid. *Pest Management Science* 73(7): 1334–1344. DOI: 10.1002/ps.4547.

Siviter, H., Richman, S. K. and Muth, F. (2021). Field-realistic neonicotinoid exposure has sub-lethal effects on non-Apis bees: a meta-analysis. *Ecology Letters* 24(12): 2586-2597. DOI: 10.1111/ele.13873.

Solomon, K. R. and Stephenson, G. L. (2017a). Quantitative weight of evidence assessment of higher-tier studies on the toxicity and risks of neonicotinoids in honeybees. *Journal of Toxicology and Environmental Health, Part B* 20(6–7): 346–364. DOI: 10.1080/10937404.2017.1388567.

Solomon, K. R. and Stephenson, G. L. (2017b). Quantitative weight of evidence assessment of higher-tier studies on the toxicity and risks of neonicotinoids in honeybees. 4. Thiamethoxam. *Journal of Toxicology and Environmental Health, Part B* 20(6–7): 365-382. DOI: 10.1080/10937404.2017.1388568.

Stadler, T., Martínez-Ginés, D. and Buteler, M. (2003). Long-term toxicity assessment of Imidacloprid to evaluate side effects on honey bees exposed to treated sunflower in Argentina. *Bulletin of Insectology* 56: 77–81.

Stephenson, G. L. and Solomon, K. R. (2017). Quantitative weight of evidence assessment of higher-tier studies on the toxicity and risks of neonicotinoids in honeybees. 2. Imidacloprid. *Journal of Toxicology and Environmental Health. Part B, Critical Reviews* 20(6–7): 330–345. DOI: 10.1080/10937404.2017.1388564.

Stevenson, J. H. (1980). Conclusions of the meeting. In International Commission for Bee Botany (ed.), Symposium on the Harmonisation of Methods for Testing the Toxicity of

Pesticides to Bees, September 23–25 1980, Wageningen, The Netherlands: Report of the Meeting: 5–13.

Stute, K. (1956). Tests for the presence of herbicides and insecticides in dead bees. *Zeitschrift für Bienenforschung* 3: 103–116.

Thompson, H. M. and Maus, Ch. (2007). The relevance of sublethal effects in honey bee testing for pesticide risk assessment. *Pest Management Science* 63(11): 1058–1061.

Thompson, H. M., Overmyer, J., Feken, M., Ruddle, N., Vaughan, S., Scorgie, E., Bocksch, S. and Hill, M. (2019). Thiamethoxam: long-term effects following honey bee colony-level exposure and implications for risk assessment. *Science of the Total Environment* 654: 60–71. DOI: 10.1016/j.scitotenv.2018.11.003.

Thompson, H. M. and Thorbahn, D. (2009). Review of honeybee pesticide poisoning incidents in Europe – evaluation of the hazard quotient approach for risk assessment. *Julius-Kühn-Archiv* 423: 103–108.

United States Environmental Protection Agency (2003). Generic ecological assessment endpoints (GEAEs) for ecological risk assessment. EPA/630/P-02/004F. . Available at: www.epa.gov/sites/default/files/2014-11/documents/generic_endpoinsts_2004.pdf.

United States Environmental Protection Agency (2012a). Ecological effects test guidelines OCSPP 850.3020: Honeybee acute contact toxicity test. EPA 712-C- 0. Available at: nepis.epa.gov/Exe/ZyPDF.cgi/P100IRFL.PDF?Dockey=P100IRFL.PDF.

United States Environmental Protection Agency (2012b). Ecological effects test guidelines OCSPP 850.3030: Honeybee toxicity of residues on foliage. EPA 712-C- 01. Available at: nepis.epa.gov/Exe/ZyPDF.cgi/P100IRF8.PDF?Dockey=P100IRF8.PDF.

United States Environmental Protection Agency (2012c). Ecological effects test guidelines OCSPP 850.3040: field testing for pollinators. EPA 712-C- 01. Available at: nepis.epa.gov/Exe/ZyPDF.cgi/P100IRF2.PDF?Dockey=P100IRF2.PDF

United States Environmental Protection Agency, Health Canada and California Department of Pesticide Regulation (2014). Guidance for assessing pesticide risks to bees. Available at: www.epa.gov/sites/default/files/2014-06/documents/pollinator_risk_assessment_guidance_06_19_14.pdf.

United States Environmental Protection Agency (2016). Guidance on exposure and effects testing for assessing risks to bees. Available at: www.epa.gov/sites/default/files/2016-07/documents/guidance-exposure-effects-testing-assessing-risks-bees.pdf.

United States Environmental Protection Agency (2021a). How we assess risks to pollinators. Available at: www.epa.gov/pollinator-protection/how-we-assess-risks-pollinators.

United States Environmental Protection Agency (2021b). Label review manual. Available at: https://www.epa.gov/pesticide-registration/label-review-manual.

van Engelsdorp, D. and Meixner, M. D. (2010). A historical review of managed honey bee populations in Europe and the United States and the factors that may affect them. *Journal of Invertebrate Pathology* 103 (Suppl. 1): S80–S95.

Végéphyl – association pour la santé des végétaux (2013). Méthode d'évaluation des effets des préparations phytopharmaceutiques sur l'abeille domestique Apis mellifera L. Available at: www.vegephyl.fr/ceb/230-methode-devaluation-des-effets-des-preparations-phytopharmaceutiques-sur-labeille-domestique-apis-mellifera-l/.

Whitehorn, P. R., O'Connor, S., Wackers, F. L. and Goulson, D. (2012). Neonicotinoid pesticide reduces bumble bee colony growth and queen production. *Science* 336(6079): 351–352, DOI: 10.1126/science.1215025.

Whitford, F., Steeger, T., Feken, M., Krupke, C., Hunt, G., Johnson, R., Gerber, C., Obermeyer, J. and Given, K. (2017). The complex life of the honey bee - purdue extension: the complex life of the honey bee. Purdue Pesticide Programs.

Woodcock, B. A., Bullock, J. M., Shore, R. F., Heard, M. S., Pereira, M. G., Redhead, J., Ridding, L., Dean, H., Sleep, D., Henrys, P., Peyton, J., Hulmes, S., Hulmes, L., Sárospataki, M., Saure, Ch., Edwards, M., Genersch, E., Knäbe, S. and Pywell, R. F. (2017). Country-specific effects of neonicotinoid pesticides on honey bees and wild bees. *Science* 356(6345): 1393–1395.

Chapter 8

Assessing the impact of alien bees on native ones

M. Zakardjian, H. Jourdan, V. Le Féon and B. Geslin, Aix-Marseille University, Avignon University, CNRS, IRD, IMBE, Marseille, France

1 Introduction

A species introduced outside of its native range, voluntarily or not, is defined as an alien species (Blackburn et al., 2011). These alien species may become invasive if their establishment and spread negatively affect biodiversity, ecosystem functioning, and/or the economy (Diagne et al., 2021). Biological invasions are among the most important threats to biodiversity worldwide (Kumschick et al., 2015). Frequently pointed out as the second cause of species extinction (Bellard et al., 2016a), the magnitude of negative effects of invasive alien species on biodiversity is nevertheless still regularly debated in the literature (Courchamp et al., 2017). With the exponential rise in global trade, the number of alien species has increased in the past few years (Bellard et al., 2016b), and projections forecast a continuous growth in the number of introductions in the following 30 years (Seebens et al., 2017, 2020).

Among alien species, insects have been less studied (Kenis et al., 2009) despite the importance of their reported negative effects on biodiversity components, especially the ones exerted by social insects such as termites or ants (Bertelsmeier, 2021). Notably, this is the case for bees, be they social or solitary, managed or wild (Russo, 2016; Geslin et al., 2017; Russo et al., 2021). Indeed, bees are generally well-perceived by the public (Sumner et al., 2018)

http://dx.doi.org/10.19103/AS.2022.0111.17

mostly because of their important role in the reproduction of wild (Ollerton et al., 2011) and cultivated plants (Klein et al., 2007; Gill et al., 2016). Therefore, because of this beneficial role, little attention is paid when an alien bee species enters a new ecosystem.

However, to date, Russo (2016) recorded 80 bee species introduced out of their native range (Fig. 1). Among these, *Apis mellifera* (L.) and *Bombus terrestris* (L.) are the two species that have been the focus of attention (Geslin et al., 2017; Mallinger et al., 2017; Goulson, 2003). These two social and domesticated species have mostly been introduced either to produce honey (for *Apis mellifera*) or to improve pollination services to crops (Aizen et al., 2020), sometimes doing more harm than good (Aizen et al., 2014, 2019; Sáez et al., 2014). Of course, not all 80 alien species described by Russo (2016) have become invasive, but the literature report many examples of deleterious consequences of the introduction of bees outside their native range (Aizen et al., 2019, 2020; Geslin et al., 2017; Mallinger et al., 2017; Traveset and Richardson, 2014; Goulson, 2003).

Hence, alien bees can directly or indirectly affect native ecosystems and therefore become invasive (Dafni et al., 2010; Russo et al., 2021, Fig. 2). They can compete with native bees for the acquisition of food resources (nectar and pollen) or nesting sites potentially affecting local populations (Geslin et al., 2020; Hung et al., 2019; Ollerton, 2017; Schaffer et al., 1983). Alien bees can transport parasites and diseases, which can, in turn, affect native populations

Figure 1 A sample of bee species introduced outside of their native range. (a) *Anthidium manicatum* (Linnaeus, 1758); Frank Vassen/Flickr/CC BY 2.0. (b) *Bombus ruderatus* (Fabricius, 1775); Dave Young/Flickr/CC BY 2.0. (c) *Megachile disjunctiformis* (Cockerell, 1911) (Franziska Maria Luthi). (d) *Osmia taurus* (Smith, 1873) Benjamin Smith/Wikimedia Commons / CC BY 2.0. (e) *Osmia cornifrons* (Radoszkowski, 1887) Flickr/CC 1.0. (f) *Xylocopa augusti* (Lepeletier, 1841) Clinton.sm/Wikimedia Commons/CC BY-SA 4.0.

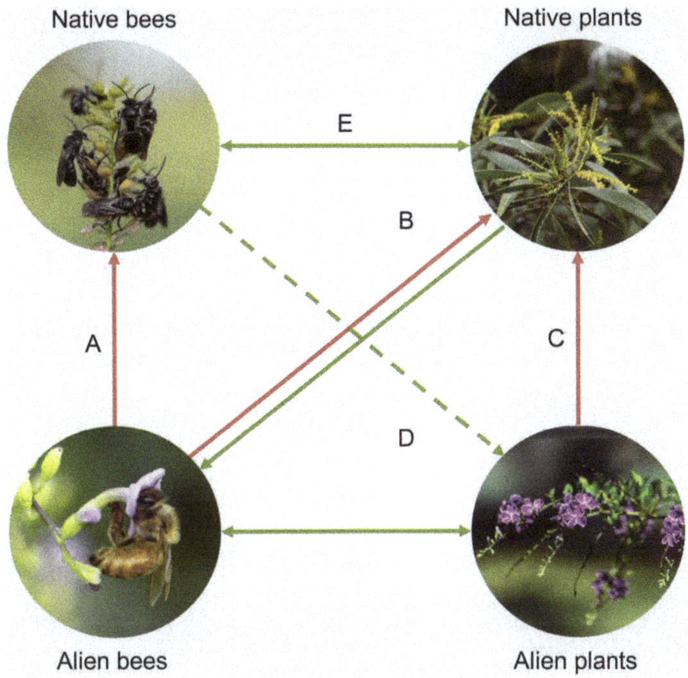

Figure 2 Positive and negative interactions (respectively green and red links) between alien and native species may disrupt native plant-pollinator systems. First, alien bees may compete for floral resources and/or nesting sites, spread parasites and pathogens and hybridize with native bees (A). Then, alien bees may also not be efficient pollinators of some native plants (e.g. due to morphological mismatch) and could thus alter their reproductive success (B). Second, alien plants may compete with native ones for soil resources (C) and corner all or part of their pollinators (D). Finally, it seems that alien plants tend to promote the presence of alien bees and in some cases alien bees visit preferentially those plant species. Thus, alien species may promote the establishment and spread of each other, potentially leading to invasional meltdowns (E). All these interactions may lead to a local decrease of abundance and richness of native species.

(Aizen et al., 2020; Goulson et al., 2015; Graystock et al., 2013). Alien bees can also disrupt plant-pollinator networks by robbing flowers, therefore decreasing their reproductive success (Dafni et al., 2010; Dohzono et al., 2008), modifying the amount or the nature of the pollen deposited (Agüero et al., 2020), or by reinforcing the pollination of particular plant species (Sanguinetti and Singer, 2014). In some cases, alien bees can even interact preferentially with alien or invasive plants, potentially enhancing their spread and negative impacts, a phenomenon called invasional meltdown (Zakardjian et al., 2020; Traveset and Richardson, 2014; Beavon and Kelly, 2012).

Despite these reported effects, the impacts of alien bees on native ecosystems are little considered on a global scale. For example, despite the devastating consequences of multiple introductions of *Bombus terrestris* worldwide (e.g. Acosta et al., 2016), the commercial trade of *Bombus* species outside of their native range continues (Aizen et al., 2019) to the point that the international trade of bees, mostly bumblebees, is nowadays an important conservation issue for global biodiversity (Sutherland et al., 2016). Therefore, it is important and timely to assess the impact of alien bees on native pollinators.

This chapter focuses on the direct and indirect impacts of alien bees on native pollinators. We will not explore the effect of alien bees on plant communities or plant-pollinator networks to focus solely on their effect on native bees. Also, because we focus on alien bees, we will not provide examples of competitive interactions between wild and domestic bees in the native range of domestic bees. In the second part, we will focus on a case study, the invasion of the alien bee *Megachile sculpturalis* in North America and Europe. We will then delineate some of the main shortcomings and directions for future studies and guide land managers to prevent future introductions.

2 Assessment of key issues related to the introduction of alien bees

2.1 Direct interactions: interference competition

Interference competition is a direct interaction, generally by aggressive contact, through which a resource that an individual should have used is made available by and for another individual (Case and Gilpin, 1974). Among alien bees, interference competition may be used to acquire food resources, nesting sites, or defend mating territories. Although rare, interference competitions between alien and native bees have been reported in other literature.

In Yucatan (Mexico), the Africanised honey bee (*Apis mellifera scutellata*) has aggressively displaced stingless bees from floral and water resources. In this region, some native stingless bees appear to have declined from anthropogenic pressures. Interference competition by honey bees could have compound effects on the most vulnerable native species (Cairns et al., 2005). In Indiana (US), honey bees have also been observed to steal pollen loads directly from the body parts of several native bees such as *Bombus impatiens*, *Megachile montivaga*, and *Melissodes desponsa* (Jean, 2005). As not all pollen loads were removed and those events remain rare, this kind of physical interaction is unlikely to severely impact native bees' fitness.

Anthidium manicatum, the most widespread unmanaged alien bee species worldwide (Strange et al., 2011), uses interference competition to defend floral territories and thus assuring mating opportunities (Starks and Reeve, 1999; Severinghaus et al., 1981). In New Zealand, *Anthidium manicatum* aggressively interferes with native *Lasioglossum* spp. (Soper and Beggs, 2013). However, in this study, native bees were less contacted than conspecific individuals and two other alien species, *Apis mellifera* and *Bombus spp*. Native *Lasioglossum* spp. shared fewer floral resources than *Apis mellifera* and *Bombus* spp. with *Anthidium manicatum*, thus reducing the probability of interference events (Case and Gilpin, 1974). Although the impact of these interactions was not tested, the authors concluded that *Anthidium manicatum* posed a minor threat to the native bee fauna of New Zealand.

In North America, *Megachile* spp. in an urban area experienced reduced visitation frequencies in the presence of *Anthidium manicatum* (Taggar et al., 2021). Unfortunately, in this study, alien and native *Megachile* spp. were not discriminated and it was therefore complicated to state whether this decrease was due to interference or exploitative competition. A native bee, *Bombus impatiens*, was shown to avoid floral enclosures where *Anthidium manicatum* was present. Even if the authors did not discriminate between interference and exploitative competition to explain their results, no effect was shown on the colony fitness of the native bumblebee (Graham et al., 2019).

In Europe and North America, several events of aggressive behaviour have been documented for *Megachile sculpturalis*. This species is likely to interfere with native species with similar nesting habits (see the case study below).

Physical interaction events appear to be uncommon between alien and native bees. Thus, interference competition is unlikely to have a deleterious impact on native bee populations. However, such behaviours could increase the potential of a species to become established or spread if coupled with other characteristics such as a high capability of exploitative competition for floral resources and nesting sites.

2.2 Exploitative competition

Alien bees may overlap with native bees in floral resource use (Fig. 3) and/or nesting sites. Exploitative competition occurs when shared resources are limiting and implies negative impacts on the fitness of native bees.

2.2.1 Exploitative competition for floral resources

Managed eusocial bees, such as *Apis mellifera* and *Bombus terrestris*, have a strong competitive potential over native wild bees. Indeed, the number of workers per managed colony compared to solitary bees and the regular renewal of populations by beekeepers (when needed) make them prone to exploit efficiently floral resources with potential impacts on native bees. Their impacts on native and invaded areas have already been reviewed by several authors (e.g. Aizen et al., 2020; Geslin et al., 2017; Mallinger et al., 2017; Osterman et al., 2021; Russo et al., 2021; Vanbergen et al., 2018), but we present here some examples of exploitative competition.

Figure 3 A honey bee and a bumblebee sharing the same floral resource (CC 1.0).

A 2-year experiment conducted in Western Australia showed that the monolectic resource used by a native bee, *Hylaeus alcyoneus*, was also used by honey bees for pollen and nectar provision. This resource overlap appears to explain the decrease in native bee fecundity (measured as the number of nests constructed) in sites with increasing honey bee density (Paini and Roberts, 2005). In this case, the monolectic behaviour of the native bee likely made it more vulnerable to exploitative competition exerted by honey bees. Another negative impact of honey bees on native bees was demonstrated in Tenerife (Canary Islands), where the number of pollinator species and that of visits by some native bees were lower after the setup of beehives (Valido et al., 2019). In this case, exploitative competition from honey bees had a negative impact on the richness and abundance of generalist native bees, with implications for pollination ecosystem services and plant reproductive success.

Conversely, some studies did not find any evidence of a detrimental effect of the exploitative competition exerted by alien honey bees on native bees. For example, the experimental increase in honey bees in a Patagonian Forest ecosystem did not affect the number of visits made by native pollinators but rather induced exploitative competition with two other invasive alien species (i.e. *Bombus ruderatus* and *Bombus terrestris*; Agüero et al., 2020). However, in this study, native pollinators were studied as a group, and contrasting responses between pollinator species may not have been highlighted. In an urban ecosystem in Montreal (Canada), an increase in honey bee capture rate did not affect the abundance of wild bees (McCune et al., 2019). The authors expected a negative effect and hypothesised that the abundance of floral resources and the very low density of beehives compared with other cities (e.g. see Ropars et al., 2019) mitigated competition. In addition, in the Ogasawara islands (Japan), *Apis mellifera* appears to have a competitive advantage over native bees (Kato et al., 1999), but the decline of native pollinators is mostly due to an invasive predator (Abe et al., 2008). Finally, in Ontario (Canada), within an agroecosystem (*Cucurbita spp.*), Willis Chan and Raine (2021) found no negative impact of the alien bee *Apis mellifera* on the wild oligolectic hoary squash bees (*Eucera Peponapis pruinosa*) foraging on *Cucurbita spp.*, likely because of an intra-day phenological difference in bees' activities (honey bees were foraging later than hoary squash bees).

Another managed eusocial bee, *Bombus terrestris*, has been massively introduced worldwide and is known to have caused disastrous impacts on native bee populations (see below, e.g. Acosta et al., 2016; Aizen et al., 2008, 2014, 2019; Morales and Aizen, 2006; Morales et al., 2013). In Japan, *Bombus terrestris* may be an efficient competitor for floral resources over some native bumblebees. For example, an experiment demonstrated that *Bombus terrestris* harvested more pollen per unit time and per unit body mass than the native *Bombus ignitus* (Inoue et al., 2010). However, the evidence of exploitative competition effects on native bees is unclear. Indeed, an experiment failed to detect a negative impact of *Bombus terrestris* on two native Japanese bumblebees (*Bombus ardens* and *Bombus hypocrita*; Nagamitsu et al., 2007a). Nevertheless, field removal of *Bombus terrestris* queens from an urban area led to an increased number of these two native bumblebee queens among five native bee species captured in the study area (Nagamitsu et al., 2009). *Bombus ardens* and *B. hypocrita* are likely to overlap for floral resource use with *Bombus terrestris* due to their tongue length (Nagamitsu et al., 2007). Thus, although the effect of competition for nesting sites could not be discriminated, the release of exploitative competition for floral resources was a major hypothesis to explain these results. Corroborating this hypothesis, *Bombus terrestris* may be responsible for the increase in robbed flowers of a native plant (*Corydalis ambigua*) and the subsequent decrease in visitation by *B. ardens* (Dohzono et al., 2008).

However, in an alpine-montane ecosystem in New Zealand, alien bees mostly composed of *Bombus terrestris* did not overlap with native bees for floral resource use (Iwasaki et al., 2018). Indeed, *Bombus terrestris* preferentially visited alien plants when native bees were mostly observed on native ones, suggesting a low potential for exploitative competition for floral resources. In addition, 3-years monitoring conducted in Japan did not show any evidence of an impact of shared plant resources between *Bombus terrestris* and the native species *B. pseudobaicalensis* (Inoue et al., 2008).

Notably, the introduction and invasion of *B. terrestris* and *B. ruderatus* in South America had notorious negative impacts on native ecosystems, including the decline and local extinction of *B. dahlbomii*, a Patagonian native bumblebee (Aizen et al., 2019; Geslin and Morales, 2015; Morales et al., 2013; Morales, 2007). In particular, *B. ruderatus* overlaps with *B. dahlbomii* for floral resources (Aizen et al., 2008; Morales and Aizen, 2006). However, Aizen et al. (2011) found no evidence of exploitative competition for floral resources between these two species, and the spread of pathogens is a favoured hypothesis to explain the decline of *B. dahlbomii* (see Section 2.3).

When first recorded in a new environment, alien solitary bees generally raise concerns about potential negative impacts on native species. Unfortunately, for most of these species, field studies following their detection rarely test for their effect on the fitness of native bee populations. However, a recent study analysing historical pan trap records in the US showed that the arrival of two alien bees, *Osmia cornifrons*, and especially *Osmia taurus*, was followed by a decline in six native *Osmia* spp. (LeCroy et al., 2020). Several factors have been hypothesised to explain this result, including exploitative competition, either for floral resources or nesting sites. Another study conducted in the US showed that urban sites with a high density of alien bees were associated with a lower abundance of native bees (Fitch et al., 2019). Once again, exploitative competition for floral resources, nesting sites, or spillover of pathogens were all hypothesised to explain these observations.

The effects of wild alien bees on native bees are also context-dependent. A study conducted in an insular system in the Southwest Pacific showed that the presence of alien bees did not affect the number of visits realised by native ones (Zakardjian et al., 2020). However, this result was explained by the generalist foraging behaviour of some insular native bees. In Brazil, a study conducted in an urban area found no impact from two alien bee species (*Apis mellifera* and *Anthidium manicatum*) on the structure of the native bee community (Graf et al., 2020), suggesting that no exploitative competition occurred. The authors proposed two hypotheses to explain this result. First, it is possible that these two alien bees, which were introduced a few decades ago, had an impact at the beginning of their establishment. Second, the urban environment may not be suitable for the species that are the most vulnerable to competition.

Most studies estimate niche overlap and the displacement of visits without measuring the impact of competition on native bee fitness. However, these metrics still reveal an effect of the competitive pressure exerted by alien bees on native ones. Mixed responses of native bees to exploitative competition by alien ones reveal that competitive pressure is context-dependent, varying according to the ecosystem (number of floral resources), the capacity of native bees to adapt their foraging preferences, and the characteristics of alien bees, such as their level of generalism and their reproductive success.

2.2.2 Exploitative competition for nesting sites

Bombus terrestris has the potential to compete with native bumblebees for nesting resources. In Japan, 3 years of monitoring studied the abundance of three native bumblebees, along with the abundance of *Bombus terrestris* queens. The results showed that a native species using the same nesting resources as *Bombus terrestris* declined with an increase in *Bombus terrestris* queens, suggesting exploitative competition for nesting sites between these two species (Inoue et al., 2008).

Concerning wild alien bees, *Megachile sculpturalis* has been reported to be a potential competitor for nesting resources (see above). In Marseille (France), near its first detection point, the presence of *Megachile sculpturalis* in bee hotels reduced the probability of occupation by native *Osmia spp*. (Geslin et al., 2020).

2.3 Apparent competition (pathogens spillover)

Parasites and pathogens may be transferred among bee species when infected individuals are abundant and share floral resources or nesting sites with other species. Thus, bees are ideal candidates for the transmission and dispersion of pathogens.

In New York State (US), the presence of the deformed wing virus (DWV; Fig. 4) in some native *Melandrena* spp. was linked to honey bee abundance (Murray et al., 2019). Even if there was no correlation with two other pathogens (black queen cell virus and sacbrood virus), these results indicate that honey bees may be a source of infection for solitary native bees, potentially increasing viral incidence for at least DWV. Still in North America, wild pollinators found close to beehives with honey bees infected with Israeli acute paralysis virus also presented this virus, contrary to wild pollinators captured close to healthy beehives (Singh et al., 2010). Similarly, Alger et al. (2019) found a strong prevalence of DWV in wild bumblebees foraging close to apiaries, whereas no DWV was found in bumblebees in the absence of honey bees.

In Argentina, *Apicystis bombi*, a parasite that especially infects bumblebees, was possibly co-introduced with *Bombus terrestris* or *B. ruderatus* (Arbetman et al., 2012; Plischuk et al., 2011) and is now also present in Colombia (Gamboa et al., 2015) and Patagonia (Arbetman et al., 2012). In Patagonia, it appears that the alien *B. ruderatus* and the native *B. dahlbomii* were free of this parasite before the invasion of *Bombus terrestris* (Arbetman et al., 2012). Thus, *A. bombi* likely arrived with and spilt over from *Bombus terrestris* to these two bumblebee species. The severe decline of *B. dahlbomii* is potentially linked to this parasite spillover (Arbetman et al., 2012).

Figure 4 Honey bee with deformed wing virus (DWV) and *Varroa destructor* (CC0 1.0).

2.4 Mating interference

In Korea, laboratory experiments demonstrated that interspecific hybridisation may happen between *Bombus terrestris* and two native bumblebees (*B. ignitus* and *B. hypocrita*). Males of the two native species mated with *Bombus terrestris* queens (Yoon et al., 2011, 2009). Thus, *Bombus terrestris* may interfere with native bumblebees for mating males, with potential negative impacts on native bees' fitness. In addition, during this experiment, *Bombus terrestris* males mated with native bumblebee queens, highlighting a risk of genetic contamination.

3 Case study: *Megachile sculpturalis*

3.1 History of introduction and spread

Megachile (*Callomegachile*) *sculpturalis* Smith (Hymenoptera: Megachilidae) - (*Msc*) is a wild bee species native to Southeast Asia (Korea, China, Taiwan, Hong Kong, and Japan; Fig. 5). We chose this species as a case study because it was the first alien bee to enter the European continent in 2008 (Vereecken and Barbier, 2009), and its spread and impacts have been carefully followed since (Le Féon et al., 2018, 2021). Before arriving in Europe, however, *Msc* colonised the North American continent. The first mention of *Msc* outside of its native range dates from 1994 in North Carolina (US; Magnum and Brooks, 1997). Since then, the species has rapidly been detected in other US states, up

Figure 5 *Megachile sculpturalis* female. Le Collet-de-Dèze, France. Picture: Frank Vassen.

to Texas in 2011 (Parys et al., 2015; Hinojosa-Díaz, 2008). Besides the US, *Msc* has also expanded its range to Canada, especially in Ontario (Paiero and Buck, 2003).

In Europe, *Msc* was first detected in Allauch, a small city close to Marseille (France; Vereecken and Barbier, 2009). Between 2008 and 2021, *Msc* has remarkably extended its range in France, with a reported presence in 259 different cities, with the most northern location being approximately 800 km from Marseille, in the northeast of Paris (near Reims, Le Féon et al., 2021). *Msc* now occupies approximately half of France, from the Atlantic coast to the Alps (Le Féon et al., 2018, 2021; Le Féon and Geslin, 2018). Besides France, *Msc* has also spread to many European countries, starting with Italy in 2009 (Quaranta et al., 2014), Switzerland in 2010 (Amiet, 2012), Germany (Westrich et al. 2015), and Hungary in 2015 (Kovàcs, 2015), Slovenia in 2016 (Gogala and Zadravec, 2018), Austria (Westrich, 2017) and Serbia (Ćetković et al., 2020) in 2017, Spain (Ortiz-Sánchez et al., 2018) and Crimean peninsula (Ivanov and Fateryga, 2019) in 2018, Croatia (inaturalist.org) and Lichtenstein (Lanner et al., 2020a) in 2019, and finally in Bosnia-Herzegovina in 2020 (inaturalist.org; Fig. 6). *Megachile sculpturalis* is therefore now widespread in Europe and occupies several habitats from very urbanised ones such as Marseille, Lyon, or Vienna (Geslin et al., 2020; Lanner et al., 2020b; Fortel et al., 2016), to montane habitats (up to 1846 m above sea level; Le Féon and Geslin, 2018).

The spread of this species is remarkably fast even if similar dispersal rates have been reported for other invasive hymenopterans such as *Bombus terrestris* in South America (Schmid-Hempel et al., 2014) or *Vespa velutina* in Europe (Robinet et al., 2017). One of the hypotheses explaining this fast spread

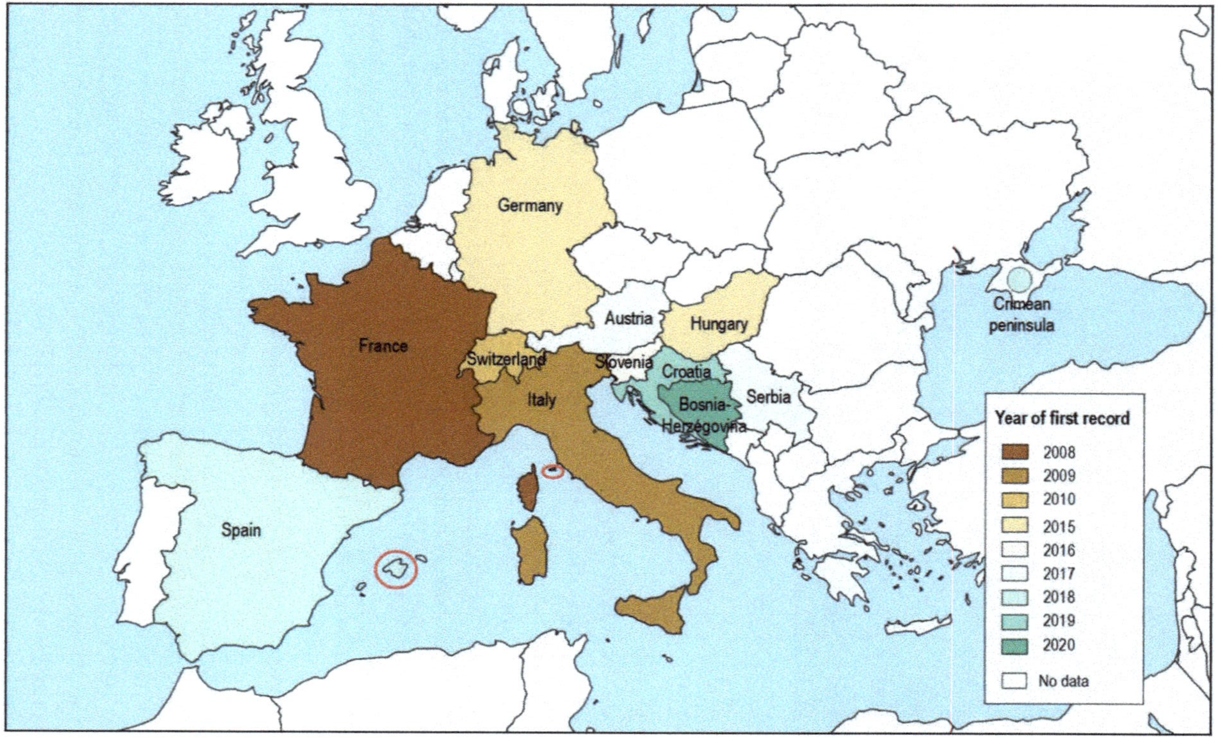

Figure 6 Year of the first record of *Megachile sculpturalis* in European countries (sources: Aguado et al., 2018; Amiet, 2012; Ćetković et al., 2020; Gogala and Zadravec, 2018; Ivanov and Fateryga, 2019; Kovács, 2015; Lanner et al., 2020b; Quaranta et al., 2014; Vereecken and Barbier, 2009; Westrich et al., 2015; Westrich, 2017). The homogeneous colouring of national territories can be misleading. For example, concerning the islands, *Msc* is so far unknown in Corsica, Sicily, and Sardinia. However, it has been recorded on the island of Elba and in Majorca, islands that we have circled in red. Map made with QGIS version 3.14. Base map downloaded from Natural Earth (naturalearthdata.com).

lies in the nesting preferences of *Msc* (in deadwood and cavities, see below), and *Msc* may have benefited from the timber trade through human-mediated transportation (e.g. Le Féon et al., 2018). Using the species distribution model (Lanner et al., 2022) recently showed that the spread of *Msc* is largely explained by anthropogenic factors and particularly facilitated by roads and maritime traffic. Through genetic analysis, Lanner et al. (2021) have also emphasised that several introductions may have occurred in Europe, partly explaining the rapid geographical expansion of *Msc*.

3.2 The morphology and ecology of Megachile sculpturalis

One of the vernacular names of *Megachile sculpturalis* is the giant resin bee. This appellation comes from the nesting biology of *Msc*, which closes its nest and brood cells with resin mainly collected from coniferous trees. *Megachile sculpturalis* is a large bee species (male 14–19 mm; female 22–27 mm; Batra, 1998), with orange to brown hairs on the thorax and infuscated wings (Fig. 5). The head is dark, with large mandibles used by females to carry mud and resin to build their nests (Fig. 7). Females carry pollen for their larvae in their ventral scopa.

Bees forage on flowers to gather pollen and nectar. While nectar is useful as a direct energy source for both adults and larvae, pollen is the principal source of protein and is gathered mostly to feed the larvae. Bees can be

Figure 7 *Megachile sculpturalis* female closing its nest with tree resin. In the background two more of her nests can be seen. Oberhofen am Thunersee, Switzerland. Picture: Christa Rohrbach.

a generalist when seeking for nectar but can be more of a specialist when foraging for a particular pollen. *Megachile sculpturalis* is a generalist species when it comes to nectar, meaning that this species can visit many different plant families. For example, in France, *Msc* has been observed foraging on 20 different plant species from 8 families, including Asteraceae (i.e. *Cirsium arvense*), Capriofoliacea (i.e. *Scabiosa columbaria*), and Lamiaceae (i.e. *Lavendula*) (Le Féon et al., 2018). However, for pollen, *Msc* appears to be more specialist with strong preferences for alien woody species also originating from Asia. In Europe, *Msc* is known for gathering pollen from alien tree species such as *Sophora japonica* (Fabaceae; Andrieu-Ponel et al., 2018) and collecting nectar from *Ligustrum lucidum* (native to Asia) and *Ligustrum vulgare* (native to Europe) (Oleaceae; Aguado et al., 2018).

Megachile sculpturalis nests in pre-existing cavities, mostly in deadwood, bamboo, or reeds. As stated earlier, *Msc* closes its nest entrances with resin, vegetal fibres, mud, and clay. Female *Msc* widely uses insect hotels (or bee hotels) to lay eggs, with much evidence in North America (MacIvor and Packer, 2015) and Europe (Díaz et al., 2021; Geslin et al., 2020; Le Féon et al., 2018; Fortel et al., 2016; Quaranta et al., 2014, Fig. 7). This propensity to nest in human-made structures may also be the reason for the rapid spread of *Msc* in Europe (Lanner et al., 2022).

3.3 Impacts on native bees

The first insight into the potential impacts of *Megachile sculpturalis* on native fauna was published in 2012 (Roulston and Malfi, 2012). The authors reported an aggressive eviction of a native carpenter bee (*Xylocopa virginica*) from its nest by *Msc* in Virginia (US). In the same year, Laport and Minckley (2012) reported the occupation of a nest of an active *X. virginica* by *Msc*. Both observations suggest potential nest competition between *Msc* and local fauna, together with aggressive territorial behaviours. Similar observations were obtained a few years ago in Europe, with (a) aggressive evictions, (b) occupation of nests, and (c) emptying of native bee nests, including *Osmia bicornis*, *O. cornuta*, and recently, *Anthidium septemspinosum* (Le Féon et al., 2018, 2021). In 2020, Geslin et al. conducted a study in urban parks of Marseille (France) to assess the impact of *Msc* on wild bees. They installed 96 bee hotels in 12 parks in the city and explored the relationship between the abundance of *Msc* and native bees emerging from bee hotels. First, they found that the most abundant bee nesting in bee hotels was *Msc*, with more than 40% of individuals. Second, they discovered a negative correlation between the presence of native bees and that of *Msc* within insect hotels, suggesting a negative impact of *Msc* on local native populations of wild bees. Two recent publications have provided new information on the impact of *Msc* on wild fauna. First, Lanner et al. (2020a) reported the killing of several *Heriades* spp. individuals, a French native species, by *Msc*, thereby documenting another direct impact of the alien bee species on native ones. Finally, Díaz et al. (2021) documented the occupation by *Msc* of *Osmia cornuta* nests without entirely emptying the nests. This means that *Msc* females laid eggs close to the cavity entrance, blocking this entrance with mud and resin. *Osmia cornuta* is active in spring, unlike *Msc*, which is active during summer (from June to September in Europe). Emerging individuals of *O. cornuta* are therefore blocked by this resin, incapable of exiting the nest, and may ultimately die from starvation or exhaustion.

Taken together, these findings show that *Msc* could have a deleterious impact on local wild bee populations (Fig. 8). However, the magnitude of these impacts remains poorly understood. For now, it does not appear easy to determine whether *Msc* could profoundly modify the structure of local bee populations, for example, by causing local extinctions of some species. The extent to which *Msc* disrupts native pollinators remains to be carefully evaluated, and future research will help us explore this question.

3.4 Other impacts

To date, there has been no documentation of other potential impacts of *Msc* on ecosystems or human economy and infrastructures. As stated earlier, *Msc*

Figure 8 *Megachile sculpturalis* female emptying the nest of a female of *Anthidium* sp. Anglet, France. Picture: Pierre-Jean Vandoorne.

appears to preferentially forage for pollen from alien plant species, potentially helping their spread. However, there is currently no evidence for such an effect. In addition, Díaz et al. (2021) showed that *Msc* could be affected by common, widespread generalist macro parasites without a substantial effect on the emerging success of *Msc*. At this stage, it appears difficult to determine whether *Msc* could reinforce parasite populations which could in turn affect native wild bees, and this remains to be explored. Finally, to the best of our knowledge, there is no reported effect of *Msc* on human economy and infrastructures. *Msc* is an opportunistic species that nests in pre-existing cavities and does not dig holes in wood by itself, reducing its potential impact on infrastructure.

Regarding the economy, a single example has been reported in the literature regarding the potential positive impact of *Msc* on agriculture: the pollination of Sunn hemp (*Crotalaria juncea* L.: Fabaceae) in Florida (Hall and Avila, 2016). *Msc* has been associated with a higher seed set for this cover crop, which contributes to soil nitrogen enhancement (Hall and Avila, 2016). Nevertheless, it is worth noting that *C. juncea* is an alien plant in the US. Moreover, this species is classified as invasive in some Pacific islands (see Hall and Avila, 2016); therefore, it is difficult to classify the interaction between *Msc* and *C. juncea* as positive or negative.

4 Conclusion

In this chapter, we have presented an overview of the known impacts of alien bees on native ones. As emphasised in the introduction, we voluntarily excluded the other known impacts of alien bees on native ecosystems, such as the disruption of native plant-pollinator networks, the population enhancement of alien plants (invasional meltdown), or the impact on crop production (except for the case study, for a comprehensive review on this subject see Aizen et al., 2020). As shown here, the potential effects exerted by alien bees on native bees are important and could, in extreme cases, be responsible for dramatic population losses of native bees (the case of *Bombus dahlbomii in South America;* see above). To date, 80 bee species have been recorded outside their native ranges (Russo, 2016). The different cases we exposed in this chapter are among the most iconic, with the strongest recorded negative impacts. Overall, however, the global impact of alien bees on native bees and ecosystems is largely unknown. Very few cases of introductions have been carefully studied, and for some of the recorded alien bees, little is known regarding their impacts on native bees and ecosystems. The case of *Msc* is an exception, being one of the single invasions that is relatively well documented (as is the case of *Bombus terrestris* in South America, Japan, and Tasmania).

Hence, we need to amplify our efforts to document new introductions, the geographic spread of alien bees, and their impacts. The number of new introductions is not expected to drop in the coming years (Seebens et al., 2017), and we need to anticipate the potentially dramatic consequences of these introductions to avoid them. For example, in 2018, Bortolotti et al. reported the presence of a second alien bee species in continental Europe, *Megachile disjunctiformis*. This species resembles *Msc* morphologically (Fig. 9) and has a similar native geographical repartition. To date, *M. disjunctiformis* has only been detected in Italy and has not yet extended its geographical range to neighbouring countries. However, we believe that this is likely to occur soon. We have little information about the ecology of this new alien species, and the consequences of its introduction should be closely observed in the future.

As stated earlier, the introduction of a new bee species in an ecosystem is often disregarded because of its role in the pollination of plants. However, we should change the way we perceive bee introductions. Like any other introduced species, alien bees can become invasive and affect native pollinators and ecosystem functioning. Introducing a new species into an environment is never a zero-sum game, and such practices must be abolished. Instead, native ecosystems should be preserved, and native bees allowed to thrive. Friendly practices favouring bees, such as the exclusion of pesticides, planting native flowers, and reduced mowing frequency (see Kawahara et al., 2021), should be encouraged. To prevent new invasions, biosecurity between countries must be reinforced and cross-border legal instruments homogenised (Diagne et al., 2021; Aizen et al., 2019; see Section 5).

Figure 9 *Megachile disjunctiformis*. Picture: Franziska Luthi.

5 Future trends in research

In this section, we have identified five major gaps that should be addressed in future research:

1. Even if some new data have been accumulated since, the first simple recommendation we can provide is to amplify research efforts on alien bees. We advocate for a new review of bees introduced worldwide 6 years after the one of Russo (2016). The cases of *M. disjunctiformis* in Italy (Bortolotti et al., 2018) or *Xylocopa pubescens* in the Canary Islands (Ruiz et al., 2020) are examples of introductions that occurred after the review of Russo (2016).
2. We should estimate the ecological impacts of alien bees more accurately, be they positive or negative. For example, Russo (2016) determined that 11 out of 80 species could positively affect native ecosystems, but no review on these effects exists to date. Of course, fieldwork and experiments to estimate impacts are costly (Goulson, 2003), but they provide valuable information for estimating impacts (e.g. see Díaz et al., 2021; Geslin et al., 2020).
3. We need to estimate the agronomic costs of alien bees. In their recent review, Aizen et al. (2020) provided convincing examples of the deleterious impacts of alien bees on agroecosystems. For example, the invasion of *Bombus terrestris* in Argentina affected the production of raspberries (Sáez et al., 2014, 2017, 2018). One underlying mechanism is the over-visitation of raspberry flowers by *Bombus terrestris*, leading to flower damage, such as style breakage (see Badano and Vergara, 2011 for an example on coffee plantations).
4. The economic cost associated with alien bees must be estimated. In addition to the costs related to agricultural production, it has been suggested that some wood-nesting bees can potentially represent a problem for wooden structures as nest excavation may generate structural weakness in lumber (Rust et al., 2004). In their recent review on the global economic costs of invasive species, Diagne et al. (2021) estimated a mean annual cost of invasive invertebrates of US$ 8.7 billion from 1970 to 2017. However, this review also highlights the critical need for more research on the economic costs of invasive insects.
5. If possible, we should provide stakeholders with management techniques to avoid the spread of alien bees. For example, Geslin et al. (2020) showed that *Megachile sculpturalis* rarely nests in bamboo with a hole diameter < 8 mm. This observation could help land managers install and build insect hotels to avoid the nesting of *Msc*.

6 How to improve biosecurity

Among the new pathways of invasions, the increase of commercial offers for the public of both managed and wild bees is of great concern (Geslin et al., 2017). Over the past 30 years, we have seen an exponential development of pollination supplementation practices in order to fight against pollinator declines and to compensate the limitation of pollination service delivered to entomophilous crops and ornamental species which may limit their yields. Since the 1970s, this has led to the first wave of global trade in social bees and bumblebees, initially for greenhouse use. As we have previously seen in this chapter, many reports of escapes of these imported bees into the wild have occurred. More recently, the second wave of the commercial trade of wild bees is emerging. In Europe or in North America, various commercial companies are selling wild solitary bees to promote better pollination service in gardens and field crops, arguing that these bees offer a much greater pollination service than domesticated social bees. Several mason bees and leaf-cutting bees have been selected for commercial purposes according to their nesting habits in cavities, offering an easy way to ship living cocoons in bamboo twigs. However, in the US and Canada, this may also promote the spread of some alien bees (e.g. *Osmia cornifrons*). These new commercial practices bring new pathways for alien pollinators to spread as the business may be done worldwide. For example, by June 2021, a wild bee package shipped from France to Seattle (US) was intercepted in New Caledonia by biosecurity services. Inside the package, we found bamboo twigs containing 23 lived cocoons of *Osmia bicornis* and two recently emerged females (H. Jourdan, *pers. obs.*). This illustrates the tremendous risk of moving large numbers of alien pollinators to a wide range of regions without any biosecurity rules implemented.

Beyond this specific pollinator trade, there are still other pathways for the involuntary spread of alien pollinators associated with global trade. The rule that the more the volume of traded commodities increases between distant countries, the more the biosecurity risk increases for alien arrivals (Hulme, 2009), is also valid for pollinators. Mason bees (including resin bees and leaf-cutting bees) and carpenter bees (including hollow stem nesting species) appear more likely to be propagated by human trade.

To deal with these new threats, enforcement of biosecurity regulations and practices against the transfer or arrival of alien bees is urgently needed. According to the Food and Agriculture Organization of the United Nations standards, biosecurity is defined as 'a strategic and integrated approach that encompasses the policy and regulatory frameworks (including instruments and activities) for analysing and managing relevant risks to human, animal and plant life and health, and associated risks to the environment'. It appears critical in the context of biological invasions (Hulme, 2010). Islands are particularly at risk and deserve special attention (Cox and Elmqvist, 2000; Moore et al., 2010). But since the late 1980s, most of the biosecurity efforts have focused on preventing the spread of parasites and pathogens of managed bees and not alien bees in general.

In recent decades, Australia has appeared as a good example of biosecurity measures enforced to protect native pollinators. Even if it was established mostly for protecting the bee industry with highly enforced regulation and surveillance against the introduction of *Varroa destructor*, tracheal mites with European (*Apis melifera*) or Asian (*Apis cerana*) beehives, it is also focusing on some bee diseases to avoid possible spillover between managed bees and wild bees (Heersink et al., 2016; Robert et al., 2017; Dalmon et al., 2021). These biosecurity efforts rely on a national network of sentinel beehives. Such a sentinel coordinated approach is also applied to the larger Pacific region (New Zealand – despite Varroa installation – New Caledonia, French Polynesia, and to a lesser extent Fiji).

Another alternative solution to reduce the unfortunate trade of alien pollinators relies on the valuing and promotion of local wild pollinators for pollination effectiveness. This implies the conservation and restoration of 'green infrastructures' across landscapes that enhance wild bee populations. In Australia, native stingless bees are promoted in a commercial way to enhance pollination service, and troubles in Tasmania linked to the European Bumblebee (*Bombus terrestris*) have led to research for local alternative pollinators. Thus, native *Amegilla* spp. were identified as highly efficient pollinators in gardens, especially for Solanaceae (McQuillan and Hingston, 1999; Hogendoorn et al., 2000, 2007). But without awareness of the public, a new biosecurity concern may develop, as people may deliberately introduce *Amegilla* spp. into neighbouring oceanian territories, mostly archipelagos, for agricultural purposes. This has been suspected of *Amegilla pulchra* in Fiji, French Polynesia, and New Caledonia (Groom et al., 2017; Zakardjian et al., 2020). However, once introduced, *Amegilla* spp. may not limit their foraging to valuable crop species and are likely to visit alien plants. If those alien plants are usually poorly visited by native pollinators, the arrival of alien bees such as *Amegilla* spp. could result in the spread of those plant species.

Finally, there is an emerging corpus of methodology and theory to better implement efficient biosecurity measures in a changing world (climate change and global trade), especially urgently needed on islands, which could be applied to pollinators (Moore et al., 2010; Aizen et al., 2019; Proesmans et al., 2021). As pointed out by Hulme (2015), pathways for alien species introductions fall into six broad classes: deliberate release; escape from captivity; contamination of a commodity; stowaway on a transport vector; via an infrastructure corridor (without which spread would not be possible) or unaided from already invaded regions. Each of them should be examined carefully in the context of alien pollinators spread, resulting in enforcement of regulation policies and *ad hoc* surveillance to prevent further biological invasions. More specifically, in the context of safeguarding pollination services, ten general rules have already been promoted as guidelines to improve governmental regulation, including an emphasis on the regulation of movement of managed pollinators and development of long-term monitoring of pollinators and pollination (Dicks et al., 2016). However, movement regulation should be extended to the wild bee trade and closest links should be developed with scientific research for robust long-term monitoring of pollinators and pollination at a global scale across a variety of landscapes. Hulme (2020) also promotes a 'one biosecurity approach' relying on multidisciplinary input from ecology, social sciences, and economics. Such a global biosecurity approach should be promoted for pollination service, which is critical to ecosystem's maintenance and human well-being in our changing world (Hulme, 2020; Proesmans et al., 2021).

7 Where to look for further information

The following articles are the key papers on the impact of alien bee species on native bees:

- Aizen, M. A., Smith-Ramírez, C., Morales, C. L., Vieli, L., Sáez, A., Barahona-Segovia, R. M., Arbetman, M. P., Montalva, J., Garibaldi, L. A., Inouye, D. W. and Harder, L. D. (2018). Coordinated species importation policies are needed to reduce serious invasions globally: The case of alien bumblebees in South America. *J. Appl. Ecol.* 1–7.
- Geslin, B., Gauzens, B., Baude, M., Dajoz, I., Fontaine, C., Henry, M., Ropars, L., Rollin, O., Thébault, E. and Vereecken, N. J. (2017). Massively introduced managed species and their consequences for plant–pollinator interactions. In: *Advances in Ecological Research*, pp. 147–199.
- Goulson, D. (2003). Effects of introduced bees on native ecosystems. *Annu. Rev. Ecol. Evol. Syst.* 34, 1–26.
- Mallinger, R. E., Gaines-Day, H. R. and Gratton, C. (2017). Do managed bees have negative effects on wild bees?: A systematic review of the literature. *PLoS ONE* 12, e0189268.
- Paini, D. R. (2004). Impact of the introduced honey bee (*Apis mellifera*) (Hymenoptera: Apidae) on native bees: A review. *Austral Ecol.* 29, 399–407.

- Russo, L. (2016). Positive and negative impacts of non-native bee species around the world. Insects 7, 101–103.

For a focus on agricultural production, we recommend:

- Aizen, M. A., Arbetman, M. P., Chacoff, N. P., Chalcoff, V. R., Feinsinger, P., Garibaldi, L. A., Harder, L. D., Morales, C. L., Sáez, A. and Vanbergen, A. J. (2020). Invasive bees and their impact on agriculture, 1st ed., *Advances in Ecological Research*. Elsevier Ltd.

We particularly recommend the work of Dr M. Aizen and his team on this thematic:

- https://sites.google.com/site/marceloaizen/home.

The open science projects following the spread of Megachile sculpturalis in Europe and in the Balkans can be found here:

- https://oabeilles.net/projets/observatoire-abeilles-exotiques.
- https://www.citizen-science.at/blog/blogger/lannerj.
- https://srbee.bio.bg.ac.rs/english.

Key journals to read articles on invasion biology are:

- Biological invasions: https://www.springer.com/journal/10530.
- Diversity and Distribution: https://onlinelibrary.wiley.com/journal/14724642.
- NeoBiota: https://neobiota.pensoft.net/.

Other interesting websites:

- https://ipbes.net/invasive-alien-species-assessment.
- http://www.issg.org/.
- https://www.gbif.org/fr/dataset/39f36f10-559b-427f-8c86-2d28afff68ca.
- https://easin.jrc.ec.europa.eu/easin.
- https://serc.si.edu/research/research-topics/biological-invasions.

Some interesting outreaches:

- https://theconversation.com/bee-battles-why-our-native-pollinators-are-losing-the-war-40620.
- https://theconversation.com/aussie-scientists-need-your-help-keeping-track-of-bees-please-128932.
- https://idtools.org/id/bees/exotic/.

8 References

Abe, T., Makino, S. and Okochi, I. (2008). Why have endemic pollinators declined on the Ogasawara Islands? *Biodivers. Conserv.* 17(6), 1465–1473.

Acosta, A. L., Giannini, T. C., Imperatriz-Fonseca, V. L. and Saraiva, A. M. (2016). Worldwide alien invasion: a methodological approach to forecast the potential spread of a highly invasive pollinator. *PLoS ONE* 11(2), e0148295.

Aguado, O., Hernández-Castellano, C., Bassols, E., Miralles, M., Navarro, D., Stefanescu, C. and Vicens, N. (2018). *Megachile* (*Callomegachile*) *sculpturalis* Smith, 1853 (Apoidea: Megachilidae): a new exotic species in the Iberian Peninsula, and some notes about its biology. *Butll. Inst. Catalana Hist. Nat.* 82, 157–162.

Agüero, J. I., Pérez-Méndez, N., Torretta, J. P. and Garibaldi, L. A. (2020). Impact of invasive bees on plant-pollinator interactions and reproductive success of plant species in mixed Nothofagus antarctica Forests. *Neotrop. Entomol.* 49(4), 557–567.

Aizen, M. A., Arbetman, M. P., Chacoff, N. P., Chalcoff, V. R., Feinsinger, P., Garibaldi, L. A., Harder, L. D., Morales, C. L., Sáez, A. and Vanbergen, A. J. (2020). Invasive bees and their impact on agriculture. *Adv. Ecol. Res.* 63, 49–92.

Aizen, M. A., Lozada, M. and Morales, C. L. (2011). Comparative nectar-foraging behaviors and efficiencies of an alien and a native bumble bee. *Biol. Invas.* 13(12), 2901–2909.

Aizen, M. A., Morales, C. L. and Morales, J. M. (2008). Invasive mutualists erode native pollination webs. *PLoS Biol.* 6(2), e31.

Aizen, M. A., Morales, C. L., Vázquez, D. P., Garibaldi, L. A., Sáez, A. and Harder, L. D. (2014). When mutualism goes bad: density-dependent impacts of introduced bees on plant reproduction. *New Phytol.* 204(2), 322–328.

Aizen, M. A., Smith-Ramírez, C., Morales, C. L., Vieli, L., Sáez, A., Barahona-Segovia, R. M., Arbetman, M. P., Montalva, J., Garibaldi, L. A., Inouye, D. W. and Harder, L. D. (2019). Coordinated species importation policies are needed to reduce serious invasions globally: the case of alien bumblebees in South America. *J. Appl. Ecol.* 56(1), 100–106.

Alger, S. A., Alexander Burnham, P. A., Boncristiani, H. F. and Brody, A. K. (2019). RNA virus spillover from managed honey bees (*Apis mellifera*) to wild bumblebees (*Bombus* spp.). *PLoS ONE* 14(6), e0217822. https://doi.org/10.1371/journal.pone.0217822.

Amiet, F. (2012). Die Blattschneiderbiene Megachile sculpturalis Smith, 1853 (Hymenoptera, Apidae) nun auch in der Schweiz. *Entomo. Helv.* 1853, 157–159.

Andrieu-Ponel, V., Ponel, P., Le Féon, V., Geslin, B. and Duvallet, G. (2018). A propos du comportement de butinage de *Megachile sculpturalis* Smith, 1853, en France méditerranéenne (Nîmes et Montpellier) (Hymenoptera, Megachilidae). *Bull. Soc. Entomol. Fr.* 123, 49–54.

Arbetman, M. P., Meeus, I., Morales, C. L., Aizen, M. A. and Smagghe, G. (2012). Alien parasite hitchhikes to Patagonia on invasive bumblebee. *Biol. Invas.* 15, 489–494.

Badano, E. I. and Vergara, C. H. (2011). Potential negative effects of exotic honey bees on the diversity of native pollinators and yield of highland coffee plantations. *Agric. For. Entomol.* 13(4), 365–372.

Batra, S. W. T. (1998). Biology of the giant resin bee, *Megachile sculpturalis* Smith, a conspicuous new immigrant in Maryland. *The Md. Nat.* 42(1-2), 1-3.

Beavon, M. A. and Kelly, D. (2012). Invasional meltdown: pollination of the invasive liana Passiflora tripartita var. mollissima (Passifloraceae) in New Zealand. *N. Z. J. Ecol.* 36, 100–107.

Bellard, C., Cassey, P. and Blackburn, T. M. (2016a). Alien species as a driver of recent extinctions. *Biol. Lett.* 12(2), 20150623.

Bellard, C., Leroy, B., Thuiller, W., Rysman, J.-F. and Courchamp, F. (2016b). Major drivers of invasion risks throughout the world. *Ecosphere* 7(3), 1–14.

Bertelsmeier, C. (2021). Globalization and the anthropogenic spread of invasive social insects. *Curr. Opin. Insect Sci.* 46, 16–23.

Bertelsmeier, C., Avril, A., Blight, O., Jourdan, H. and Courchamp, F. (2015). Discovery-dominance trade-off among widespread invasive ant species. *Ecol. Evol.* 5(13), 2673–2683.

Blackburn, T. M., Pysek, P., Bacher, S., Carlton, J. T., Duncan, R. P., Jarosık, V., Wilson, J. R. U. and Richardson, D. M. (2011). A proposed unified framework for biological invasions. *Trends Ecol. Evol.* 26(7), 333–339.

Bortolotti, L., Luthi, F., Flaminio, S., Bogo, G. and Sgolastra, F. (2018). First record of the Asiatic bee *Megachile disjunctiformis* in Europe. *Bull. Insectology* 71, 143–149.

Cairns, C. E., Villanueva-Gutiérrez, R., Koptur, S. and Bray, D. B. (2005). Bee populations, forest disturbance, and Africanization in Mexico. *Biotropica* 37(4), 686–692.

Case, T. J. and Gilpin, M. E. (1974). Interference competition and niche theory. *Proc. Natl. Acad. Sci. U. S. A.* 71(8), 3073–3077.

Ćetković, A., Stanisavljević, L., Pleća, M., Raičević, J., Žikić, V., Glavendekić, M. and Bila Dubaić, J. (2020). Project: monitoring the spread of sculptured resin bee (*Megachile sculpturalis*) in Serbia. Centar za biologiju pčela, Biološki fakultet, Univerzitet u Beogradu.

Courchamp, F., Fournier, A., Bellard, C., Bertelsmeier, C., Bonnaud, E., Jeschke, J. M. and Russell, J. C. (2017). Invasion biology: specific problems and possible solutions. *Trends Ecol. Evol.* 32(1), 13–22.

Cox, P. A. and Elmqvist, T. (2000). Pollinator extinction in the Pacific Islands. *Conserv. Biol.* 14(5), 1237–1239.

Dafni, A., Kevan, P., Gross, C. L. and Goka, K. (2010). *Bombus terrestris*, pollinator, invasive and pest: an assessment of problems associated with its widespread introductions for commercial purposes. *Appl. Entomol. Zool.* 45(1), 101–113.

Dalmon, A., Diévart, V., Thomasson, M., Fouque, R., Vaissière, B. E., Guilbaud, L., Le Conte, Y. and Henry, M., (2021). Possible spillover of pathogens between bee communities foraging on the same floral resource. *Insects* 12, 122.

Diagne, C., Leroy, B., Vaissière, A. C., Gozlan, R. E., Roiz, D., Jarić, I., Salles, J. M., Bradshaw, C. J. A. and Courchamp, F. (2021). High and rising economic costs of biological invasions worldwide. *Nature* 592(7855), 571–576.

Díaz, S. S., Carisio, L., Manino, A., Biella, P. and Porporato, M. (2021). Nesting, sex ratio and natural enemies of the Giant Resin Bee in relation to native species in Europe. *Insects*, 12 (6): 545.

Dicks, L. V., Viana, B., Bommarco, R., Brosi, B., del Coro Arizmendi, M. D., Cunningham, S. A., Galetto, L., Hill, R., Lopes, A. V., Pires, C., Taki, H. and Potts, S. G. (2016). Ten policies for pollinators. *Science* 354(6315), 975–976.

Dohzono, I., Kunitake, Y. K., Yokoyama, J. and Goka, K. (2008). Alien bumble bee affects native plant reproduction through interactions with native bumble bees. *Ecology* 89(11), 3082–3092.

Fitch, G., Wilson, C. J., Glaum, P., Vaidya, C., Simao, M. C. and Jamieson, M. A. (2019). Does urbanization favour exotic bee species? Implications for the conservation of native bees in cities. *Biol. Lett.* 15(12), 20190574.

Fortel, L., Henry, M., Guilbaud, L., Mouret, H., Vaissière, B. E. (2016). Use of human-made nesting structures by wild bees in an urban environment. *J. Insect Conserv.* 20, 239–253.

Gamboa, V., Ravoet, J., Brunain, M., Smagghe, G., Meeus, I., Figueroa, J., Riaño, D. and de Graaf, D. C. (2015). Bee pathogens found in *Bombus atratus* from Colombia: a case study. *J. Invertebr. Pathol.* 129, 36–39.

Geslin, B., Gachet, S., Deschamps-Cottin, M., Flacher, F., Ignace, B., Knoploch, C., Meineri, É., Robles, C., Ropars, L., Schurr, L. and Le Féon, V. (2020). Bee hotels host a high abundance of exotic bees in an urban context. *Acta Oecol.* 105, 103556.

Geslin, B., Gauzens, B., Baude, M., Dajoz, I., Fontaine, C., Henry, M., Ropars, L., Rollin, O., Thébault, E. and Vereecken, N. J. (2017). Massively introduced managed species and their consequences for plant-pollinator interactions. *Adv. Ecol. Res.* 57, 147–199.

Geslin, B. and Morales, C. L. (2015). New records reveal rapid geographic expansion of *Bombus terrestris* Linnaeus, 1758 (Hymenoptera: Apidae), an invasive species in Argentina. *Check List* 11(3), 3–5.

Gill, R. J., Baldock, K. C. R., Brown, M. J. F., Cresswell, J. E., Dicks, L. V., Fountain, M. T., Garratt, M. P. D., Gough, L. A., Heard, M. S., Holland, J. M., Ollerton, J., Stone, G. N., Tang, C. Q., Vanbergen, A. J., Vogler, A. P., Woodward, G., Arce, A. N., Boatman, N. D., Brand-Hardy, R., Breeze, T. D., Green, M., Hartfield, C. M., O'Connor, R. S., Osborne, J. L., Phillips, J., Sutton, P. B. and Potts, S. G. (2016). Protecting an ecosystem service. *Adv. Ecol. Res.* 53, 135–206.

Gogala, A. and Zadravec, B. (2018). First record of *Megachile sculpturalis* smith in Slovenia (hymenoptera: Megachilidae). *Acta Entomol. Slov.* 26, 79–82.

Goulson, D. (2003). Effects of introduced bees on native ecosystems. *Annu. Rev. Ecol. Evol. Syst.* 34(1), 1–26.

Goulson, D., Nicholls, E., Botias, C. and Rotheray, E. L. (2015). Bee declines driven by combined stress from parasites, pesticides, and lack of flowers. *Science* 347(6229), 1255957.

Graf, L. V., Zenni, R. D. and Gonçalves, R. B. (2020). Ecological impact and population status of non-native bees in a Brazilian urban environment. *Rev. Bras. Entomol.* 64(2), e20200006.

Graham, K. K., Eaton, K., Obrien, I. and Starks, P. T. (2019). *Anthidium manicatum*, an invasive bee, excludes a native bumble bee, *Bombus impatiens*, from floral resources. *Biol. Invas.* 21(4), 1089–1099.

Graystock, P., Yates, K., Evison, S. E. F., Darvill, B., Goulson, D. and Hughes, W. O. H. (2013). The Trojan hives: pollinator pathogens, imported and distributed in bumblebee colonies. *J. Appl. Ecol.* 50(5), 1207–1215.

Groom, S. V. C., Stevens, M. I., Ramage, T. and Schwarz, M. P. (2017). Origins and implications of apid bees (Hymentopera: Apidae) in French Polynesia. *Entomol. Sci.* 20(1), 65–75.

Hall, H. G. and Avila, L. (2016). *Megachile sculpturalis*, the Giant resin Bee, overcomes the blossom structure of Sunn Hemp (*Crotalaria juncea*) that impedes pollination. *J. Melittology* 65, 1–11.

Heersink, D. K., Caley, P., Paini, D. R. and Barry, S. C. (2016). Quantifying the establishment likelihood of invasive alien species introductions through ports with application to honey bees in Australia. *Risk Anal.* 36(5), 892–903.

Hinojosa-Díaz, I. (2008). The giant resin bee making its way west: first record in Kansas (Hymenoptera: Megachilidae). *ZooKeys* 1, 67–71.

Hogendoorn, K., Coventry, S. and Keller, M. A. (2007). Foraging behaviour of a blue banded bee, *Amegilla chlorocyanea* in greenhouses: implications for use as tomato pollinators. *Apidologie* 38(1), 86–92.

Hogendoorn, K., Steen, Z. and Schwarz, M. P. (2000). Native Australian carpenter bees as a potential alternative to introducing bumble bees for tomato pollination in greenhouses. *J. Apic. Res.* 39(1–2), 67–74.

Hulme, P. E. (2009). Trade, transport and trouble: managing invasive species pathways in an era of globalization. *J. Appl. Ecol.* 46(1), 10–18.

Hulme, P. E. (2010). Biosecurity: the changing face of invasion biology, in M. R. Richardson (Ed), *Fifty Years of Invasion Ecology – the Legacy of Charles Elton*, pp. 301–314, Oxford: Blackwells.

Hulme, P. E. (2015). Invasion pathways at a crossroad: policy and research challenges for managing alien species introductions. *J. Appl. Ecol.* 52(6), 1418–1424.

Hulme, P. E. (2020). One biosecurity: a unified concept to integrate human, animal, plant, and environmental health. *Emerg. Top. Life Sci.* 4(5), 539–549.

Hung, K. J., Kingston, J. M., Lee, A., Holway, D. A. and Kohn, J. R. (2019). Non-native honey bees disproportionately dominate the most abundant floral resources in a biodiversity hotspot. *Proc. Biol. Sci.* 286(1897), 20182901.

Inoue, M. N., Makino, T. T., Yokoyama, J. and Sakai, S. (2010). Is *Bombus terrestris* (Hymenoptera: Apidae) a stronger competitor against native Japanese species? A comparison of foraging efficiency. *Appl. Entomol. Zool.* 45(1), 71–75.

Inoue, M. N., Yokoyama, J. and Washitani, I. (2008). Displacement of Japanese native bumblebees by the recently introduced *Bombus terrestris* (L.) (Hymenoptera: Apidae). *J. Insect Conserv.* 12(2), 135–146.

Ivanov, S. P. and Fateryga, A. V. (2019). First record of the invasive giant resin bee *Megachile* (callomegachile) *sculpturalis* smith, 1853 (hymenoptera: Megachilidae) in the crimea. *Far East. Entomol.* 395, 7–13.

Iwasaki, J. M., Dickinson, K. J. M., Barratt, B. I. P., Mercer, A. R., Jowett, T. W. D. and Lord, J. M. (2018). Floral usage partitioning and competition between social (*Apis mellifera*, *Bombus terrestris*) and solitary bees in New Zealand: niche partitioning via floral preferences? *Austral Ecol.* 43(8), 937–948.

Jean, R. P. (2005). Quantifying a rare event: pollen theft by honey bees from bumble bees and other bees (Apoidea: Apidae, Megachilidae) foraging at flowers. *J. Kans. Entomol. Soc.* 78(2), 172–175.

Kato, M., Shibata, A., Yasui, T. and Nagamasu, H. (1999). Impact of introduced honey bees, *Apis mellifera*, upon native bee communities in the Bonin (Ogasawara) Islands. *Popul. Ecol.* 41(2), 217–228.

Kawahara, A. Y., Reeves, L. E., Barber, J. R. and Black, S. H. (2021). Eight simple actions that individuals can take to save insects from global declines. *Proc. Natl. Acad. Sci. U. S. A.* 118(2), 2–7.

Kenis, M., Auger-Rozenberg, M.-A., Roques, A., Timms, L., Péré, C., Cock, M. J. W., Settele, J., Augustin, S. and Lopez-Vaamonde, C. (2009). Ecological effects of invasive alien insects. *Biol. Invasions* 11, 21–45.

Klein, A. M., Vaissière, B. E., Cane, J. H., Steffan-Dewenter, I., Cunningham, S. A., Kremen, C. and Tscharntke, T. (2007). Importance of pollinators in changing landscapes for world crops. *Proc. Biol. Sci.* 274(1608), 303–313. https://doi.org/10.1098/rspb.2006.3721.

Kovàcs, T. (2015). *Megachile sculpturalis* Smith, 1853. In: *Hungary (Hymenoptera, Megachilidae). Folia Historico-Naturalia Musei Matraensis* (vol. 39), 73–76. https://

matramuzeum.nhmus.hu/sites/default/files/nhmusfiles/kiadvanyok/folia/vol39/08_Megachile_Kovacs_39.pdf.

Kumschick, S., Gaertner, M., Vila, M., Essl, F., Jeschke, J. M., Pysek, P., Ricciardi, A., Bacher, S., Blackburn, T. M., Dick, J. T. A., Evans, T., Hulme, P. E., Kühn, I., Mrugala, A., Pergl, J., Rabitsch, W., Richardson, D. M., Sendek, A. and Winter, M. (2015). Ecological impacts of alien species: quantification, scope, caveats, and recommendations. *BioScience* 65(1), 55–63.

Lanner, J., Dubos, N., Geslin, B., Leroy, B., Hernández-Castellano, C., Dubaić, J. B., Bortolotti, L., Calafat, J. D., Ćetković, A., Flaminio, S., Le Féon, V., Margalef-Marrase, J., Orr, M., Pachinger, B., Ruzzier, E., Smagghe, G., Tuerlings, T., Vereecken, N. J. and Meimberg, H. (2022). On the road: anthropogenic factors drive the invasion risk of a wild solitary bee species. *Sci. Total Environ.* 827, 154246. https://doi.org/10.1016/j.scitotenv.2022.154246.

Lanner, J., Gstöttenmayer, F., Curto, M., Geslin, B., Huchler, K., Orr, M. C., Pachinger, B., Sedivy, C. and Meimberg, H. (2021). Evidence for multiple introductions of an invasive wild bee species currently under rapid range expansion in Europe. *BMC Ecol. Evol.* 21(1), 17.

Lanner, J., Meyer, P., Harmetzky, F., Meimberg, H. and Pachinger, B. (2020a). Die Asiatische Mörtelbiene (Hymenoptera: *Megachile sculpturalis* Smith, 1853) – eine neue Bienenart für Österreich. *Beiträge zur Entomofaunistik* 21, 87–95.

Lanner, J., Huchler, K., Pachinger, B., Sedivy, C. and Meimberg, H. (2020b). Dispersal patterns of an introduced wild bee, *Megachile sculpturalis* Smith, 1853 (Hymenoptera: Megachilidae) in European alpine countries. *PLoS ONE* 15(7), e0236042.

Laport, R. G. and Minckley, R. L. (2012). Occupation of active Xylocopa virginica nests by the recently invasive *Megachile sculpturalis* in upstate New York. *J. Kans. Entomol. Soc.* 85(4), 384–386.

LeCroy, K. A., Savoy-Burke, G., Carr, D. E., Delaney, D. A. and Roulston, T. H. (2020). Decline of six native mason bee species following the arrival of an exotic congener. *Sci. Rep.* 10(1), 18745.

Le Féon, V., Aubert, M., Genoud, D., Andrieu-Ponel, V., Westrich, P. and Geslin, B. (2018). Range expansion of the Asian native giant resin bee *Megachile sculpturalis* (Hymenoptera, Apoidea, Megachilidae) in France. *Ecol. Evol.* 8(3), 1534–1542.

Le Féon, V., Genoud, D. and Geslin, B. (2021). Actualisation des connaissances sur l'abeille *Megachile sculpturalis* Smith 1853 en Europe. *Osmia* 9, 25–36.

Le Féon, V. and Geslin, B. (2018). Écologie et distribution de l'abeille originaire d'Asie *Megachile sculpturalis* Smith 1853 (Apoidea - Megachilidae - Megachilini) : un état des connaissances dix ans après sa première observation en Europe. *Osmia* 7, 31–39.

MacIvor, J. S. and Packer, L. (2015). "Bee hotels" as tools for native pollinator conservation: a premature verdict? *PLoS ONE* 10(3), e0122126.

Magnum, W. A. and Brooks, R. W. (1997). First records of *Megachile* (Callomegachile) *sculpturalis* Smith (Hymenop- tera: Megachilidae) in the Continental United States. *J. Kans. Entomol. Soc.* 70(2), 140–142.

Mallinger, R. E., Gaines-Day, H. R. and Gratton, C. (2017). Do managed bees have negative effects on wild bees?: a systematic review of the literature. *PLoS ONE* 12(12), e0189268.

McCune, F., Normandin, É., Mazerolle, M. J. and Fournier, V. (2019). Response of wild bee communities to beekeeping, urbanization, and flower availability. *Urban Ecosyst.* 23(1), 39–54.

McQuillan, P. B. and Hingston, A. B. (1999). Displacement of Tasmanian native megachilid bees by the recently introduced bumblebee *Bombus terrestris* (Linnaeus, 1758) (Hymenoptera: Apidae). *Aust. J. Zool.* 47(1), 59-65.

Moore, J. L., Rout, T. M., Hauser, C. E., Moro, D., Jones, M., Wilcox, C. and Possingham, H. P. (2010). Protecting islands from pest invasion: optimal allocation of biosecurity resources between quarantine and surveillance. *Biol. Conserv.* 143(5), 1068-1078.

Morales, C. L. (2007). Introducción de abejorros (*Bombus*) no nativos: causas, consecuencias ecológicas y perspectivas. *Ecol. Austral* 17, 51-65.

Morales, C. L. and Aizen, M. A. (2006). Invasive mutualisms and the structure of plant-pollinator interactions in the temperate forests of north-west Patagonia, Argentina. *J. Ecol.* 94(1), 171-180.

Morales, C. L., Arbetman, M. P., Cameron, S. A. and Aizen, M. A. (2013). Rapid ecological replacement of a native bumble bee by invasive species. *Front. Ecol. Environ.* 11(10), 529-534.

Murray, E. A., Burand, J., Trikoz, N., Schnabel, J., Grab, H. and Danforth, B. N. (2019). Viral transmission in honey bees and native bees, supported by a global black queen cell virus phylogeny. *Environ. Microbiol.* 21(3), 972-983.

Nagamitsu, T., Kenta, T., Inari, N., Horita, H., Goka, K. and Hiura, T. (2007a). Foraging interactions between native and exotic bumblebees: enclosure experiments using native flowering plants. *J. Insect Conserv.* 11(2), 123-130.

Nagamitsu, T., Kenta, T., Inari, N., Kato, E. and Hiura, T. (2007b). Abundance, body size, and morphology of bumblebees in an area where an exotic species, *Bombus terrestris*, has colonized in Japan. *Ecol. Res.* 22(2), 331-341.

Nagamitsu, T., Yamagishi, H., Kenta, T., Inari, N. and Kato, E. (2009). Competitive effects of the exotic *Bombus terrestris* on native bumble bees revealed by a field removal experiment. *Popul. Ecol.* 52(1), 123-136.

Ollerton, J. (2017). Pollinator diversity: distribution, ecological function, and conservation. *Annu. Rev. Ecol. Evol. Syst.* 48(1), 353-376.

Ollerton, J., Winfree, R. and Tarrant, S. (2011). How many flowering plants are pollinated by animals? *Oikos* 120(3), 321-326.

Ortiz-Sánchez, F. J., Navarro, J. F. and Taeger, U. (2018). *Megachile* (Callomegachile) *sculpturalis* Smith, 1853, nueva especie para la fauna ibérica (Hymenoptera, Megachilidae). *Bol. Soc. Entom. Aragonesa* 63, 259 -261.

Osterman, J., Aizen, M. A., Biesmeijer, J. C., Bosch, J., Howlett, B. G., Inouye, D. W., Jung, C., Martins, D. J., Medel, R., Pauw, A., Seymour, C. L. and Paxton, R. J. (2021). Global trends in the number and diversity of managed pollinator species. *Agric. Ecosyst. Environ.* 322, 107653.

Paiero, S. M. and Buck, M. (2003). The giant resin bee, *Megachile sculpturalis* Smith, and other newly introduced and newly recorded native Megachlidae and Andrenidae (Apoidea) from ontario. *J. Entomol. Soc. Ont.* 134, 141-143.

Paini, D. R. and Roberts, J. D. (2005). Commercial honey bees (*Apis mellifera*) reduce the fecundity of an Australian native bee (*Hylaeus alcyoneus*). *Biol. Conserv.* 123(1), 103-112.

Parys, K. A., Tripodi, A. D. and Sampson, B. J. (2015). The Giant resin Bee, Megachile sculpturalis Smith: new Distributional Records for the Mid- and Gulf-south USA. *Biodivers. Data J.* 3(3), e6733.

Plischuk, S., Meeus, I., Smagghe, G. and Lange, C. E. (2011). *Apicystis bombi* (Apicomplexa: Neogregarinorida) parasitizing *Apis mellifera* and *Bombus terrestris* (Hymenoptera: Apidae) in Argentina. *Environ. Microbiol. Rep.* 3(5), 565-568.

Proesmans, W., Albrecht, M., Gajda, A., Neumann, P., Paxton, R. J., Pioz, M., Polzin, C., Schweiger, O., Settele, J., Szentgyörgyi, H., Thulke, H. H. and Vanbergen, A. J. (2021). Pathways for novel epidemiology: plant-pollinator-pathogen networks and global change. *Trends Ecol. Evol.* 36(7), 623-636.

Quaranta, M., Sommaruga, A., Balzarini, P. and Felicioli, A. (2014). A new species for the bee fauna of italy: Megachile sculpturalis continues its colonization of Europe. *Bull. Insectology* 67, 287-293.

Roberts, J. M. K., Anderson, D. L. and Durr, P. A. (2017). Absence of deformed wing virus and Varroa destructor in Australia provides unique perspectives on honeybee viral landscapes and colony losses. *Scientific reports*, 7(1), 1-11.

Robinet, C., Suppo, C. and Darrouzet, E. (2017). Rapid spread of the invasive yellow-legged hornet in France: the role of human-mediated dispersal and the effects of control measures. *J. Appl. Ecol.* 54(1), 205-215.

Ropars, L., Dajoz, I., Fontaine, C., Muratet, A. and Geslin, B. (2019). Wild pollinator activities negatively related to honey bee colony densities in urban context. *PLoS ONE* 14(9), 667725.

Roulston, T. and Malfi, R. (2012). Aggressive eviction of the eastern carpenter bee (*Xylocopa virginica* (Linnaeus)) from its nest by the giant resin bee (*Megachile sculpturalis* Smith). *J. Kans. Entomol. Soc.* 85(4), 387-388.

Ruiz, C., Suárez, D., Naranjo, M. and De la Rúa, P., (2020). First record of the carpenter bee Xylocopa pubescens (Hymenoptera, Apidae) in the Canary Islands confirmed by DNA barcoding. *J. Hymenopt. Res.* 80, 169-175.

Russo, L. (2016). Positive and negative impacts of non-native bee species around the world. *Insects* 7(4), 101-103.

Russo, L., de Keyzer, C. W., Harmon-Threatt, A. N., LeCroy, K. A. and MacIvor, J. S. (2021). The managed-to-invasive species continuum in social and solitary bees and impacts on native bee conservation. *Curr. Opin. Insect Sci.* 46, 43-49. https://doi.org/10.1016/j.cois.2021.01.001.

Rust, R. W., Cambon, G., Grossa, J. T. and Vaissière, B. E. (2004). Nesting biology and foraging ecology of the wood-boring bee Lithurgus Chrysurus (Hymenoptera: Megachilidae). *J. Kans. Entomol. Soc.* 77(3), 269-279.

Sáez, A., Morales, C. L., Garibaldi, L. A. and Aizen, M. A. (2017). Invasive bumble bees reduce nectar availability for honey bees by robbing raspberry flower buds. *Basic Appl. Ecol.* 19, 26-35.

Sáez, A., Morales, C. L., Ramos, L. Y. and Aizen, M. A. (2014). Extremely frequent bee visits increase pollen deposition but reduce drupelet set in raspberry. *J. Appl. Ecol.* 51(6), 1603-1612.

Sáez, A., Morales, J. M., Morales, C. L., Harder, L. D. and Aizen, M. A. (2018). The costs and benefits of pollinator dependence: empirically-based simulations predict raspberry fruit quality. *Ecol. Appl.* 28(5), 1215-1222.

Sanguinetti, A. and Singer, R. B. (2014). Invasive bees promote high reproductive success in Andean orchids. *Biol. Conserv.* 175, 10-20.

Schaffer, W. M., Zeh, D. W., Buchmann, S. L., Kleinhans, S., Valentine Schaffer, M., Antrim, J. (1983). Competition for nectar between introduced honeybees and native North American bees and ants. *Ecology* 64, 564-577.

Schmid-Hempel, R., Eckhardt, M., Goulson, D., Heinzmann, D., Lange, C., Plischuk, S., Escudero, L. R., Salathé, R., Scriven, J. J. and Schmid-Hempel, P. (2014). The invasion of southern South America by imported bumblebees and associated parasites. *J. Anim. Ecol.* 83(4), 823–837.

Seebens, H., Bacher, S., Blackburn, T. M., Capinha, C., Dawson, W., Dullinger, S., Genovesi, P., Hulme, P. E., van Kleunen, M., Kühn, I., Jeschke, J. M., Lenzner, B., Liebhold, A. M., Pattison, Z., Pergl, J., Pyšek, P., Winter, M. and Essl, F. (2020). Projecting the continental accumulation of alien species through to 2050. *Glob. Chang. Biol.* 27, 970–982.

Seebens, H., Blackburn, T. M., Dyer, E. E., Genovesi, P., Hulme, P. E., Jeschke, J. M., Pagad, S., Pyšek, P., Winter, M., Arianoutsou, M., Bacher, S., Blasius, B., Brundu, G., Capinha, C., Celesti-Grapow, L., Dawson, W., Dullinger, S., Fuentes, N., Jäger, H., Kartesz, J., Kenis, M., Kreft, H., Kühn, I., Lenzner, B., Liebhold, A., Mosena, A., Moser, D., Nishino, M., Pearman, D., Pergl, J., Rabitsch, W., Rojas-Sandoval, J., Roques, A., Rorke, S., Rossinelli, S., Roy, H. E., Scalera, R., Schindler, S., Štajerová, K., Tokarska-Guzik, B., Van Kleunen, M., Walker, K., Weigelt, P., Yamanaka, T. and Essl, F. (2017). No saturation in the accumulation of alien species worldwide. *Nat. Commun.* 8, 14435.

Severinghaus, L. L., Kurtak, B. H. and Eickwort, G. C. (1981). The reproductive Behavior of *Anthidium manicatum* (Hymenoptera: Megachilidae) and the Significance of Size for Territorial Males. *Behav. Ecol. Sociobiol.* 9(1), 51–58.

Singh, R., Levitt, A. L., Rajotte, E. G., Holmes, E. C., Ostiguy, N., Vanengelsdorp, D., Lipkin, W. I., Depamphilis, C. W., Toth, A. L. and Cox-Foster, D. L. (2010). RNA viruses in hymenopteran pollinators: evidence of inter-Taxa virus transmission via pollen and potential impact on non-Apis hymenopteran species. *PLoS ONE* 5(12), e14357.

Soper, J. and Beggs, J. (2013). Assessing the impact of an introduced bee, *Anthidium manicatum*, on pollinator communities in New Zealand. New Zeal. *J. Bot.* 51, 213–228.

Starks, P. T. and Reeve, H. K. (1999). Condition-based alternative reproductive tactics in the wool-carder bee, *Anthidium manicatum*. *Ethol. Ecol. Evol.* 11(1), 71–75.

Strange, J. P., Koch, J. B., Gonzalez, V. H., Nemelka, L. and Griswold, T. (2011). Global invasion by *Anthidium manicatum* (Linnaeus) (Hymenoptera: Megachilidae): assessing potential distribution in North America and beyond. *Biol. Invas.* 13(9), 2115–2133.

Sumner, S., Law, G. and Cini, A. (2018). Why we love bees and hate wasps. *Ecol. Entomol.* 43(6), 836–845.

Sutherland, W. J., Broad, S., Caine, J., Clout, M., Dicks, L. V., Doran, H., Entwistle, A. C., Fleishman, E., Gibbons, D. W., Keim, B., LeAnstey, B., Lickorish, F. A., Markillie, P., Monk, K. A., Mortimer, D., Ockendon, N., Pearce-Higgins, J. W., Peck, L. S., Pretty, J., Rockström, J., Spalding, M. D., Tonneijck, F. H., Wintle, B. C. and Wright, K. E. (2016). A horizon scan of global conservation issues for 2016. *Trends Ecol. Evol.* 31(1), 44–53.

Taggar, A. K., McGrath, E. and Despland, E. (2021). Competition between a native and introduced pollinator in unmanaged urban meadows. *Biol. Invas.* 23(6), 1697–1705.

Traveset, A. and Richardson, D. M. (2014). Mutualistic interactions and biological invasions. *Annu. Rev. Ecol. Evol. Syst.* 45(1), 89–113.

Valido, A., Rodríguez-Rodríguez, M. C. and Jordano, P. (2019). Honey bees disrupt the structure and functionality of plant-pollinator networks. *Sci. Rep.* 9(1), 4711.

Vanbergen, A. J., Espíndola, A. and Aizen, M. A. (2018). Risks to pollinators and pollination from invasive alien species. *Nat. Ecol. Evol.* 2, 16–25.

Vereecken, N. J. and Barbier, É. (2009). Premières données sur la présence de l'abeille asiatique *Megachile* (Callomegachile) *sculpturalis* SMITH (Hymenoptera, Megachilidae) en Europe. *Osmia* 3, 4–6.

Westrich, P. (2017). Faszination Wildbienen, Forschungsprojekte: *Megachile sculpturalis*. Available at: https://www.wildbienen.info/forschung/projekte_17.php (consulté le 01 juin 2021).

Westrich, P., Knapp, A. and Berney, I. (2015). *Megachile sculpturalis* Smith 1853 (Hymenoptera, Apidae), a new species for the bee fauna of Germany, now north of the Alps. *Eucera* 9, 3–10.

Willis Chan, D. S. and Raine, N. E. (2021). Hoary squash bees (*Eucera pruinosa*: hymenoptera: Apidae) provide abundant and reliable pollination services to Cucurbita Crops in Ontario (Canada). *Environ. Entomol.* 50(4), 968–981. https://doi.org/10.1093/ee/nvab045.

Yoon, H. J., Kim, S. Y., Lee, K. Y., Lee, S. B., Park, I. G. and Kim, I. (2009). Interspecific hybridization of the bumblebees *Bombus ignitus* and *B. terrestris*. *Int. J. Ind. Entomol.* 18, 41–48.

Yoon, H.-J., Park, I.-G., Lee, K.-Y., Kim, M.-A. and Jin, B.-R. (2011). Interspecific hybridization of the Korean native bumblebee *Bombus hypocrita sapporoensia* and the European bumblebee *B. terrestris*. *Int. J. Ind. Entomol.* 23(1), 167–174.

Zakardjian, M., Geslin, B., Mitran, V., Franquet, E. and Jourdan, H. (2020). Effects of urbanization on plant-pollinator interactions in the tropics: an experimental approach using exotic plants. *Insects* 11(11), 1–16.

Part 3

Promoting pollinators and pollination

Chapter 9

The role of habitat conservation and restoration in protecting pollinators in agricultural landscapes

Stephen Buchmann, University of Arizona, USA

1 Introduction

The world's biomes (temperate and tropical forests, deserts, tundra, boreal forests, grasslands and savannas), and the many local ecosystems and habitats they contain, are rapidly losing biodiversity. This has been termed the sixth mass extinction during the Holocene epoch (sometimes referred to as the Anthropocene) due to the results of planet-wide human activity. Ongoing extinction rates for various species are estimated to be about 100 times greater than background extinction rates during the past million years or so (Pimm et al., 2014).

Especially alarming, due to many keystone species involved and the ecosystem services they provide, are losses of invertebrates. These include pollinators of both wild plants and about 1400 species of crop plants grown within agricultural ecosystems (Roubik, 2018; Wagner et al., 2021). The majority of the world's approximately 350 000 flowering plant species (angiosperms) are engaged in largely mutualistic relationships with animal pollinators. Estimates suggest that 82% of angiosperm species are pollinated by insects (mostly bees), while 6% are pollinated by vertebrates (birds, bats and non-flying mammals) and about 12% are pollinated by wind dispersal (Winfree, 2008; Winfree et al., 2008, 2011; Ollerton et al., 2011; Aslan et al., 2013). A new

http://dx.doi.org/10.19103/AS.2022.0111.18

meta-analysis of 1174 angiosperm species indicates that, without pollinators, one-third of flowering plant species would produce no seeds, while about half would suffer a drastic and unsustainable 80% reduction in fertility (Rodger et al., 2021). Pollinating animals are thus essential for the maintenance of healthy flowering plant communities and the myriad species that depend upon plants for food, reproduction and shelter.

Factors such as climate change, deforestation, desertification, pollution, pesticide use, invasive species, habitat modification and loss have all contributed to the recent dramatic losses of insect, bird, bat and other mammalian pollinators (Potts et al., 2010). A review of over 70 studies of insect decline suggests widespread insect losses (Sánchez-Bayo and Wyckhuys, 2019). Since humans are dependent upon insect-pollinated crops for around one-third of their diets, this is a crucial issue for all of us. Many global pollinator initiatives have called for immediate action to better understand and address these alarming declines (Buchmann and Ascher, 2005; Harvey et al., 2020; Marselle et al., 2020).

2 Bees as pollinators

There are 20000 bee species worldwide, the majority of which are pollinators (Danforth et al., 2019). Eusocial bees live in populous colonies, each with a reproductive queen, a small number of drones (males) and many sterile female workers. Eusociality can be found in many species in the family Apidae (honeybees, bumblebees, carpenter bees and stingless bees) and the family Halictidae (sweat bees). Among eusocial bees, the western honeybee (*Apis mellifera*) is the most widely managed pollinator species used globally to improve crop yields (Whitfield et al., 2006; Klein et al., 2007). Commercialized bumblebees (e.g. *Bombus terrestris and B. impatiens*) are employed in the greenhouse production of fruits and vegetables such as tomatoes, strawberries and cucurbits (Velthuis and van Doorn, 2006). While wild honeybees nest above ground (e.g. in hollow tress), bumblebee species vary, with some nesting belowground, e.g. in ditch banks, hillsides or old mammal burrows such as mouse nests, while others nest above ground, e.g. in grasses or trees (Fussell and Corbet, 1992a). A UK survey found that bumblebee nest densities were highest in gardens (36 nests/ha), lower in countryside habitats (fence lines, hedgerows and woodland ecotones (edges)) (20–37 nests/ha) and lowest in woodlands and grassland areas (11–15 nests/ha) (Osborne et al., 2007).

However, in contrast to better-known eusocial species such as honeybees, over 90% of bee species are solitary. Among solitary bees, all females are reproductive, building their own nests and foraging for floral resources for their offspring without the help of other individuals (Danforth et al., 2019). Commonly observed solitary genera in the USA include leaf-cutter (*Megachile*), mason (*Osmia*), squash (*Eucera*), metallic green sweat (*Agapostemon*), *Lasioglossum* spp. and small carpenter bees (*Ceratina*), among others. Nest sites used by solitary bees include holes, burrows or tunnels in the soil, hollow plant stalks and gaps in wood or mortar structures such as walls (Danforth et al., 2019). Solitary bee species often provide pollination services in specialty crop production (e.g. for alfalfa seed, pumpkins, tomato, eggplant and blueberries). The leaf-cutter bee (*Megachile rotunda*) is used for pollination in alfalfa and clover seed production as well as blueberry production, while mason bees (*Osmia* spp.) are managed for a variety of orchard crops, including almonds and cherries (Delaplane and Mayer, 2000). The alkali bee (*Nomia melanderi*) is used to pollinate alfalfa for seed production (Cane, 2008a,b). Squash bees (*Eucera pruinosa*) are important pollinators of cucurbit crops, especially pumpkin, squash and watermelon (Willis Chan and Raine, 2021). Alkali and squash bees are unusual among solitary bee species in that they can form dense nesting aggregations with thousands of individual nests in close proximity to each other and to crops and some solitary bees locate their nests within cropping areas (e.g. squash bees and *Dieunomia triangulifera*) (Minckley et al., 1994; Willis Chan and Raine, 2021). In addition to pollination, solitary bees perform other ecosystem services such as soil bioturbation and water infiltration and nutrient cycling and also serve as food for species such as insect predators, spiders, birds, lizards and mammals (Christmann, 2020).

It is generally recognized that pollination and other ecosystem services are optimized by promoting a diverse range of bee species. For pollination-dependent crops, wild bees can complement services provided by managed pollinators, improving production and yield (Morandin and Winston, 2005; Greenleaf and Kremen, 2006; Winfree et al., 2007; Garibaldi et al., 2013; Mallinger et al., 2019; Nicholson et al., 2019). Promoting a diverse assemblage of wild pollinators serves the dual benefit of preserving biodiversity and increasing the efficiency of overall pollinator services (Burkle et al., 2017; Christmann, 2022).

However, the wide range of nesting habits and habitats used by bees can complicate attempts at habitat conservation or restoration, e.g. in reestablishing lost populations of wild ground-nesting solitary bees. There have been examples of moving colonies of bumblebees or highly social stingless bees near or within crops to promote pollination (Button and Elle, 2014). However, these are temporary measures designed to provide pollinators only while crops are blooming. These social bee domiciles are removed following the bloom of the crop and are not intended to conserve or restore populations of bees that formerly lived in the area. The example and case history of the USA alkali bee industry are examined as one success story for the active/managed reintroduction of a native bee to its former rangelands, although as a pollinator of an introduced highly valuable crop, alfalfa, which originated in the Mediterranean basin.

3 Challenges in habitat restoration

Attempts at habitat restoration must consider the range of threats facing bee species. Habitat loss due to urban development or agricultural intensification has frequently been highlighted as a leading cause of bee species decline (Wilcove et al., 1998; Venter et al., 2006). One meta-analysis found that fragmentation and outright loss of habitat negatively affected both species richness and the overall abundance of wild bees (Winfree et al., 2009). One relatively neglected aspect of habitat loss is the impact of roads and highways (Bennett, 1991). By 2019, the highway network in the USA had a total length of about 4.2 million miles. Such road networks kill millions of insects, including bees, every year (Muñoz et al., 2014; Moller, 2019). Roads have other negative effects such as pollution from traffic as well as acting as barriers to the movement of bees and other species (Forman, 2003).

However, it is important to note that not all habitat changes are detrimental. It has been shown that native solitary bee populations can survive in gardens, parks and other urban settings (Hall et al., 2017). In Tucson (in the US state of Arizona), native bee visitors were studied at urban sites colonized by the creosote bush (*Larrea tridentata*). Perennial creosote bushes bloom during both the spring and summer and are a prolific floral host for about 125 species of native bees, occurring as habitat islands as small as one hectare. Even the smallest patches of creosote bushes supported healthy populations of native ground-nesting desert bees (Cane et al., 2005, 2006).

Research suggests that wild bee density can be affected by tillage, with farms practicing no-till techniques having significantly higher populations of ground-nesting bees (Schuler et al., 2005). It has also been suggested that agricultural activities such as tilling and grazing may have a limited impact on some ground-nesting bee species (Winfree et al., 2008). Tilling, for example, may remove stems used by cavity-nesting bees or non-crop weeds or native plants that provide pollen and nectar sources. However, while tillage may reduce belowground brood cells for many bees, deeper-nesting species may not be affected and some species (i.e. squash bees) may prefer to nest in tilled soil on farms (Ullmann et al., 2016).

Invasive species also impact the availability of nesting sites and flowering plants for pollen and nectar. Weedy plants can 'smother' native and agricultural habitats with dense mixed or unifloral stands of non-native flowering plants, obstructing the ability of native vegetation to flower and thrive. Examples from the USA include yellow star thistle (*Centaurea solstitialis*) in southern and central California (also considered a prime honey source by beekeepers), various knapweeds including *C. stoebe* and leafy spurge (*Euphorbia esula*). Invasive grasses include downy brome or cheatgrass (*Bromus tectorum*) and foxtail buffalo grass (*Cenchrus ciliaris*), the latter presenting an ever-increasing problem in Arizona, USA and Mexico (Di Tomaso, 2000; Lyons et al., 2017). Invasive grasses such as foxtail buffalo grass (which forms dense mats) can also increase the risk of wildfires in arid regions such as the Sonoran Desert in Arizona (Wilder et al., 2021). Effects on ground-nesting bees of wildfires are mixed: bee brood cells less than 5 cm deep are at high mortality risk from fires, those from 5 cm to 10 cm deep are at low risk and those bees nesting below 10 cm are essentially at no risk from fires passing overhead (Cane and Neff, 2011).

There are about 23 species of alien bees that have been accidentally or purposefully introduced into the USA (Ascher, 2001; Cane, 2003; Sheffield et al., 2013). Introduced bees in the USA include species of *Andrena*, *Anthidium*, *Chelostoma*, *Hoplitis*, *Hylaeus* and especially *Megachile (*leaf-cutter bees). Most accidentally introduced bees in the USA are stem-nesting bees (*Hylaeinae*, *Megachilidae* and *Xylocopinae*), probably introduced via imported plant materials. Although introduced bees only make up about 0.5% of the North American bee fauna (Ascher, 2001), they can cause significant harm. If they become established, introduced non-native bees can compete with native bees (Cane, 2003). An example is the introduced giant resin bee (*Megachile sculpturalis*) which competes with the eastern carpenter bee (*Xylocopa virginica*) in the Eastern United States for nesting sites and floral resources (Laport and Minckley, 2012). Honeybees may also compete with native bees and other pollinators for limited floral resources and nesting sites (Goulson, 2003; Paini, 2004). Invasive species have also been found to cause mating disruption in some anthidiine solitary bees in the Eastern United States (Stout and Morales, 2009). Additionally, there are native cuckoo (cleptoparasitic) bees (genera including *Nomada* and *Triepeolus*) that lay their eggs in the nests of other bees, with their larvae then consuming the host eggs or larvae and the food meant for the host larvae (Danforth et al., 2019). Other common parasites of many ground-nesting bees are rhipiphorid beetles, shiny black blister beetles (Meloidae) and velvet ants (wingless female wasps in the family *Mutillidae*).

Both native and introduced parasitic organisms create further problems for the health and well-being of native bees. Bees are routinely attacked by at least 124 species of entomophagous fungi (Batra, 1984; Omuse et al., 2021), at least four kinds of nematodes (Quinn, pers. comm.), many bacteria and viruses, at least three species of parasitic mites on honeybees and many more on solitary bees (Sammatraro et al., 2000). Common diseases of social and some solitary bees include chalkbrood (*Ascosphaera apis*), American foulbrood (*Paenibacillus larvae*) and sacbrood virus. Many of the disease organisms associated with solitary ground-nesting bees have not yet been identified. A particular threat is pathogen spillover. Honeybees and commercialized bumblebees can harbor microorganisms and diseases, which then may infect other wild bees (Colla and Packer, 2008; Colla et al., 2005). This occurs when microbial pathogens (especially *Nosema bombi* and *Crithidia*) infecting honey and commercialized bumblebees are deposited on flowers by the bees (Cordes et al., 2011; Murray et al., 2013). These pathogens can easily infect female and male solitary bees as well as wild bumblebees visiting these blooms (Colla and Packer, 2008; Otterstatter and Thomson, 2008; Cameron et al., 2011; Szabo et al., 2012). This infection route has been linked to the demise of several US bumblebee species (e.g. *Bombus affinis*, *B. terricola*, *B. pennsylvanicus* and possibly *B. occidentalis* and *B. franklini*) (Szabo et al., 2012). It is also important to note that there are many bacteria, archaea and yeasts that inhabit the guts of bee adults and larvae (Danforth et al., 2019). Many now appear to be mutualists which promote digestion and combat coinhabiting pathogenic microbes, especially in solitary bees (Dharampal et al., 2019).

Changes in global temperature and precipitation are likely to have drastic effects on bees, their metamorphosis, sociality, emergence and interactions with other bees and other species (Radmacher and Strohm, 2010; Davison and Field, 2018; Bartomeus et al., 2011; Straka et al., 2014). Most studies examining the effects of climate change on bees have been done in trap nest blocks or laboratory colonies with few direct measurements in nature for ground-nesting bees (Forrest and Chisholm, 2017; Vinchesi et al., 2013). Many researchers have suggested that the early arrival of spring, which affects the blooming phenology of plants, is already causing a mismatch between when flowers bloom and when bees, especially specialist bees, emerge (Forrest and Chisholm, 2017). Increases in air temperature and changes in winter conditions will affect soil temperatures and the flowering of plants that provide bee food and nesting materials (Anderson and Harmon-Threat, 2016). A recent study of 66 bumblebee (Bombus) species across the UK and North America, which compared the period 1901–1974 with the period 2000–2014, concluded that bumblebee species richness declined in areas particularly affected by climate change (Soroye et al., 2020). They suggest that climate change has been the cause of greater declines of European bumblebee species than was previously assumed.

However, there is evidence that certain ground-nesting bees may be resilient to some effects of climate change. Some bee species have been found to be resilient to hurricanes (Cane, 1997) and may survive complete inundation of nesting sites (Stephen, 1966; Visscher et al., 1994), fires (Cane and Neff, 2011) and drought conditions (Minckley et al., 2013). To date, extreme and prolonged cold is one factor that has received limited study. Warming environments in northern regions may encourage range expansions and concomitant increasing species richness (Deutsch et al., 2008; Sirois-Delisle and Kerr, 2018). It has been found that ground-nesting bees typically nest at depths of 9 cm to 23 cm (Cane and Neff, 2011; Harmon-Threatt, 2020). One way bees have adapted to adverse climatic conditions (e.g. drought) is to remain underground and in diapause, in one case for up to 10 years (Torchio, 1976). Some ground-nesting bee species are able to tolerate high temperatures (40°C) at the depths at which they nest (Visscher and Danforth, 1993).

4 Assessing the effectiveness of habitat restoration

There are a range of approaches to foraging habitat restoration that may benefit wild bee populations. These include the use of burning and grazing to improve floral diversity in grasslands, planting of hedgerows, creating field margins as well as the use of mowing, weeding of invasive species and reseeding to create more ecologically diverse and rich landscapes. Some studies have highlighted how quickly bee populations can recolonize restored sites, though factors such as proximity to existing wild bee communities are important to recolonization success (Exeler et al., 2009; Richards et al., 2011; Kutby, 2013).

A study by Winfree (2010) found that agri-environmental restoration projects in the European Union using such techniques significantly benefitted bee communities in both species richness and abundance compared with conventionally cultivated sites. However, the author also highlighted the range of challenges in delivering successful outcomes for wild bee populations. These included gaps in our understanding of the complex range of wild bee needs, a corresponding focus of many projects on restoring floral resources (often with limited information on what floral resources might be appropriate) with less focus on nesting sites, challenges in establishing appropriate benchmarks to measure restoration outcomes, issues in accurately measuring changes in bee numbers and behaviors in response to changes in management, limited monitoring of the long-term impacts of some projects as well as challenges in isolating what factors in a particular restoration project had the greatest impact. The challenges of the effective study of wild bee populations and of assessing the effectiveness of restoration projects have been highlighted by Colla et al. (2012) and Menz et al. (2011).

A more recent meta-analysis has examined the impact of habitat enhancement in agricultural settings for wild bee populations (Tonietto and Larkin, 2018). The authors analyzed 28 studies across 11 habitat types and 7 restoration techniques. Overall, they also found strong evidence that ecological restoration projects significantly advanced wild bee conservation in improving native bee species richness and abundance across habitat types. However, among other challenges, they also highlighted a continued focus of restoration projects on enhancing floral resources to improve overall biodiversity, with few studies considering wild bees specifically or issues such as habitat (nesting site) requirements for ground- or cavity-nesting bees.

Similar results have been obtained when looking at the restoration of particular habitat types such as prairie ecosystems. A study of restored tallgrass prairie sites in Ontario (Canada) found they supported twice as many bee species and three times more bees than unrestored prairies, an improvement the authors associated with a wider diversity of floral resources (Paterson et al., 2019). Another study of eight restored prairie sites in Minnesota (USA) also found they provided sufficient pollen resources to maintain wild bee populations despite the continued use of intensive cultivation in adjacent fields (Ritchie et al., 2020). A study of restored grassland-wetland complexes in North American prairies also found that bee diversity increased rapidly following restoration compared to reference sites, with bee numbers and diversity of species linked to floral species composition (Purvis et al., 2020). However, a recent study of 14 restored tallgrass prairie sites across North America found that wild bees responded best to soil disturbance by grazing bison and the overall amount of prairie and forest across the landscape rather than to the greater availability of local floral resources in restored sites (Griffin et al., 2020). They also found a high sensitivity of bees to agricultural land use (adjacent row-crop agriculture).

One approach to ecological restoration in North America and particularly in Europe has been to reverse the destruction of hedgerows (a centuries-old feature in many European rural landscapes) resulting from agricultural intensification, a loss which has been linked to pollinator decline (Park, 1988; Osborne et al., 1991). Hedgerows provide safe resting and feeding areas for numerous species of insects and vertebrate wildlife, including wild bees (Kremen et al., 2018; Bentrup et al., 2019). Perennial plants in the UK hedgerows that are important for bees include hawthorn (*Crataegus*), brambles (*Rubus*) and lime (*Tilia*). As a result, there has been a range of initiatives to restore or replant hedgerows to create more biodiverse agricultural landscapes (Hopwood, 2013; Garratt et al., 2017; Kremen and Merenlender, 2018).

Recently, land managers, state and federal agencies in the United States have expressed an interest in monitoring the effects of electrical transmission line rights of way on honey bee and unmanaged native bees and other pollinators. A review of these concerns and studies is found here (see Wojcik and Buchmann, 2012).

Analyzing the results of over 300 surveys of 15 Californian sites across an 8-year period, M'Gonigle et al. (2015) concluded that hedgerow restoration promoted enhanced colonization and persistence of wild bee populations compared to unrestored field edges. Another study of Californian hedgerows by Morandin and Kremen (2013) found that they played a significant role in promoting less common native bee species, confirmed by Ponisio et al. (2016), who highlighted the role of hedgerows in promoting bee species diversity. Research by Kremen et al. (2018) found that pollinator numbers and species diversity responded most to the growth in floral diversity as hedgerows matured and that the surrounding natural habitat also played a key role in promoting pollinators. However, a recent study of field margins and hedgerows in Canada by Clausen et al. (2022) concluded that, overall, grass margins contributed more to wild bee abundance and diversity than remnant or planted hedgerows, findings supported by other European studies (e.g. Pywell at al., 2005, 2012; Haaland et al., 2011). A study of riparian forest restoration projects by Williams (2010) also found that they promoted diverse and abundant wild bee communities, though community structure differed compared to remnant sites.

One potential challenge that potentially limits the effectiveness of restoration projects is the selection of the right floral and other resources for particular wild bee species. Bees need flowering plants to survive. Even the approximately 10% of bees that are cleptoparasites ultimately depend upon the nutrients ingested by the hosts on which they depend. Angiosperms provide pollen, nectar, floral oils and other nutritive substances. Certain bee species (leaf-cutter bees, orchid bees and stingless bees) also require leaves, saps, resins and volatiles from other plants as nest-building materials or precursors for mating attractant blends. Long-lived bee colonies or individuals need a mixed array of flowering plants on a daily and seasonal basis. This is especially true for honeybees (*Apis*), bumblebees (*Bombus*), stingless bees (*Melipona*, *Trigona* and other genera) and carpenter bees (*Xylocopa*) (Lawson et al., 2021). Consuming a mixed pollen diet offsets unfavorable properties of any one pollen source including low protein content or outright toxicity (Eckhart et al., 2014). Restoration ecologists, conservationists and gardeners agree that using a diversity of different annuals and perennials with overlapping bloom times during the season is beneficial to providing a regular and nutritious food supply (Majewska and Altizer, 2020). Many solitary bee species are specialist foragers, preferring to collect pollen from a narrow group of plants (Cane and Sipes, 2006; Wright, 2018; Fowler and Droege, 2020; Fowler, 2020a,b). Restoring habitat for these species requires planting specific plants that meet the needs of their pollen specialization. For example, the squash bees only collect pollen from wild or cultivated plants in the genus *Cucurbita*, and between 25% and 35% of bee species in the USA are pollen specialists, many on species of goldenrod (Asteraceae) (Fowler and Droege, 2020; Fowler, 2020a,b).

Selecting the right plants for a garden or large-scale restorative planting depends upon a number of factors (Menz et al., 2011). Firstly, in garden and urban landscapes, it is important to select and use native plants that are adapted to local edaphic and climatic conditions so that less watering, fertilization and maintenance are required. Secondly, it is important to choose heirloom varieties and non-hybridized flowers. This is because the hybrids of many flowers have been selected by plant breeders to appeal to human visual sensibilities or to reduce messiness in cut flowers (i.e. pollenless sunflowers) rather than pollinator dietary and other requirements. Pollinator attractiveness tests for garden flower varieties have been conducted in the UK and elsewhere (Corbet et al., 1984).

The importance of the correct selection can be seen in the activities of the US Bureau of Land Management (BLM) whose responsibilities include the planting of seed mixes to control erosion and restore floristic diversity to rangelands after over-grazing or fires. One such area is the American sagebrush steppe which stretches across the Great Basin and adjoining region of the US Intermountain West, covering much of Nevada, Utah and Oregon, and which is at increased risk of wildfires due to climate change. The bulk seed purchases made by the BLM have included non-native Eurasian species together with species such as western yarrow (*Achillea millefolium*) and Lewis' flax (*Linum lewisii*), which do not provide ideal resources for wild bees. Recognizing this problem, the White House issued a National Pollinator Health Strategy in 2015 which required federal land management agencies to adapt programs to target the protection of managed honey and native bee populations.

In response to the new strategy, the BLM conducted walking scans of native bees, and their densities, across 219 locations together with surveying 17 wildflower species within the Intermountain West to determine the most important bee plants that should be used for reseeding (Cane and Love, 2016). Over 35 000 plants were sampled, and 3500 bees were collected in this study. Few bees were observed on the traditionally used yarrow and flax. Instead, most bees were observed on and collected from plants including *Phacelia*, *Lomatium*, *Sphaeralcea*, *Balsamorhiza* and *Dalea* (Cane and Love, 2016). These findings are being used to redesign seed mixes to better target and benefit wild bee populations.

5 Case studies

*5.1 Case study: managing alkali bees (***Nomia melanderi***)*

Fewer than a dozen species of bees are managed commercially as crop pollinators. These are honeybees (*A. mellifera* and *A. cerana*) typically for field crops and bumblebees (*Bombus* spp.) mainly for the pollination of greenhouse crops like tomatoes and strawberries. The alfalfa leaf-cutter bee (*M. rotundata*) is used for pollination of alfalfa (lucerne) in the USA and Canada. The only native US solitary bee that has been managed commercially for pollination services is the alkali bee (*Nomia melanderi*) (Cane, 2021). The alkali bee is native to the Pacific Northwest, especially within the intermountain basin regions of Washington and Oregon but extending into California's Central Valley. The species is a gregarious ground-nesting solitary bee. Females nest in close proximity (hundreds per square meter). Each female digs and provisions her own nest.

In the 1940s, after noticing this species collecting pollen from alfalfa blossom more actively than other bees, some farmers began experimenting with locating bee-nesting beds in alkaline flats and then planting alfalfa crops adjacent to them for pollination (Stephen, 2003). Farmers have found that while yields using honeybees for pollination achieved seed yields of 200 kg/ha, use of alkali bees has increased seed yields to 1300 kg/ha or more (Olmstead and Wooten, 1987; Cane, 2021). In the Walla Walla region of Washington state, at least 17 million alkali bees are managed in dozens of artificially created nesting sites, the oldest established 60 years ago and the largest home to some 5 million bees (Cane, 2008a).

The first commercial alkali nesting beds were long narrow beds dug about 1 m deep and lined with plastic. A layer of gravel was placed over the plastic, and then about a meter of silty-loam soil added. Plastic standpipes were inserted at intervals into the soil bed. A surface treatment of rock salt was added to cap the bee-nesting site. The salt draws water up by capillary action in a way that mimics the wild nesting sites of this halictid bee. Today subsurface drip irrigation tubing performs this role. New bee beds are 'inoculated' with brood cells (containing dormant pupae) from other artificial beds or natural nesting sites. Backhoes with metal attachments punch into the soil and remove cubes of soil containing brood cells. These are placed in cardboard inserts then carefully set into waiting holes. An operator can remove and transplant up to 1000 bee cubes (Cane, 2021). Some growers use spiked rollers across beds to create starter nest holes which are adopted by nesting females (Cane, 2008b).

Surprisingly, given their dense nesting aggregations, alkali bees are not subject to parisitization by cleptoparasitic bees. Instead, they are predated by bee flies (Bombyliidae) along with meloid (blister) beetles and mutillid wasps. The greatest mortality can occur from spoiled pollen ball provisions of nectar and pollen upon which eggs are laid, with up to 32% of nest cells affected in this way (Batra et al., 1973). Forty-six species of fungi (including yeasts) have been isolated from alkali bee brood cells, though it is still not clear which are pathogenic, saprophytic or beneficial mutualists (Danforth et al., 2019).

Alkali bees are an unusual case of a 'domesticated' native solitary ground-nesting bee used for agricultural pollination. No other ground-nesting bees anywhere in the world have been managed due to the difficulty of managing underground brood cells. Although alkali bees are not endangered and provide commercial crop pollination services, this example shows the potential for supporting other solitary, ground-nesting bee species once their nesting biology is better understood.

5.2 Case study: promoting stem- or cavity-nesting bees

About 10% of the nearly 4000 species of native bees in the USA are stem- or cavity-nesting bees. Commonly observed genera include leaf-cutter (*Megachile*), mason (*Osmia*) and carpenter bees (*Xylocopa* and *Ceratina*).

Many practitioners have also advocated for installing drilled board or straw-containing wooden boxes to attract and house nesting leaf-cutter and mason bees (Frankie et al., 1998; MacIvor and Packer, 2015).

These bees can be supported by providing nesting wood substrates such as leaving dead trees (snags) or uncut dead branches as potential nest sites. Dead trees and branches provide habitats for beetles, bees, birds and other wildlife. When the larvae of many beetle species chew through dead wood, they create galleries leading to the surface. Once beetles pupate and emerge, these empty tunnels provide nest sites for many megachilid bees since stem-nesting species cannot excavate burrows of their own. These beetle burrow nests may be reused over many years by multiple generations of a diverse assemblage of bees in this beetle-bee ecosystem. Evidence from leaving dead specimens of Indian White Teak (*Gmelina arborea*) after teak logging in Costa Rica suggests their potential as a habitat for bees (Galbraith et al., 2015). Research on cavity-nesting bee and wasp species in southern Amazonia looked at the effects of different types of forest regeneration (pastures, teak reforestation, fig and mixed native species, naturally regenerated secondary forest) compared to sites with intact primary forest (Araujo et al., 2020), concluding that the most 'complex' vegetation structure supported the most diverse bee and wasp communities.

Holes can also be drilled into dead trees, upright wooden fence posts, wooden outbuildings or small towers mimicking dead trees. These holes should be spaced about 10–20 cm apart and drilled to a depth of 5–10 cm if possible. To encourage the greatest diversity of twig-nesting bee genera and species, hole diameters should include holes from 2 mm up to 8 mm in diameter. Paper straw liners can also be added. However, a recent work by MacIvor and Packer (2015) in Ontario, Canada, has shown that artificial nests such as these may attract more wasps and exotic bee species than native bee species.

5.3 Case study: providing underground-nesting boxes for bumblebees

Globally, there are around 250 species of bumblebees (*Bombus*), with about a dozen species that have been reared within insectaries and used for the pollination of agricultural crops, especially tomatoes, peppers and cucumbers (Goulson, 2003, 2010; Kevan et al., 1990). They are especially valued for their ability to vibrate flowers with pored anthers, also known as buzz pollination (Buchmann, 1983: De Luca and Vallejo-Marin, 2013). Agricultural crops requiring buzz pollination include tomatoes, eggplant, peppers, blueberries, cranberries and kiwi fruit. Honeybees are sometimes used to pollinate blueberry or cranberry crops but are generally less effective than bumblebees (Buchmann, 1983; Cane and Payne, 1988; DeLuca and Vallejo-Marin, 2013).

Bumblebees typically nest underground, especially in ditch banks or hillsides, in old mammal burrows, especially mouse nests (Fussell and Corbet, 1992a,b). In areas where bee colonies are lacking, it is possible to encourage wild bumblebee queens to make their nests inside wood or plastic boxes provided for them at the edges of fields situated in ditch banks. Bumblebee boxes have been developed and tested in several countries (Lye et al., 2011). An early study developed polystyrene nesting boxes for four species of introduced bumblebees in New Zealand for crop pollination (Donovan and Wier, 1978). Over a period of 3 years, for 340 nest boxes placed, 84 (24.7%) were occupied by all four introduced bumblebee species (*B. hortorum*, *B. uderatus*, *B. subterraneus* and *B. terrestris*). Other initiatives have experienced low levels of occupancy. Since bumblebee species have declined drastically across the UK (Goulson et al., 2008; 2015), partly due to habitat loss from intensive farming, artificial nest boxes have been used to help reverse the decline. However, one study of underground *Bombus* boxes in garden settings found occupancy as low as 1.5% (Fussell and Corbet, 1992a). Another study found that only 23 of 736 domiciles deployed (3.1%) were occupied by queens (Lye et al., 2011).

6 Conclusion

Bee species, especially ground-nesting ones, can vary in their resilience to human-caused and other disturbances. Deeper nesting species, for example, are relatively immune to the effects of wildfires, insulated by the soil above their brood cells, whereas cavity-nesting bees living at or with a few meters of the ground are at greater risk. Habitat loss and degradation are the most severe threat to native bee populations, followed by invasive plants and bee species. Parasites, diseases and predators take their toll but are generally at sufficiently low levels to allow local bee populations to be sustained. A major exception is pathogen spillover from commercial to wild bee species affecting five at-risk species of bumblebees within the USA and a similar suite of endangered bumblebee species in Great Britain and other European countries. Agrochemicals including insecticides and herbicides are an omnipresent threat to all bees. Systemic insecticides are especially dangerous because they are secreted in nectar used by bees. Bees that have 'wax furniture' (honeybees, stingless bees and bumblebees) are at risk of falling prey to something akin to sick building syndrome since these persistent toxic chemicals become incorporated into these waxes and can persist for years. Attention needs to be paid to invasive bees like the Giant Resin Bee, which competes for floral resources with many native bees and for nesting holes with native carpenter bees. Urbanization effects seem to have helped certain bee species (e.g. *Megachile*) by providing abundant holes and cracks in human structures along with floral resources available the year-round. Habitat islands within cities have, for example, shown to be refugia for Sonoran Desert bee species that visit patches of creosote bush (Larrea) within the city limits of Tucson (in the state of Arizona, USA).

Habitat restoration for plant communities is at a more advanced stage than for the restoration of wild bee communities in all habitats. Some success has been achieved, but long-term studies and careful bee monitoring are needed to properly gauge the permanent effects of restoration programs for wild bee populations. Importantly, land managers and restoration ecologists should become familiar with the entirely of bees' needs, not just flowers. In the case of solitary bee species, they should make attempts to provide bare ground, vertical banks, water and mud sources along with plants that provide stems for cavity nests and resins and leaf pieces as nest building materials.

7 Where to look for further information

Students, professors, land managers and conservation biologists are encouraged to search for additional and more modern references in the primary scientific literature. One excellent way of doing this is simply to do keyword searches in Google Scholar (https://scholar.google.com) or JSTOR (a digital library founded in New York City in 1995). Advanced key word searches can be easily made in its online database (www.jstor.org/action/showAdvancedSearch). Looking in the literature citations of recent review papers (e.g. in Annual Reviews of Entomology or Ecology and Systematics: in Annual Reviews: https://www.annualreviews.org) is another excellent way to mine the historical and modern literature to find more articles on topics of your choice.

8 References

Anderson, N. and Harmon-Threatt, A. 2016. 7:The effects of seed Mix diversity on soil conditions and nesting of bees in prairie restorations. *North American Prairie Conference Proceedings* (vol. 17). Available at: https://ir.library.illinoisstate.edu/napc/17.

Araujo, G. J., Stork-Tonon, D. and Izzo, T. J. 2020. Temporal stability of cavity-nesting bee and wasp communities in different types of reforestation in southeastern Amazonia. *Restoration Ecology* 28(6):1528–1540. https://doi.org/10.1111/rec.13250.

Ascher, J. S. 2001. Hylaeus Hyalinatus Smith, a European bee new to North America, with notes on other adventive bees (Hymenoptera: Apoidea). *An*. Paper 49. Available at: https://digitalcommons.usu.edu/bee_lab_an/49.

Aslan, C. E., Zavaleta, E. S., Tershy, B. and Croll, D. 2013. Mutualism disruption threatens global plant biodiversity: a systematic review. *PLoS ONE* 8(6):e66993. https://doi.org/10.1371/journal.pone.0066993.

Bartomeus, I., Ascher, J. S., Wagner, D., Danforth, B. N., Colla, S., Kornbluth, S. and Winfree, R. 2011. Climate-associated phenological advances in bee pollinators and bee-pollinated plants. *Proceedings of the National Academy of Sciences of the United States of America* 108(51):20645–20649.

Batra, L. R., Batra, S. W. T. and Bohart, G. E. 1973. The mycoflora of domesticated and wild bees (Apoidea). *Mycopathologia et Mycologia Applicata* 49(1):13–44.

Batra, S. W. T. 1984. Solitary bees. *Scientific American* 250(2):120–127.

Bennett, A. F. 1991. Roads, roadsides and wildlife conservation: a review. In: *Conservation 2: The Role of Corridors*, Saunders, D. A. and Hobbs, R. J. (Eds). Nature. Surrey Beatty & Sons, Chipping Norton, pp. 99–118.

Bentrup, G., Hopwood, J., Adamson, N. L. and Vaughan, M. 2019.Temperate agroforestry systems and insect pollinators: a review. *Forests* 10(11):981. https://doi.org/10.3390/f10110981.

Buchmann, S. and Ascher, J. S. 2005. The plight of pollinating bees. *Bee World* 86(3):71–74, Https://doi.org/10.1080/0005772X.2005.11417316.

Buchmann, S. L. 1983. Buzz pollination in angiosperms. In: *Handbook of Experimental Pollination Biology*, Jones, C. E. and Little, R. J. (Eds). Van Nostrand Reinhold Company, New York, NY, pp. 73–113.

Burkle, L. A., Delphia, C. M. and O'Neill, K. M. 2017. A dual role for farmlands: food security and pollinator conservation. *Journal of Ecology* 105(4):890–899.

Button, L. and Elle, E. 2014. Wild bumble bees reduce pollination deficits in a crop mostly visited by managed honey bees. *Agriculture, Ecosystems and Environment* 197:255–263. https://doi.org/10.1016/j.agee.2014.08.004.

Cameron, S. A., Lozier, J. D., Strange, J. P., Koch, J. B., Cordes, N., Solter, L. F. and Griswold, T. L. 2011. Patterns of widespread decline in North American bumble bees. *Proceedings of the National Academy of Sciences of the United States of America* 108(2):662–667. https://doi.org/10.1073/pnas.1014743108.

Cane, J. H. 1997. Violent weather and bees; populations of the barrier island endemic, *Hesperapis oraria* (Hymenoptera: Melittidae) survive a Category 3 hurricane. *Journal of the Kansas Entomological Society* 70(1):73–75.

Cane, J. H. 2003. Exotic non-social bees (Hymenoptera: Apoidea) in North America: ecological implications. In: *For Nonnative Crops, Whence Pollinator for the Future?*, Strickler, K. and Cane, J. H. (Eds), Thomas Say Publications in Entomology, Lanham, MD, pp. 113–126.

Cane, J. H. 2008a. A native ground-nesting bee (*Nomia melanderi*) sustainably managed to pollinate alfalfa across an intensively agricultural landscape. *Apidologie* 39(3):315–323. Available at: https://doi-org.subzero.lib.uoguelph.ca/10.1051/apido:2008013.

Cane, J. H. 2008b. Pollinating bees crucial to farming wildflower seed for U.S. Habitat restoration. In: *Bee Pollination in Agricultural Ecosystems*, James, R. R. and Pitts-Singer, T. (Eds). Oxford University Press, Oxford, pp. 49–64.

Cane, J. H. 2021. Global warming, advancing bloom and evidence for pollinator plasticity from long-term bee emergence monitoring. *Insects* 12(5):457.

Cane, J. H. and Love, B. 2016. Floral guilds of bees in sagebrush steppe: comparing bee usage of wildflowers available for postfire Restoration. *Natural Areas Journal* 36(4):377–391.

Cane, J. H., Minckley, R., Kervin, L. and Roulston, T. 2005. Temporally persistent patterns of incidence and abundance in a pollinator guild at annual and decadal scales: the bees of *Larrea tridentata*. *Biological Journal of the Linnean Society* 85(3):319–329. https://doi.org/10.1111/j.1095-8312.2005.00502.x.

Cane, J. H., Minckley, R. L., Kervin, L. J., Roulston, T. H. and Williams, N. M. 2006. Complex responses within a desert bee guild (Hymenoptera: Apiformes) to urban habitat fragmentation. *Ecological Applications* 16(2):632–644.

Cane, J. H. and Neff, J. L. 2011. Predicted fates of ground-nesting bees in soil heated by wildfire: thermal tolerances of life stages and a survey of nesting depths. *Biological Conservation* 144(11):2631–2636. https://doi.org/10.1016/j.biocon.2011.07.019.

Cane, J. H. and Payne, J. A. 1988. Foraging ecology of the bee Habropoda laboriosa (Hymenoptera: Anthophoridae), an Oligolege of Blueberries (Ericaceae: Vaccinium) in the Southeastern United States. *Annals of the Entomological Society of America* 81(3):419–427.

Cane, J. H. and Sipes, S. 2006. Floral specialization by bees: analytical methodologies and a revised lexicon for oligolecty. In: *Plant-Pollinator Interactions: From Specialization to Generalization*, Waser, N. and Ollerton, J. (Eds). University of Chicago Press, Chicago,Illinois, pp. 99–122.

Christmann, G. B. 2020. Introduction: struggling with innovations. Social innovations and conflicts in urban development and planning. *European Planning Studies* 28(3):423–433.

Christmann, S. 2022. Regard and protect ground-nesting pollinators as part of soil biodiversity. *Ecological Applications* 32(3):e2564. https://doi.org/10.1002/eap.2564.

Clausen, M. A., Elle, E. and Smukler, S. M. 2022. Evaluating hedgerows for wild bee conservation in intensively managed agricultural landscapes. *Agriculture, Ecosystems and Environment* 326:107814.

Colla, S. R., Ascher, J. S., Arduser, M., Cane, J., Deyrup, M., Droege, S., Givvs, J., Griswold, T., Hall, H. G., Henne, C., Neff, J., Jean, R. P., Rightmyer, M. G., Sheffield, C., Veit, M. and Wolf, A. 2012. Documenting Persistence of most eastern North American bee species (Hymenoptera: Apoidea, Anthophila) to 1990–2009. *Journal of the Kansas Entomological Society* 85(1):14–22. http://doi.org/10.2317/JKES110726.1.

Colla, S. R., Otterstatter, M. C., Gegear, R. J. and Thomson, J. D. 2005. Plight of the bumble bee: pathogen spillover from commercial to wild populations. *Biological Conservation* 129(4):461–467. https://doi.org/10.1016/j.biocon.2005.11.013.

Colla, S. R. and Packer, L. 2008. Evidence for decline in eastern North American bumblebees (Hymenoptera: Apidae), with special focus on Bombus affinis Cresson. *Biodiversity and Conservation* 17(6):1379–1391. https://doi.org/10.1007/s10531-008-9340-5.

Corbet, S. A., Kerslake, C. J. C., Brown, D. and Morland, N. E. 1984. Can bees select nectar-rich flowers in a patch? *Journal of Apicultural Research* 23(4):234–242, https://doi.org/10.1080/00218839.1984.11100638.

Cordes, N., Huang, W.-F., Strange, J. P., Cameron, S. A., Griswold, T. L., Lozier, J. D. and Solter, L. F. 2011. Interspecific geographic distribution and variation of the pathogens Nosema bombi and Crithidia species in United States bumble bee populations. *Journal of Invertebrate Pathology* 109(2):209–216. https://doi.org/10.1016/j.jip.2011.11.005.

Danforth, B. N., Minckley, R. L. and Neff, J. L. 2019. *The Solitary Bees: Biology, Evolution, Conservation*. Princeton University Press, Princeton and Oxford, 472 pp.

Davison, P. J. and Field, J. 2018. Environmental barriers to sociality in an obligate eusocial sweat bee. *Insectes Sociaux* 65(4):549–559. https://doi.org/10.1007/s00040-018-0642-7.

Delaplane, K. S. and Mayer, D. F. 2000. *Crop Pollination by Bees*. CABI Publishing.

DeLuca, P. A. and Vallejo-Marin, M. 2013. What's the "buzz" about? The ecology and evolutionary significance of buzz-pollination. *Current Opinion in Plant Biology* 16(4):429–435. https://doi.org/10.1016/j.pbi.2013.05.002.

Deutsch, C. A., Tewksbury, J. J., Huey, R. B., Sheldon, K. S., Ghalambor, C. K., Haak, D. C. and Martin, P. R. 2008. Impacts of climate warming on terrestrial ectotherms across latitude. *Proceedings of the National Academy of Sciences of the United States of America* 105(18):6668–6672. https://doi.org/10.1073/pnas.0709472105.

Dharampal, P. S., Carlson, C., Currie, C. R. and Steffan, S. A. 2019. Pollen-borne microbes shape bee fitness. *Proceedings. Biological Sciences* 286(1904):20182894. http://doi.org/10.1098/rspb.2018.2894.

Di Tomaso, J. M. 2000. Invasive weeds in rangelands: species, impacts, and management. *Weed Science* 48(2):255–265.

Donovan, B. J. and Wier, S. S. 1978. Development of hives for field population increase, and studies on the life cycles of the four introduced bumble bees in New Zealand. *New Zealand Journal of Agricultural Research* 21(4):733–756. https://doi.org/10.1080/00288233.1978.10427476.

Eckhart, M., Haider, M., Dorn, S. and Muller, A. 2014. Pollen mixing in pollen generalist solitary bees: a possible strategy to complement or mitigate unfavourable pollen properties? *Journal of Animal Ecology* 83(3):588–597. https://doi.org/10.1111/1365-2656.12168.

Exeler, N., Kratochwil, A. and Hochkirch, A. 2009. Restoration of riverine inland sand dune complexes: implications for the conservation of wild bees. *Journal of Applied Ecology* 46(5):1097–1105. https://doi.org/10.1111/j.1365-2664.2009.01701.x.

Forman, R. T. T. 2003. *Road Ecology: Science and Solutions*. Island Press, Washington, DC.

Forrest, J. R. K. and Chisholm, S. P. M. 2017. Direct benefits and indirect costs of warm temperatures for high-elevation populations of a solitary bee. *Ecology* 98(2):359–369. Available at: https://doi.org/10.1002/ecy.1655.

Fowler, J. 2020a. Pollen specialist bees of the Western United States. Available at: https://jarrodfowler.com/pollen_specialist.html.

Fowler, J. 2020b. Pollen specialist bees of the Central United States. Available at: https://jarrodfowler.com/bees_pollen.html.

Fowler, J. and Droege, S. 2020. Pollen specialist bees of the Eastern United States. Available at: https://jarrodfowler.com/specialist_bees.html.

Frankie, G. W., Thorp, R. W., Newstrom-Lloyd, L. E., Rizzardi, M. A., Barthell, J. F., Griswold, T. L., Kim, J. and Kappagoda, S. 1998. Monitoring solitary bees in modified wildland habitats: implications for bee ecology and conservation. *Environmental Entomology* 27(5):1137–1148. https://doi.org/10.1093/ee/27.5.1137.

Fussell, M. and Corbet, S. A. 1992a. Flower usage by bumble-bees: a basis for forage plant management. *Journal of Applied Ecology* 29(2):451–465. https://doi.org/10.2307/2404513.

Fussell, M. and Corbet, S. A. 1992b. The nesting places of some British bumble bees. *Journal of Apicultural Research* 31(1):32–41. https://doi.org/10.1080/00218839.1992.11101258.

Galbraith, S. M., Vierling, L. A. and Bosque-Perez, N. A. 2015. Remote sensing and ecosystem services; current status and future opportunities for the study of bees and pollination-related services. *Current Forestry Reports* 1(4):261–274.

Garibaldi, L. A., Steffan-Dewenter, I., Winfree, R., Aizen, M. A., Bommarco, R., Cunningham, S. A., Kremen, C., Carvalheiro, L. G., Harder, L. D., Afik, O., Bartomeus, I., Benjamin, F., Boreux, V., Cariveau, D., Chacoff, N. P., Dudenhöffer, J. H., Freitas, B. M., Ghazoul, J., Greenleaf, S., Hipólito, J., Holzschuh, A., Howlett, B., Isaacs, R., Javorek, S. K., Kennedy, C. M., Krewenka, K. M., Krishnan, S., Mandelik, Y., Mayfield, M. M., Motzke, I., Munyuli, T., Nault, B. A., Otieno, M., Petersen, J., Pisanty, G., Potts, S. G., Rader, R., Ricketts, T. H., Rundlöf, M., Seymour, C. L., Schüepp, C., Szentgyörgyi, H., Taki, H., Tscharntke, T., Vergara, C. H., Viana, B. F., Wanger, T. C., Westphal, C., Williams, N. and Klein, A. M. 2013. Wild pollinators enhance fruit set of crops regardless of honey bee abundance. *Science* 339(6127):1608–1611.

Garratt, M. P. D., Senapathi, D., Coston, D. J., Mortimer, S. R. and Potts, S. G. 2017. The benefits of hedgerows for pollinators and natural enemies depends on hedge quality and landscape context. *Agriculture, Ecosystems and Environment* 247:363-370. https://doi.org/10.1016/j.agee.2017.06.048.

Goulson, D. 2003. Effects of introduced bees on native ecosystems. *Annual Review of Ecology, Evolution, and Systematics* 34(1):1-26.

Goulson, D. 2010. Impacts of non-native bumblebees in Western Europe and North America. *Applied Entomology and Zoology* 45(1):7-12.

Goulson, D., Lye, G. C. and Darvill, B. 2008. Decline and conservation of bumble bees. *Annual Review of Entomology* 53:191–208. https://doi.org/10.1146/annurev.ento.53.103106.093454.

Goulson, D., Nicholls, E., Botias, C. and Rotheray, E. L. 2015. Bee declines driven by combined stress from parasites, pesticides, and lack of flowers. *Science* 347(6229):1255957–1255957.

Greenleaf, S. S. and Kremen, C. 2006. Wild bees enhance honey bees' pollination of hybrid sunflower. *Proceedings of the National Academy of Sciences of the United States of America* 103(37), 13890–13895.

Griffin, S. R., Bruninga-Socolar, B. and Gibbs, J. 2020. Bee communities in restored prairies are structured by landscape and management, not local floral resources. *Basic and Applied Ecology* 50:144–154. https://doi.org/10.1016/j.baae.2020.12.004.

Haaland, C., Naisbit, R. E. and Bersier, L. 2011. Sown wildflower strips and insect conservation: a review. *Insect Conservation and Diversity* 4(1):60–80.

Hall, D. M., Camilo, G. R., Tonietto, R. K., Ollerton, J., Ahrné, K., Arduser, M., Ascher, J. S., Baldock, K. C., Fowler, R., Frankie, G., Goulson, D., Gunnarsson, B., Hanley, M. E., Jackson, J. I., Langellotto, G., Lowenstein, D., Minor, E. S., Philpott, S. M., Potts, S. G., Sirohi, M. H., Spevak, E. M., Stone, G. N. and Threlfall, C. G. 2017.The city as a refuge for insect pollinators. *Conservation Biology* 31(1):24–29.

Harmon-Threatt, A. 2020. Influence of nesting characteristics on health of wild bee communities. *Annual Review of Entomology* 65(1):39–56. Available at: https://doi.org/10.1146/annurev-ento-011019-024955.

Harvey, J. A., Heinen, R., Gols, R. and Thakur, M. P. 2020. Climate change-mediated temperature extremes and insects: from outbreaks to breakdowns. *Global Change Biology* 26(12):6685–6701. https://doi.org/10.1111/gcb.15377.

Hopwood, J. L. 2013. Roadsides as habitat for pollinators: management to support bees and butterflies. *Proceedings of the 2013 International Conference on Ecology and Transportation (ICOET 2013).*

Kerr, J. T., Pindar, A., Galpern, P., Packer, L., Potts, S. G., Roberts, S. M., Rasmont, P., Schweiger, O., Colla, S. R., Richardson, L. L., Wagner, D. L., Gall, L. F., Sikes, D. S. and Pantoja, A. 2015. Climate change impacts on bumblebees converge across continents. *Science* 349(6244):177–180. https://doi.org/10.1126/science.aaa7031.

Kevan, P. G., Clark, E. A. and Thomas, V. G. 1990. Insect pollinators and sustainable agriculture. *American Journal of Alternative Agriculture* 5(1):13–22. https://doi.org/10.1017/S0889189300003179.

Klein, A. M., Vaissiere, B. E., Cane, J. H., Steffan-Dewenter, I., Cunningham, S. A., Kremen, C. and Tscharntke, T. 2007. Importance of pollinators in changing landscapes for world crops. *Proceedings. Biological Sciences* 274(1608):303–313.

Kremen, C. and Merenlender, A. M. 2018. Landscapes that work for biodiversity and people. *Science* 362(6412):6020.

Kremen, C., M'Gonigle, L. K. and Ponisio, L. C. 2018. Pollinator community assembly tracks changes in floral resources as restored hedgerows mature in agricultural landscapes. *Frontiers in Ecology and Evolution* 6:Article 170. https://doi.org/10.3389/fevo.2018.00170.

Kutby, R. A. 2013. Bee communities in restored landfill sites of Niagara region. Unpublished MSc. Thesis. Brock University. Available at: http://hdl.handle.net/10464/5124.

Laport, R. G. and Minckley, R. L. 2012. Occupation of active Xylocopa virginica nests by the recently invasive Megachile sculpturalis in upstate New York. *Journal of the Kansas Entomological Society* 85(4):384–386. https://doi.org/10.2317/0022-8567-85.4.384.

Lawson, S. P., Kennedy, K. B. and Rehan, S. M. 2021. Pollen composition significantly impacts the development and survival of the native small carpenter bee, Ceratina Calcarata. *Ecological Entomology* 46(2):232–239. https://doi.org/10.1111/een.12955.

Lye, G. C., Jennings, S. N., Osborne, J. L. and Goulson, D. 2011. Impacts of the use of nonnative commercial bumble bees for pollinator supplementation in raspberry. *Journal of Economic Entomology* 104(1):107–114. https://doi.org/10.1603/EC10092.

Lyons, K. G., Maldonado-Leal, B. G. and Owen, G. 2017. Community ecosystem effects of buffelgrass (*Pennisetum ciliare*) and nitrogen deposition in the Sonoran Desert. *Invasive Plant Science and Management* 6(1):65–78. https://doi.org/10.1614/IPSM-D-11-00071.1.

MacIvor, J. S. and Packer, L. 2015. 'Bee Hotels' as tools for native pollinator conservation: a premature verdict? *PLoS ONE* 10(3):e0122126. https://doi.org/10.1371/journal.pone.0122126.

Majewska, A. A. and Altizer, S. 2020. Planting gardens to support insect pollinators. *Conservation Biology: The Journal of the Society for Conservation Biology* 34(1):15–25.

Mallinger, R. E., Franco, J. G., Prischmann-Voldseth, D. A. and Prasifka, J. R. 2019. Annual cover crops for managed and wild bees: optimal plant mixtures depend on pollinator enhancement goals. *Agriculture, Ecosystems and Environment* 273:107–116.

Marsele, M. R., Turbe, A., Shwartz, A., Bonn, A. and Colleony, A. 2020. Addressing behavior in pollinator conservation policies to combat the implementation gap. *Conservation Biology* 35(2):610–622. https://doi.org/10.1111/cobi.13581.

Menz, M. H. M., Phillips, R. D., Winfree, R., Kremen, C., Aizen, M. A., Johnson, S. D. and Dixon, K. W. 2011. Reconnecrting plants and pollinators: challenges in the restoration of pollination mutualisms. *Trends in Plant Science* 16(1):4–12. https://doi.org/10.1016/j.tplants.2010.09.006.

M'Gonigle, L. K., Ponisio, L. C., Cutler, K. and Kremen, C. 2015. Habitat restoration promotes pollinator persistence and colonization in intensively managed agriculture. *Ecological Applications* 25(6):1557–1565.

Minckley, R. L., Roulston, T. H. and Williams, N. M. 2013. Resource assurance predicts specialist and generalist bee activity in drought. *Proceedings. Biological Sciences* 280(1759):20122703. http://doi.org/10.1098/rspb.2012.2703.

Minckley, R. L., Wcislo, W. T., Yanega, D. and Buchmann, S. L. 1994. Behavior and phenology of a specialist bee (Dieunomia) and sunflower (Helianthus) pollen availability. *Ecology* 75(5):1406–1419. https://www.jstor.org/stable/1937464.

Moller, A. P. 2019. Parallel declines in abundance of insects and insectivorous birds in Denmark over 22 years. *Ecology and Evolution* 9(11):6581–6587. https://doi.org/10.1002/ece3.5236.

Morandin, L. A. and Kremen, C. 2013. Hedgerow restoration promotes pollinator populations and exports native bees to adjacent fields. *Ecological Applications* 23(4):829–839.

Morandin, L. A. and Winston, M. L. 2005. Wild bee abundance and seed production in conventional, organic, and genetically modified canola. *Ecological Applications* 15(3):871–881.

Muñoz, P. T., Tamayo, P., Torres, A. G., Pascual, F., and Megías, A. G. 2014. Effects of roads on insects: a review. *Biodiversity and Conservation* 24(3):659–682. https://doi.org/10.1007/s10531-014-0831-2.

Murray, T. E., Coffey, M. F., Kehoe, E. and Horgan, F. G. 2013. Pathogen prevalence in commercially reared bumble bees and evidence of spillover in conspecific populations. *Biological Conservation* 159:269–276. https://doi.org/10.1016/j.biocon.2012.10.021.

Nicholson, C. C., Ricketts, T. H., Koh, I., Smith, H. G., Lonsdorf, E. V. and Olsson, O. 2019. Flowering resources distract pollinators from crops: model predictions from landscape simulations. *Journal of Applied Ecology* 56(3):618–628.

Ollerton, J., Winfree, R. and Tarrant, S. 2011. How many flowering plants are pollinated by animals? *Oikos* 120(3):321–326. https://doi.org/10.1111/j.1600-0706.2010.18644.x.

Olmstead, A. L. and Wooten, D. B. 1987. Bee pollination and productivity growth: the case of alfalfa. *American Journal of Agricultural Economics* 69(1):56–63. https://doi.org/10.2307/1241306.

Omuse, E., Niassy, S., Wagacha, J. M., Ong'amo, G. O., Lattorff, H. M. G., Kiatoko, N., Mohamed, S. A., Subramanian, S., Akutse, K. S. and Dubois, T. 2021. Susceptibility of the western honey bee *Apis mellifera* and the African stingless bee *Meliponula ferruginea* (Hymenoptera: Apidae) to the entomopathogenic fungi *Metarhizium anisopliae* and *Beauveria bassiana*. *Journal of Economic Entomology* 115(1):46–55. https://doi.org/10.1093/jee/toab211.

Osborne, J. L., Martin, A. P., Shortall, C. R., Todd, A. D., Goulson, D., Knight, M. E., Hale, R. J. and Sanderson, R. A. 2007. Quantifying and comparing bumblebee nest densities in gardens and countryside habitats. *Journal of Applied Ecology* 45(3):784–792.

Osborne, J. L., Williams, I. H. and Corbet, S. A. 1991. Bees, pollination and habitat change in the European Community. *Bee World* 72(3):99–116. DOI: 10.1080/0005772X.1991.11099088.

Otterstatter, M. C. and Thomson, J. D. 2008. Does pathogen spillover from commercially reared bumble bees threaten wild pollinators? *PLoS ONE* 3(7):e2771. https://doi.org/10.1371/journal.pone.0002771.

Paini, D. R. 2004. Impact of the introduced honey bee (*Apis mellifera*) (Hymenoptera: Apidae) on native bees: a review. *Austral Ecology* 29(4):399–407.

Park, J. R. (Ed.) 1988. Environmental management in agriculture: European perspectives. *Proceedings of the a Workshop Held from 14 to 17 July 1987 at Bristol, United Kingdom, Under the Aegis of the CEC Land and Water Use and Management Committee (vol. 11169)*. Burns & Oates.

Paterson, C., Cottenie, K. and MacDougall, A. S. 2019. Restored native prairie supports abundant and species-rich native bee communities on conventional farms. *Restoration Ecology* 27(6):1291–1299. https://doi.org/10.1111/rec.12987.

Pimm, S. L., Jenkins, C. N., Abell, R., Brooks, T. M., Gittleman, J. L., Joppa, L. N., Raven, P. H., Roberts, C. M. and Sexton, J. O. 2014. The biodiversity of species and their rates of extinction, distribution, and protection. *Science* 344(6187):1246752. https://doi.org/10.1126/science.1246752.

Ponisio, L. C., M'Gonigle, L. K. and Kremen, C. 2016. On-farm habitat restoration counters biotic homogenization in intensively managed agriculture. *Global Change Biology* 22(2):704–715.

Potts, S. G., Biesmeijer, J. C., Kremen, C., Neumann, P., Schweiger, O. and Kunin, W. E. 2010. Global pollinator declines: trends, impacts and drivers. *Trends in Ecology and Evolution* 25(6):345–353. https://doi.org/10.1016/j.tree.2010.01.007.

Purvis, E. E. N., Vickruck, J. L., Best, L. R., Devries, J. H. and Galpern, P. 2020. Wild bee community recovery in restored grassland-wetland complexes of prairie North America. *Biological Conservation* 252: https://doi.org/10.1016/j.biocon.2020.108829.

Pywell, R. F., Heard, M. S., Bradbury, R. B., Hinsley, S., Nowakowski, M., Walker, K. J. and Bullock, J. M. 2012. Wildlife-friendly farming benefits rare birds, bees and plants. *Biology Letters* 8(5):772–775.

Pywell, R. F., Warman, E. A., Carvell, C., Sparks, T. H., Dicks, L. V., Bennett, D., Wright, A., Critchley, C. N. R. and Sherwood, A. 2005. Providing foraging resources for bumblebees in intensively farmed landscapes. *Biological Conservation* 121(4):479–494.

Radmacher, S. and Strohm, E. 2010. Factors affecting offspring body size in the solitary bee Osmia bicornis (Hymenoptera, Megachilidae). *Apidologie* 41(2):169–177. https://doi.org/10.1051/apido/2009064.

Richards, M. H., Rutgers-Kelly, A., Gibbs, J., Vickruck, J. L., Rehan, S. M. and Sheffield, C. S. 2011. Bee diversity in naturalizing patches of Carolinian grasslands in southern Ontario, Canada. *The Canadian Entomologist* 143(3):279–299. https://doi.org/10.4039/n11-010.

Ritchie, A. D., Lane, I. G. and Cariveau, D. P. 2020. Pollination of a bee-dependent forb in restored prairie: no evidence of pollen limitation in landscapes dominated by row crop agriculture. *Restoration Ecology* 28(4):919–926. https://doi.org/10.1111/rec.13157.

Rodger, J. G., Bennett, J. M., Razanajatovo, M., Knkight, T. M., van Kleunen, M., Ashman, T. L., Steets, J. A., Hui, C., Arceo-Gomez, G., Burd, M., Burkle, L. A., Burns, J. H., Durka, W., Freitas, L., Kemp, J. E., Li, J., Pauw, A., Vamosi, J. C., Wolowski, M., Xia, J. and Ellis, A. G. 2021. Widespread vulnerability of flowering plant seed production to pollinator declines. *Science Advances* 7(42):eabd3524.

Roubik, D. W. (Eds.) 2018. *The Pollination of Cultivated Plants. A Compendium for Practitioners* (vols. 1 and 2). Food and Agriculture Organization of the United Nations, Rome.

Sammatraro, D., Gerson, U. and Needham, G. 2000. Parasitic mites of honey bees: life history, implications, and impact. *Annual Review of Entomology* 45:519–548. https://doi.org/10.1146/annurev.ento.45.1.519.

Sánchez-Bayo, F. and Wyckhuys, K. A. G. 2019. Worldwide decline of the entomofauna: a review of its drivers. *Biological Conservation* 232:8–27. https://doi.org/10.1016/j.biocon.2019.01.020.

Schuler, R. E., Roulston, T. H. and Farris, G. E. 2005. Farming practices influence wild pollinator populations on squash and pumpkin. *Journal of Economic Entomology* 98(3):790–795.

Sheffield, C. S., Pindar, A., Packer, L. and Kevan, P. G. 2013. The potential of cleptoparasitic bees as indicator for assessing bee communities. *Apidologie* 44(5):501–510 https://doi.org/10.1007/s13592-013-0200-2.

Sirois-Delisle, C. and Kerr, J. T. 2018. Climate change-driven range losses among bumblebee species are poised to accelerate. *Scientific Reports* 8(1):14464 https://doi.org/10.1038/s41598-018-32665-y.

Soroye, P., Newbold, T. and Kerr, J. 2020. Climate change contributes to widespread declines among bumble bees across continents. *Science* 367(6478):685–688. https://doi.org/10.1126/science.aax8591.

Stephen, W. P. 1966. Native bees- An untapped pollinator resources. *Bee World (Suppl.)* 47:191–194.

Stephen, W. P. 2003. Solitary bees in North America: a perspective. In: *For Non-native Crops, Whence Pollinators of the Future?*, Peter, G. Kevan (Ed), Thomas Say Publications, Entomological Society of America, Annapolis, MD, pp. 41–66.

Stout, J. C. and Morales, C. L. 2009. Ecological impacts of invasive alien species on bees. *Apidologie* 40(3):388–409. https://doi.org/10.1051/apido/2009023.

Straka, J., Cerna, K., Machackova, L., Zemenova, M. and Keil, P. 2014. Life span in the wild: the role of activity and climate in natural populations of bees. *Functional Ecology* 28(5):1235–1244. https://doi.org/10.1111/1365-2435.12261.

Szabo, N. D., Colla, S. R., Wagner, D. L., Gall, L. F. and Kerr, J. T. 2012. Do pathogen spillover, pesticide use, or habitat loss explain recent North American bumblebee declines? *Conservation Letters* 5(3):232–239. https://doi.org/10.1111/j.1755-263X.2012.00234.x.

Tonietto, R. K. and Larkin, D. J. 2018. Habitat restoration benefits wild bees: a meta-analysis. *Journal of Applied Ecology* 55(2):582–590. https://doi.org/10.1111/1365-2664.13012.

Torchio, P. F. 1976. Use of Osmia lignaria ay (Hymenjoptera: Apoidea, Megachilidae) as. Pollinator in an apple and prune Orchard. *Journal of the Kansas Entomological Society* 49(4):475–482. Available at: http://www.jstor.org/stable/25082856.

Ullmann, K. S., Meisner, M. H. and Williams, N. M. 2016. Impact of tillage on the crop pollinating, ground-nesting bee Peponapis pruinosa in California. *Agriculture,*

Ecosystems and Environment 232:240–246. https://doi.org/10.1016/j.agee.2016.08.002.

Velthuis, H. H. W. and Van Doorn, A. 2006. A century of advances in bumblebee domestication and the economic and environmental aspects of its commercialization for pollination. *Apidologie* 37(4):421–451.

Venter, O., Brodeur, N. N., Nemiroff, L., Belland, B., Dolinsek, I. J. and Grant, J. W. A. 2006. Threats to endangered species in Canada. *BioScience* 56(11):903–910.

Vinchesi, A., Cobos, D., Lavine, L. and Walsh, D. 2013. Manipulation of soil temperatures to influence brood emergence in the alkali bee (Nomia melanderi). *Apidologie* 44(3):286–294. https://doi.org/10.1007/s13592-012-0180-7.

Visscher, P. K. and Danforth, B. N. 1993. Biology of Calliopsis pugionis (Hymenoptera: Andrenidae): nesting, foraging, and investment sex ratio. *Annals of the Entomological Society of America* 86(6):822–832. Available at: https://doi.org/10.1093/aesa/86.6.822.

Visscher, P. K., Vetter, R. S. and Orth, R. 1994. Benthic bees? Emergence Phenology of Calliopsis pugionis (Hymenoptera: Andrenidae) at a seasonally flooded site. *Annals of the Entomological Society of America* 87(6):941–945. https://doi.org/10.1093/aesa/87.6.941.

Wagner, D. L., Grames, E. M., Forister, M. L., Berenbaum, M. R. and Stopak, D. 2021. Insect decline in the Anthropocene: death by a thousand cuts. *Proceedings of the National Academy of Sciences of the United States of America* 118(2):e2023989118; https://doi.org/10.1073/pnas.2023989118.

Whitfield, C. W., Behura, S. K., Berlocher, S. H., Clark, A. G., Johnston, J. S., Sheppard, W. S., Smith, D. R., Suarez, A. V., Weaver, D. and Tsutsui, N. D. 2006. Thrice out of Africa: ancient and recent expansions of the honey bee, Apis mellifera. *Science* 314(5799):642–645.

Wilcove, D. S., Rothstein, D., Dubow, J., Phillips, A. and Losos, E. 1998. Quantifying threats to imperiled species in the United States. *BioScience* 48(8):607–615. Available at: http://links.jstor.org/sici?sici=0006-3568%28199808%2948%3A8%3C607%3AQTTISI%3E2.0.CO%3B2-B.

Wilder, B. T., Jarnevich, C. S., Baldwin, E., Black, J. S., Franklin, K. A., Grissom, P., Hovanes, K. A., Olsson, A., Malusa, J., Kibria, A. S. M. G., Li, Y. M., Lien, A. M., Ponce, A., Rowe, J. A., Soto, J. R., Stahl, M. R., Young, N. E. and Betancourt, J. L. 2021. Grassification and fast-evolving fire connectivity and risk in the Sonoran Desert, United States. *Frontiers in Ecology and Evolution* 9:655561. https://doi.org/10.3389/fevo.2021.655561.

Williams, N. M. 2010. Restoration of nontarget species: bee communities and pollination function in riparian forests. *Restoration Ecology* 19(4):450–459. https://doi.org/10.1111/j.1526-100X.2010.00707.x.

Willis Chan, D. S. and Raine, N. E. 2021. Hoary squash bees (Eucera pruinosa: Hymenoptera: Apidae) provide abundant and reliable pollination services to cucurbita crops in Ontario (Canada). *Environmental Entomology* 50(4):968–981, https://doi.org/10.1093/ee/nvab045.

Winfree, R. 2008. Pollinator-dependent crops: an increasingly risky business. *Current Biology* 18(20):R968–R969. https://doi.org/10.1016/j.cub.2008.09.010.

Winfree, R. 2010. The conservation and restoration of wild bees. *Annals of the New York Academy of Sciences* 1195:169–197. https://doi.org/10.1111/j.1749-6632.2010.05449.x.

Winfree, R., Aguilar, R., Vázquez, D. P., LeBuhn, G. and Aizen, M. A. 2009. A meta-analysis of bees responses to anthropogenic disturbance. *Ecology* 90(8):2068–2076.

Winfree, R., Bartomerus, I. and Cariveau, D. P. 2011. Native pollinators in anthropogenic habitats. *Annual Review of Ecology, Evolution, and Systematics* 42(1):1–22. https://doi.org/10.1146/annurev-ecolsys-102710-145042.

Winfree, R., Williams, N. M., Dushoff, J. and Kremen, C. 2007. Native bees provide insurance against ongoing honey bee losses. *Ecology Letters* 10(11):1105–1113.

Winfree, R., Williams, N. M., Gaines, H., Ascher, J. S. and Kremen, C. 2008. Wild bee pollinators provide the majority of crop visitation across land-use gradients in New Jersey and Pennsylvania, USA. *Journal of Applied Ecology* 45(3):793–802. DOI: 10.1111/j.1365-2664.2007.01418.x.

Wojcik, V. A. and Buchmann, S. 2012. Pollinator conservation and management on electrical transmission and roadside rights-of-way: a review. *Journal of Pollination Ecology* 7:16–26.

Wright, K. W. 2018. The evolution of diet breadth in Melissodes bees (Apidae: Eucerini). Available at: https://digitalrepository.unm.edu/biol_etds/301.

Chapter 10

Altering crop management practices to promote pollinators

Jose G. Franco, USDA ARS, USA; and Rachel E. Mallinger, University of Florida, USA

1 Introduction

Agricultural intensification, or the increase in crop production per unit land area, has resulted in high-input, large-scale monoculture cropping and overall declines in crop diversity in the USA (Aguilar et al., 2015) and greater homogeneity in the crops grown across regions at a global scale (Martin et al., 2019). This has often resulted in agricultural landscapes devoid of the habitat and floral resources needed to sustain pollinators. In addition to a reduction in crop diversity, the intensified use of agrochemical inputs reduces populations of flowering weeds that can provide floral resources for pollinators and introduces compounds into the environment found to be toxic to sensitive pollinator species (van der Slujis et al., 2013). Modern crop production systems are, thus, thought to contribute to declines in pollinator populations that we rely on for the pollination of economically important crops and wild plants.

To address this pressing issue, agriculture must develop practices that diversify the agricultural landscape to enhance the availability of floral resources and habitat within the crop production space. This approach, known as land sharing, is in contrast to the land sparing paradigm where natural lands are set aside for conservation purposes and are distinct management units from agricultural lands (Phalan et al., 2011). The land-sharing paradigm, however, is thought to be a more realistic approach given the demands a growing

http://dx.doi.org/10.19103/AS.2022.0111.20

population has placed on agricultural lands and the benefits that increased biodiversity has on agroecosystems and agricultural production (Tscharntke et al., 2005, 2012).

Another chapter discusses in more detail the management of field margins, wildflower strips, or other adjacent and surrounding habitat to crop fields that contribute to landscape diversity and enhanced pollinator resources. In this chapter, the focus is on approaches to enhance plant diversity within the crop production space. Approaches by which this can be achieved include in-field practices such as the planting of flowering cover crops between crop rotations, intercropped with cash crops, or in place of a cash crop, planting of diverse perennial forages that also provide floral resources, and maintaining flowering weed communities if they pose little threat to crop production.

2 Approaches to managing production space for pollinators

2.1 Cover crops

Cover crops play a critical role in providing habitat and floral resources to promote pollinators and pollination within the crop production space or field, i.e. not only field edges or margins as discussed in a previous chapter. Integrating flowering cover crops is challenging due to the need to reduce competition for space and other resources with the primary cash crop. However, there are a number of ways in which this can be achieved, including planting overwintering (cold-hardy) cover crops that provide early-season floral resources prior to a cash crop, planting full-season cover crops in lieu of a cash crop or fallow ground (bare ground not planted to a cash crop), or strip or companion plantings between crop rows or on orchard floors.

Planting overwintering cover crops in the fall for early spring flowering can be an effective way to reduce competition with the primary cash crop while providing floral resources at a time when few are available across the landscape (Bretagnolle and Gaba, 2015). Examples of cold-hardy species that also flower early the subsequent year, which is particularly important in northern growing zones, are winter camelina (*Camelina sativa* L.), winter canola (*Brassica napus* L.), and pennycress (*Thlaspi arvense* L.) (Eberle et al., 2015). Other cover crops that overwinter, though to varying extents depending on planting date and climate, include red clover (*Trifolium pratense* L.) and Austrian winter pea (*Pisum sativum* subsp. *arvense* L.; Ellis and Barbercheck, 2015). In warmer climates or southern growing zones, additional cover crops will likely overwinter, but the timing of flowering may limit their use if impacts on the subsequent cash crop are a primary consideration. Further, if the cover crop is to double as an economic crop (winter camelina and pennycress, for instance, are being developed as oilseed crops; Cubins et al., 2019), adjustments to the crop rotation and agronomic management would be necessary to accommodate cover crop seed harvest for a relay- or double-crop system (Patel et al., 2021; Ott et al., 2019; Johnson et al., 2017; Gesch et al., 2014).

Cover crops may be planted as full-season crops for a variety of reasons. They may be planted in fields that would otherwise remain fallow or in fields where adverse conditions such as drought or excess moisture result in a failure to plant an insured crop by a certain date (also known as prevented planting; 2013 Crop insurance handbook, USDA-RMA, 2013). Though not as common, full-season cover crop mixtures may also be integrated into a crop rotation to achieve soil health benefits and provide a forage source for livestock (Housman et al., 2021; Sanderson et al., 2018). With the inclusion of flowering species in full-season cover crop mixtures, season-long floral resources that support both managed and wild pollinators may be achieved, though cover crop species selection and mixture composition will influence the species of pollinators attracted (Mallinger et al., 2019).

Companion planting with flowering cover crops in intercropping systems is another approach that can sustain pollinator communities within crop production fields (Kordbacheh et al., 2020; Norris et al., 2018). In some instances, intercropping with flowering species does not only attract beneficial insects strictly for pollination services, but, as is the case with alyssum (*Lobularia maritima* (L.) Desv.) in organic lettuce (*Lactuca sativa* L.) and broccoli (*Brassica oleracea* L. var. *italica* Plenck) production fields, it may also attract predatory insects such as hoverflies that reduce pest populations (Brennan, 2013, 2016). Intercropping with non-flowering ground covers that release volatile organic compounds such as peppermint (*Mentha* × *piperita*) can also support pollinators and other beneficial insects (Gowton et al., 2021). Further, orchard production systems with cover crop ground covers can also act as a companion planting system to provide a number of benefits including habitat and floral resources for pollinators (Van Sambeek, 2017; Saunders et al., 2013). Companion or intercropping systems should, however, be managed with a holistic approach that reduces reliance on commercial inputs, as pesticide exposure, which is a well-documented source of bee declines, is an ever-present risk to pollinators (Nicholls and Altieri, 2013).

2.2 Flowering weeds for pollinators

Within the production space, floral resources for pollinators can be provided by commercial crops, cover crops, intercrops, and even flowering weeds. Where flowering weeds do not significantly compete with and reduce crop yields, they can be an effective and affordable way to provide pollen and nectar for pollinators. Flowering weeds can supplement the crop plant during crop bloom, as an alternative source of nectar and pollen, or, more importantly, extend the flowering season to provide floral resources for pollinators outside of crop bloom. Weeds often flower in high density and require few inputs, thus offering a cost-effective way to enhance or conserve pollinator communities.

Weed communities within crop fields have shifted with increasing agricultural intensification to favor weeds that are resistant to herbicides or germinate late to escape herbicides, with resulting on-farm weed community simplification (Bretagnolle and Gaba, 2015). Nonetheless, on-farm weeds can provide equal or even better floral resources for pollinators as compared to adjacent natural/semi-natural habitat as the open canopy of crop fields facilitates high flower density (Gemmill-Herren and Ochieng, 2008). Furthermore, weeds are often top nectar and pollen producers, outproducing intentionally grown plant species that are typically present in lower densities (Hicks et al., 2016). For honey bees, in particular, flowering weeds have been found to be a major source of food (Requier et al., 2015).

Flowering weeds can enhance pollination services through increasing crop flower visitation rates via a spillover effect in which pollinators attracted to flowering weeds move over to pollinate the crop plant, or they can enhance pollinator abundance and/or diversity via increased season-long resource availability. In the case of the former, effects on crop pollination could be realized immediately, while in the case of the latter, effects are typically seen 1–3 years following resource enhancement. Examples from sunflower crop fields have illustrated the benefit of weeds for pollinator communities and pollination services. For example, higher on-farm weed diversity in sunflower fields was correlated with a higher diversity of pollinators visiting sunflowers, and seed mass increased with flowering weed species richness. In a different example, the diversity of weed flowers had a positive impact on the species richness of pollinators visiting sunflowers and on sunflower seed set (Carvalheiro et al., 2011). Additionally, the presence of flowering weeds within sunflower fields mitigated some of the negative effects of reduced natural habitat in the surrounding landscape on pollinator diversity (Carvalheiro et al., 2011). These examples from sunflowers illustrate how flowering weeds can benefit not only pollinator communities but pollination services and crop yields as well.

However, weeds can compete with the crop plant for space, nutrients, and water, thereby reducing crop plant densities and overall yields (Sidemo-Holm et al., 2021). They can also compete for pollinators if bloom time overlaps and if they are more attractive than the crop plant. To reduce competition for pollinators, weeds can be mowed or spot treated during bloom, but allowed to flower outside of bloom. Another cost is greater pesticide risk to pollinators; the presence of flowering weeds could increase pollinator exposure to pesticides applied outside of crop bloom with significant negative effects on pollinator populations (Larson et al., 2013). To prevent exposure to pesticides, flowering weeds should be mowed or treated prior to insecticide applications when possible.

2.3 Flowering crops

Large homogeneous row crop systems dominated by corn and soybean, often representative of agriculture in the US Midwest, do not provide the nectar and pollen resources needed to support many pollinator populations. Declines in crop diversity in US cropland (Aguilar et al., 2015) have resulted in fewer flowering crops that sustain pollinator resources. Therefore, diversification strategies are needed to restore plant diversity and re-introduce flowering economic crops into crop rotations that can improve habitat and foraging resources for pollinators (Nicholls and Altieri, 2013).

Crops such as sunflower (*Helianthus annuus* L.), buckwheat (*Fagopyrum esculentum* Moench), flax or linseed (*Linum usitatissimum* L.), canola, and other flowering (forbs) crops can provide floral resources that attract diverse species of pollinators at varying times of the growing season (Mallinger et al., 2019; Campbell et al., 2016; Eberle et al., 2015) and, in some instances, may also benefit from insect pollination (Mallinger et al., 2018; Witter et al., 2015). Forage crops such as alfalfa (*Medicago sativa* L.) and clovers (*Trifolium* spp.) have also been shown to provide floral resources attractive to some pollinator species (Butters et al., 2022; Bryan et al., 2021), though these crops are typically harvested prior to mass bloom events.

Other lesser known but emerging forage crops such as silflower (*Silphium integrifolium* Michx.) and cup plant (*Silphium perfoliatum* L.) support a high abundance of diverse bees when planted as crop borders (Butters et al., 2022), but more work is needed to develop breeding and agronomic management strategies to successfully integrate them into crop production systems (Van Tassel et al., 2017; Lehmkuhler et al., 2007). It is likewise worthy to note the potential for extreme climate scenarios, i.e. drought, to reduce nectar production of crops such as buckwheat and cup plant, thereby reducing the benefit of these flowering crops for pollinators (Rering et al., 2020; Mueller et al., 2020).

2.4 Crop rotations

Rotating crops across growing seasons and years can mean temporally inconsistent resources for pollinators, especially when nectar and pollen-rich crops, such as melons, are rotated with less attractive and resourceful crop such as grains. However, crop rotations can be designed to allow for pollinator populations to persist in heterogeneous and disturbed environments. It may be particularly important to consider how crop rotations affect pollinator population persistence when animal-pollinated crops are used in rotation with non-flowering crops or crops that are wind-pollinated. Examples include oilseed crops such as sunflowers, oilseed rape, and flax, which are typically grown on 3–4-year rotations with wind-pollinated crops including wheat or barley, or with fallow fields. Additionally, pollinator-dependent melons, squash, and other cucurbits are grown on a 3+-year rotation with a variety of crops including many that do not provide significant floral resources for pollinators (e.g. greens, root vegetables, legumes, oats, and wheat) or with fallow fields.

Rotated crops pollinated by specialist pollinators are at particular risk of experiencing reduced pollination services due to inconsistent and small pollinator populations. For example, sunflowers in North America receive significant pollination services from specialist bees including long-horned bees in the tribe Eucerini that rely on sunflowers or other Asteraceae species (Mallinger et. al., 2018). Squash is also pollinated by a few species of specialist squash bees that rely on squash and related Cucurbitaceae plants (Tepedino, 1981). For specialist sunflower and squash bees, the absence of their preferred or required plant hosts in some years may limit population growth and persistence. Unlike with generalist pollinators, these specialists cannot take advantage of other crops in the rotation or flowering weeds in the crop fields. Thus, particular attention needs to be given to designing crop rotations to ensure some resource availability within the specialist pollinator's foraging range.

Limited research on crop rotation effects on pollinator populations has been conducted, and therefore recommendations are also limited. Rotating crops at the level of individual fields, and staggering rotation cycles across adjacent fields, can achieve the benefits of crop rotations for disease pressure and soil health while still ensuring that the pollinator resources provided by certain crops are present in the broader landscape. Larger-bodied pollinators, including some specialist sunflower and squash bees, may forage a few hundred meters and up to 1 km from their nests (Gathmann and Tscharntke, 2002; Greenleaf et al., 2007), though other studies suggest most bees forage locally within a small radius (Sardiñas et al., 2016). Thus, rotations would need to be designed to ensure resource availability within a few hundred meters, which may not be feasible in many landscapes. However, some studies suggest that consistent and high presence of the focal pollinator-dependent crop in the landscape does not improve pollination services to the focal crop (Andersson et al., 2014). Alternatively, including some flowering forbs in the landscape each year or growing season, especially for generalist pollinators that are able to take advantage of a diversity of floral resources, can reduce negative effects of crop rotations. These floral resources could be provided by other crops in the rotation or by stable, flowering field margins and flower strips. In some studies, generalist pollinator abundance was affected by crop rotation schemes, but stable crop boundary features mitigated the negative effects of crop rotations on floral resource availability and pollinator populations. Thus, having larger and more stable areas of pollinator habitat near crop fields may be the best way to ensure temporally consistent resources in rotational landscapes (Marja et al., 2018; Gardner et al., 2021).

2.5 Providing protected nesting areas within the production space

Many pollinators need nesting habitat in addition to floral resources. For example, the majority of bees nest belowground and require soil of suitable texture, slope, and accessibility, and with minimal disturbance (Danforth et al., 2019). For bees that nest aboveground, the majority are considered renters and need hollowed cavities, while a minority will excavate nests in a substrate such as wood (Danforth et al., 2019). While the majority of research, particularly for wild bees, has focused on nest habitat adjacent to the field within field margins or hedgerows, in this section we cover agronomic practices to provide or enhance nest habitat within the production space. In fact, some wild bees, such as specialist squash bees, preferentially nest within crop fields over areas adjacent to the field (Julier and Roulston, 2009).

Agronomic practices that can affect nesting habitat for bee pollinators include irrigation and tillage, inclusion of trap nests for aboveground nesting bees, and manipulation of soil to intentionally create belowground nesting habitat. Squash bees, important solitary bee pollinators of *Cucurbita* spp., have been shown to prefer irrigated soils (Julier and Roulston, 2009), and thus regular irrigation within these fields may enhance their nesting rates. However, little is known about how irrigation affects other bee species. The effects of tillage are better documented, though the focus is largely on squash bees, and results are mixed. In some contexts, squash bee abundance did not vary across farms differing in tillage intensity (Julier and Roulston, 2009), but in other studies, populations were more abundant in no or reduced tillage systems (Shuler et al., 2005; Appenfeller et al., 2020). In a more controlled study, tillage moderately reduced and delayed offspring emergence but did not completely eliminate emergence or destroy nests (Ullmann et al., 2016). However, squash bees also seem to prefer nesting in tilled soils (Skidmore et al., 2019) despite the impacts that tillage may have on future offspring emergence. And, in the lone example conducted in other agricultural systems (perennial vineyards), alternating tillage systems had a positive effect on bee pollinator abundance and diversity as compared to no tillage, suggesting that some amount of soil disturbance can increase nesting rates and bee populations (Kratschmer et al., 2019). These studies suggest overall that there may be benefits to reduced tillage, especially for squash bees and within agroecosystems with significant tillage intensity, though soil accessibility is also important for bee nesting rates and can be achieved through occasional or alternating tillage practices.

Intentionally creating nesting space for wild bees within the production space is not common and limited to a few specific agroecosystems. Perhaps the most well-known example of creating ground-nesting habitat is with the alkali bee, an important pollinator of alfalfa in western North America. While this bee is solitary, it nests in aggregations underground within soil of sufficient alkalinity, texture, and moisture (Danforth et al., 2019). These soil attributes have been successfully manipulated in nest boxes or artificial bee beds at ground level through the addition of soil, salt, and irrigation, and have been shown to increase local populations (Stephen, 1960; Cane, 2008). Another opportunity for providing ground-nesting habitat is for the Southeastern blueberry bee *Habropoda laboriosa*, an important pollinator of blueberries in the Southeastern USA. It has been shown to readily nest in manmade holes in the ground, suggesting that nest habitat could be created but is also highly sensitive to changes in visual cues associated with the nest site (Cane, 1994). Finally, the use of trap nests within the crop field to provide nesting habitat for both wild and managed aboveground cavity nesters has been done in alfalfa for the managed alfalfa leafcutter bee (Pitts-Singer and Cane, 2011), in almonds for the managed orchard mason bee (Boyle and Pitts-Singer, 2019), and in oilseed rape fields for wild bees (Dainese et al., 2018).

3 Case studies

3.1 Annual row crop systems

Increasing plant diversity within row crop production fields can be achieved by integrating cover crops or native species within a crop rotation using four primary strategies: (1) intercropping (mixed, strip, or otherwise) or companion cropping with cover crops, forages, or native species that provide floral resources and habitat for pollinators; (2) strategic placement of flowering cash crops (including forages) across the landscape; (3) replacement of a monoculture cash crop, fallow fields, or in prevented plantings scenarios with cover crops in the rotation; and (4) adding overwintering cover crops that provide floral resources early in the season and do not compete with cash crops. The examples provided in this section will focus on the two latter strategies.

In North Dakota, USA, annual cover crops planted specifically for the purpose of providing full-season ground cover and floral resources were shown to support diverse bee populations (Mallinger et al., 2019). Flowering cover crops such as buckwheat, sunflower, and phacelia were planted in single-species stands, i.e. monocultures, and in mixtures of two, three, and six species (with cowpea, hairy vetch, and kale/mustard). Low-diversity mixtures with phacelia and buckwheat provided season-long floral resources at high densities that supported honey bees and generalist bumble bees. Adding sunflower into the mixture increased visitation by more rare wild bees but reduced visitation by managed and generalist bees due to lower floral density of buckwheat and phacelia flowers. Thus, more diverse mixtures increased bee species richness but resulted in reduced visitation by more common species. In addition, the cover crops evaluated in this study attracted several bee species considered to be in decline, many of which were also recorded in adjacent rangeland (minimally managed grazing areas).

Providing early-season floral resources in crop production fields lacking in plant diversity to sustain pollinators during this critical time of year is also crucial. Planting cold-hardy, overwintering cover crops may play an important role in filling this niche. In Pennsylvania, USA, increasing cover crop diversity to include flowering crops such as canola (an early flowering crop, April–May) can support diverse populations of pollinators including honey bees, native bees, and Syrphid flies (Ellis and Barbercheck, 2015). However, crop rotations restrict the flowering window and effectiveness of other flowering service crops such as red clover and Austrian winter pea that flower later in the season (May–July). This study also demonstrated the importance of fall planting date on winter survival and spring flowering, with cover crops planted earlier (August–September) generally resulting in increased overwintering as well as earlier flowering in the spring. Work in South Dakota and Minnesota, USA, also found variability in winter survival of fall-planted flowering cover crops (Eberle et al., 2015). In this study, pennycress, winter canola, and, in particular, winter camelina provided floral and nectar resources early in the season to support pollinators in a corn-soybean crop rotation. The extent to which these floral resources benefit insect pollinators, however, may be impacted by weather variability, particularly in the spring when conditions are often cool and wet in this region of the USA and may have variable impacts on pollinator populations (Forcella et al., 2020).

3.2 Annual specialty crop systems

Within annual cropping systems, intercropping and on-farm crop diversification are the primary ways in which the production system can be modified for pollinators and pollination services. As compared to perennial systems and annual row crops, there are fewer examples of altering crop management practices within the production space for annual specialty crops, likely because their high value discourages modifications within the production space in favor of modifications adjacent to the crop field (i.e. field margins). Additionally, there is minimal vertical structure to manipulate as in perennial orchard crops, and many specialty annual crops (e.g. vegetable crops) are not pollinator-dependent. Here, we present three examples from annual specialty crops.

In strawberries, intercropping with *Coriandrum sativum* or *Mentha arvensis* increased flower visitor frequency overall within the field but reduced pollinator visitation specifically to strawberry flowers, suggesting that the intercrops competed with the strawberry flowers for pollinators (Hodgkiss et al., 2019). Treatments in which strawberries were intercropped with *C. sativum* or *M. arvensis* did not show improved pollination or yields. However, in a different system, intercropping bell peppers with basil increased pollinator visits to the bell peppers specifically, with resulting increases in pepper fruit and seed production (Pereira et al., 2015). Combined, the research to date suggests mixed effects of intercropping on pollination of annual specialty crops likely dependent on both the attractiveness of the specialty crop and the intercrop. Intercrops should be attractive enough to draw in or enhance pollinator populations, but not significantly more attractive than the specialty crop so as to compete for pollinators. Intercrops with peak bloom time slightly before the peak bloom of the specialty crop may be most effective for facilitating pollination. Finally, in an example of crop diversification, more bees were found on diversified polyculture farms than on monoculture farms, including more specialized bees such as the squash bee, but there was no assessment of how these higher bee abundances affected crop yields (Guzman et al., 2019).

3.3 Perennial orchard systems

While within annual specialty crops, intercropping and on-farm crop diversification were the primary management strategies examined, for perennial orchard systems, research focuses primarily on cover crops or living ground cover with examples from almonds, vineyards, and apples. Flowering ground cover within almond orchards can increase pollinator abundance and richness as well as fruit set. On small farms in Egypt, more abundant and species-rich ground cover provided by both crops and wild plants was associated with a greater abundance and diversity of pollinators within the orchard (Norfolk et al., 2016). Importantly, the abundance and species richness of ground vegetation were also directly associated with increased almond fruit set (Norfolk et al., 2016). In Spanish almond orchards, flower abundance on the orchard floor positively predicted wild pollinator visitation to almonds and pollinator species diversity and was associated with increased almond set both via direct effects and indirect effects mediated through increasing pollinator diversity (Alomar et al., 2018). Finally, in Australia, the richness of ground cover plants in almond orchards was associated with greater native bee abundance (Saunders et al., 2013), though effects on yield were not examined. Additionally, no examples from the primary almond production region (California, USA) link within-orchard vegetation to pollinators, pollination, or yield.

Within European vineyards, greater interrow floral resource availability and percent vegetation cover increased wild pollinator abundance and diversity (Kratschmer et al., 2019), but effects on crop production were not evaluated. Finally, while the aforementioned studies all examined existing gradients in ground cover floral resources, one study found that planting cover crops within vineyards, including a mix of *Phacelia tancetifolia*, *Ammi majus*, and *Daccus carota*, resulted in higher abundance and diversity of wild bees as compared to control plots (Wilson et al., 2018). The abundances of numerous individual wild bee taxa, including *A. mellifera*, *Bombus* spp., *Halictus* spp., and *Lasioglossum* spp., were all higher in vineyard blocks with cover crops as compared to control, though again effects on crop production were not examined (Wilson et al., 2018). Yields in these vineyard studies were likely not examined in part because wine grapes have overall minimal dependence on insect pollinators (Klein et al., 2007), though in some cases insect pollination can improve yields and grape quality (Sampson et al., 2001).

3.4 Perennial forage systems (pastures and rangelands)

Perennial forage systems offer perhaps the greatest opportunity for improving agroecosystems for pollinators due to the large land area that they comprise and the diverse plant communities that they can contain. Perennial forage systems, including pasture and rangeland, are a major land use in North America and globally. Pastures and rangelands constitute approximately 22–26% of global ice-free land surface area and represent more land than is in forest or cropland (Ellis et al., 2010; Ellis and Ramankutty, 2008; Klein Goldewijk and Ramankutty, 2004). Therefore, changes to perennial forage systems could have significant impacts on pollinator communities and pollination services.

Perennial forage systems can be improved for pollinators through plant diversification and alteration of management practices including grazing intensities and regimes. Adding flowering plants to forage systems has been shown to increase the diversity of bees and other pollinators, and with consequences for pollination services provided to wild and crop plants. For example, in a study comparing forage systems with (1) bluestem grass only, (2) bluestem grass mixed with alfalfa, (3) alfalfa only, and (4) mixed native grasses, there was a higher insect pollinator abundance and richness in the plots with alfalfa and native grasses as compared to bluestem only plots, and this trend was driven largely by native sweat bees (Bhandari et al., 2018). Additionally, on working grasslands, increasing plant richness by adding legumes and forbs resulted in increased functional diversity of pollinator communities (Orford et al., 2016). Higher pollinator functional diversity in turn reduced variation in flower visitation rates to two of three sentinel crop species, including strawberry and red campion, that were placed within the pastures (Orford et al., 2016). These results conclusively show the benefits of increased forage plant diversity on pollinator abundance and diversity, with potential benefits for the pollination of adjacent crop fields.

However, one limitation to the adoption of plant diversification for pollinators in perennial forage systems is the perceived lack of benefits for the farmer given that forage system productivity is not generally thought to be pollinator-dependent. There is also limited information on how diverse forage plant communities specifically designed for pollinators could improve animal production or the health and productivity of the pasture. Increased pollinator activity and diversity could improve pollination services to flowering forage species, thereby reducing the need to reseed forage each year, but this has not been well studied. If benefits for either animal production or long-term forage quantity and quality could be demonstrated, the adoption of plant diversification for pollinators within perennial forage systems may be increased.

Along with plant diversification, altering the grazing intensity and stocking levels of pastures can influence floral resource availability and pollinator communities. However, the direction of the relationships between grazing intensity, floral resources, and pollinators is variable. While some studies show that high-intensity grazing can have negative effects on floral resources and bee abundance and diversity (Kearns and Oliveras, 2009; Kruess and Tscharntke, 2002; Xie et al., 2008), other studies show no or even positive effects of high-intensity grazing on flowers and bees (Vulliamy et al., 2006; Wallen, 2010). Some of this variability may be due to variable responses to grazing intensity across bee taxa, with bumble bees showing greater sensitivity to increased grazing intensity, particularly in the early season, as compared to sweat bees (Kimoto et al., 2012). Additionally, the relationship between grazing intensity and floral resources may be hump-shaped, suggesting that intermediate levels of grazing may be optimal for maintaining abundant and diverse pollinator communities (Lázaro et al., 2016). As with plant diversification, modifying pasture management for pollinators is unlikely to be widely adopted unless it can show benefits for animal production or system sustainability.

4 Assessing efficacy of alternative agronomic practices

Assessing the impact of agronomic practices on pollinators and farm productivity is crucial for identifying best management practices and encouraging adoption. However, most studies examined the responses of pollinator abundance and diversity to agronomic practices but did not include crop productivity and yield responses. In notable exceptions to this trend, Norfolk et al. (2016) and Alomar et al. (2018) were able to link increased ground cover diversity and flower abundance to increased almond production via direct effects and indirect effects on increased pollinator abundance and diversity. Other work including yield measures found contrasting conclusions with intercropping having no effect on strawberry production (Hodgkiss et al., 2019) but a positive effect on bell pepper production via increased pollinator activity (Pereira et al., 2015). In all these cases, the focal crop was a pollinator-dependent crop. Examples of agronomic practices for pollinators benefiting crop productivity and yields are more limited for crops perceived to have little to no benefit from insect pollinators, including vineyards (Kratschmer et al., 2019; Wilson et al., 2018) and perennial forage systems. For crops with little to no dependence on pollinators, farmer motivation to conserve or enhance pollinators within the production space may be out of general conservation principles or because such practices provide other tangible benefits to production.

Additionally, the scale at which these agronomic practices were assessed may influence the applicability of study conclusions to real-world agroecosystems. Many of the previously-described studies were conducted in small plots (Hodgkiss et al., 2019; Pereira et al., 2015; Orford et al., 2016; Wilson et al., 2018; Bhandari et al., 2018), and those that were done at the farm scale relied on existing gradients in floral resources, including weeds and intentionally planted species (Norfolk et al., 2016; Saunders et al., 2013; Guzman et al., 2019; Kratschmer et al., 2019). Few to no studies manipulated agronomic practices at the farm scale. These limitations can make it challenging to draw conclusions or make predictions for working farms. When using existing gradients, treatments are not randomly assigned, and variables of interest such as floral resource availability and diversity are likely correlated with other factors that vary across farms. For example, farms with greater cover crop diversity may also have reduced chemical inputs or a greater diversity of crop plants, which could additionally impact pollinator abundance and diversity. In small plot studies, treatments are typically randomly assigned and other variables are better controlled, but these results may not scale up the level of the whole farm. Thus, analyzing the scale at which the study was conducted, and correlations among different agronomic practices, is crucial for developing management recommendations from study results.

5 Conclusion

To summarize, agronomic strategies that seek to diversify crop and forage production to support pollinators and the services they provide continue to be evaluated, but we must be intentional and strategic in how we integrate them. Overall, given the large land area occupied by perennial pasture and rangeland along with their relatively more diverse plant communities (as compared to crop production systems), these areas may provide the greatest opportunity for altering management practices to support pollinators. However, acreage of row crops such as corn and soybean, which typically are characterized by large-scale,

Published by Burleigh Dodds Science Publishing Limited, 2023.

monoculture, input-intensive practices, has steadily increased over the last year 30 years (USDA NASS, 2020). In 2021, corn and soybean, along with all types of wheat, were grown on 227 million acres in the USA alone (USDA NASS, 2021), and additional acreage is dedicated to other row crops, specialty crops, and orchard production. Thus, agricultural lands overall represent opportunities to not only reduce adverse impacts on pollinators but also support and sustain pollinator populations through strategic design and management modifications.

One challenge associated with these approaches, particularly where agrochemicals are heavily used, is the exposure of sensitive pollinator species to toxic pesticide residues. Therefore, for these strategies to be effective within the crop production space, use of agrochemical inputs would need to be reduced in order to prevent the agroecosystem from functioning as a population sink. Integrated pest management (IPM) and holistic approaches would need to be implemented more broadly to reduce adverse impacts on diverse pollinator populations. Conversely, IPM approaches that reduce insecticide applications may support greater populations of wild pollinators and maintain or enhance crop yields (Pecenka et al., 2021), contributing to both ecosystem and agronomic resilience.

Certainly, as is the case with other ecosystem services such as carbon markets and water quality credits, payments for pollinator provisioning and habitat conservation services may be one of several approaches integrated into a system that further incentivizes many of the practices discussed in this chapter. Globally, payments for ecosystem services have increased, with hundreds of programs now in existence, but with biodiversity PES lagging behind other sectors (Boetzl et al., 2020; Geppert et al., 2020; Salzman et al., 2018). Programs and structures that support biodiversity would, by default, also promote pollinators. Finally, if crop rotations are to include more flowering, pollinator-dependent crops, crop diversity must simultaneously be maintained to support greater biodiversity, sustained pollination services, and the long-term sustainability of agriculture (Aizen et al., 2019).

6 Future trends in research

Altering agronomic management practices, particularly within the crop production space, is challenging and poses more questions than we currently have answers to. For example, the literature is scant on data that quantifies the agronomic (e.g. yield, return on investment) and ecosystem (biodiversity) tradeoffs involved when implementing the approaches described in this chapter. Can we achieve an optimal state whereby both food production and farm economics are balanced with pollinator habitat and floral resource provisioning? What are the long-term impacts of these practices compared to business as usual (current) farming practices and how do they impact agroecosystem functioning? How do fluctuations in crop commodity prices and the development of new or emerging markets impact farmer willingness to alter their crop management practices? What other ecosystem services can be bundled with pollinator services to increase the use of conservation-focused agronomic practices across the agricultural landscape?

Future research should address these and other relevant questions if we are to transition from conventional production practices to more agroecologically based approaches. Additionally, within the context of different cropping systems and farming practices, research has yet to address how different pollinator populations will respond to these changes under current and future climate scenarios, and how they respond to landscape-level changes beyond the crop production space. These uncertainties can only be addressed and implemented at a transformative level by more transdisciplinary research that bridges disciplines in the biophysical and social sciences and invites co-production of research with farmers and other stakeholders.

7 Where to look for further information

The following articles and guides provide a good overview:

- Sustainable Agriculture Research and Education. (2013). *Managing Cover Crops Profitably: Third Edition*. Available at https://www.sare.org/resources/managing-cover-crops-profitably-3rd-edition/.
- Green, R. E., Cornell, S. J., Scharlemann, J. P. W. and Balmford, A. (2005). Farming and the Fate of Wild Nature, *Science* 307(5709), 550–555.
- Kremen, C. and Merenlender, A. M. (2018). Landscapes That Work for Biodiversity and People, *Science* 362(6412).
- Nicholls, C. A. and Altieri, M. A. (2013). Plant biodiversity enhances bees and other insect pollinators in agroecosystems. A review, *Agronomy for Sustainable Development*, 33, 257–274.
- Phalan, B., Onial, M., Balmford, A. and Green, R. E. (2011). Reconciling food production and biodiversity conservation: Land sharing and land sparing compared, *Science* 333(6047), 1289–1291.
- Potts, S. G., Biesmeijer, J. C., Kremen, C., Neumann, P., Schweiger, O. and Kunin, W. E. (2010). Global pollinator declines: trends, impacts and drivers, *Trends in Ecology & Evolution* 25(6), 345–353.

Key research and resources in this area can be found at the following organizations:

- Food and Agriculture Organization of the United Nations (https://www.fao.org/pollination/en/).

- United States Department of Agriculture (USDA) (www.usda.gov).
- Xerces Society for Invertebrate Conservation (https://www.xerces.org/pollinator-conservation).
- Pollinator Partnership (https://www.pollinator.org/international).

8 References

Aguilar, J., Gramig, G. G., Hendrickson, J. R., Archer, D. W., Forcella, F. and Liebig, M. A. (2015). Crop species diversity changes in the United States: 1978–2012, *PLoS ONE* 10(8), e0136580.

Aizen, M. A., Aguiar, S., Biesmeijer, J. C., Garibaldi, L. A., Inouye, D. W., Jung, C., Martins, D. J., Medel, R., Morales, C. L., Ngo, H., Pauw, A., Paxton, R. J., Sáez, A. and Seymour, C. L. (2019). Global productivity is threatened by increasing pollinator dependence without a parallel increase in crop diversification, *Global Change Biology* 25(10), 3516–3527. doi:10.1111/gcb.14736.

Alomar, D., González-Estévez, M. A., Traveset, A. and Lázaro, A. (2018). The intertwined effects of natural vegetation, local flower community, and pollinator diversity on the production of almond trees, *Agriculture, Ecosystems and Environment* 264, 34–43. doi:10.1016/j.agee.2018.05.004.

Andersson, G. K. S., Ekroos, J., Stjernman, M., Rundlöf, M. and Smith, H. G. (2014). Effects of farming intensity, crop rotation and landscape heterogeneity on field bean pollination, *Agriculture, Ecosystems and Environment* 184, 145–148. doi:10.1016/j.agee.2013.12.002.

Appenfeller, L. R., Lloyd, S. and Szendrei, Z. (2020). Citizen science improves our understanding of the impact of soil management on wild pollinator abundance in agroecosystems, *PLoS ONE* 15(3), e0230007. doi:10.1371/journal.pone.0230007.

Bhandari, K. B., West, C. P., Longing, S. D., Brown, C. P., Green, P. E. and Barkowsky, E. (2018). Pollinator abundance in semiarid pastures as affected by forage species, *Crop Science* 58(6), 2665–2671.

Boetzl, F. A., Krauss, J., Heinze, J., Hoffmann, H., Juffa, J., König, S., Krimmer, E., Prante, M., Martin, E. A., Holzschuh, A. and Steffan-Dewenter, I. (2020). A multitaxa assessment of the effectiveness of agri-environmental schemes for biodiversity management, *Proceedings of the National Academy of Sciences of the United States of America* 118(10), e2016038118. doi:10.1073/pnas.2016038118.

Boyle, N. K. and Pitts-Singer, T. L. (2019). Assessing blue orchard bee (Osmia lignaria) propagation and pollination services in the presence of honey bees (Apis mellifera) in Utah tart cherries (Osmia lignaria), *PeerJ* 7, e7639. doi:10.7717/peerj.7639.

Brennan, E. B. (2013). Agronomic aspects of strip intercropping lettuce with alyssum for biological control of aphids, *Biological Control* 65(3), 302–311.

Brennan, E. B. (2016). Agronomy of strip interropping broccoli with alyssum for biological control of aphids, *Biological Control* 97, 109–119.

Bretagnolle, V. and Gaba, S. (2015). Weeds for bees? A review, *Agronomy for Sustainable Development* 35(3), 891–909.

Bryan, C. J., Sipes, S. D., Arduser, M., Kassim, L., Gibson, D. J., Scott, D. A. and Gage, K. L. (2021). Efficacy of cover crops for pollinator habitat provision and weed suppression, *Environmental Entomology* 50(1), 208–221.

Butters, J., Murrell, E., Spiesman, B. J. and Kim, T. N. (2022). Native flowering border crops attract high pollinator abundance and diversity, providing growers the opportunity to enhance pollination services, *Environmental Entomology* 51(2), 492–504.

Campbell, J. W., Irvin, A., Irvin, H., Stanley-Stahr, C. and Ellis, J. D. (2016). Insect visitors to flowering buckwheat, *Fagopyrum esculentum* (Polygonales: Polygonaceae), in North-Central Florida, *Florida Entomologist* 99(2), 264–268.

Cane, J. H. (1994). Nesting biology and mating behavior of the southeastern blueberry bee, Habropoda laboriosa (Hymenoptera: Apoidea), *Journal of the Kansas Entomological Society* 67(3), 236–241.

Cane, J. H. (2008). A native ground-nesting bee (Nomia melanderi) sustainably managed to pollinate alfalfa across an intensively agricultural landscape, *Apidologie* 39(3), 315–323. doi:10.1051/apido:2008013.

Carvalheiro, L. G., Veldtman, R., Shenkute, A. G., Tesfay, G. B., Pirk, C. W. W., Donaldson, J. S. and Nicolson, S. W. (2011). Natural and within-farmland biodiversity enhances crop productivity, *Ecology Letters* 14(3), 251–259. doi:10.1111/j.1461-0248.2010.01579.x.

Dainese, M., Riedinger, V., Holzschuh, A., Kleijn, D., Scheper, J. and Steffan-Dewenter, I. (2018). Managing trap-nesting bees as crop pollinators: spatiotemporal effects of floral resources and antagonists, *Journal of Applied Ecology* 55(1), 195–204. doi:10.1111/1365-2664.12930.

Danforth, B. N., Minckley, R. L., Neff, J. L. and Fawcett, F. (2019). *The Solitary Bees: Biology, Evolution, Conservation*. Princeton University Press. doi:10.2307/j.ctvd1c929.

Cubins, J. A., Wells, M. S., Frels, K., Ott, M. A., Forcella, F., Johnson, G. A., Walia, M. K., Becker, R. L. and Gesch, R. W. (2019). Management of pennycress as a winter annual cash cover crop: a review, *Agronomy for Sustainable Development* 39(5), 46.

Eberle, C. A., Thom, M. D., Nemec, K. T., Forcella, F., Lundgren, J. G., Gesch, R. W., Riedell, W. E., Papiernik, S. K., Wagner, A., Peterson, D. H. and Eklund, J. J. (2015). Using pennycress, Camelina, and canola cash cover crops to provision pollinators, *Industrial Crops and Products* 75, 20–25.

Ellis, K. E. and Barbercheck, M. E. (2015). Management of overwintering cover crops influences floral resources and visitation by native bees, *Environmental Entomology* 44(4), 999–1010.

Ellis, E. C., Klein Goldewijk, K., Siebert, S., Lightman, D. and Ramankutty, N. (2010). Anthropogenic transformation of the biomes, 1700 to 2000, *Global Ecology and Biogeography* 19.

Ellis, E. C. and Ramankutty, N. (2008). Putting people in the map: anthropogenic biomes of the world, *Frontiers in Ecology and the Environment* 6(8), 439–447.

Forcella, F., Patel, S., Lenssen, A. W., Hoerning, C., Wells, M. S., Gesch, R. W. and Berti, M. T. (2020). Weather and landscape influences on pollinator visitation of flowering winter oilseeds (field pennycress and winter Camelina), *Journal of Applied Entomology* 145(4), 286–294.

Gardner, E., Breeze, T. D., Clough, Y., Smith, H. G., Baldock, K. C. R., Campbell, A., Garratt, M. P. D., Gillespie, M. A. K., Kunin, W. E., McKerchar, M., Potts, S. G., Senapathi, D., Stone, G. N., Wäckers, F., Westbury, D. B., Wilby, A. and Oliver, T. H. (2021). Field boundary features can stabilise bee populations and the pollination of mass-flowering crops in rotational systems, *Journal of Applied Ecology* 58(10), 2287–2304. doi:10.1111/1365-2664.13948.

Gathmann, A. and Tscharntke, T. (2002). Foraging ranges of solitary bees, *Journal of Animal Ecology* 71(5), 757–764. doi:10.1046/j.1365-2656.2002.00641.x.

Gemmill-Herren, B. and Ochieng', A. O. (2008) Role of native bees and natural habitats in eggplant (Solanum melongena) pollination in Kenya, *Agriculture, Ecosystems and Environment* 127(1-2), 31-36. doi:10.1016/j.agee.2008.02.002.

Geppert, C., Hass, A., Földesi, R., Donkó, B., Akter, A., Tscharntke, T. and Batáry, P. (2020). Agri-environment schemes enhance pollinator richness and abundance but bumblebee reproduction depends on field size, *Journal of Applied Ecology* 57(9), 1818-1828.

Gesch, R. W., Archer, D. W. and Berti, M. T. (2014). Dual cropping winter Camelina with soybean in the northern corn belt, *Agronomy Journal* 106(5), 1735-1745.

Gowton, C. M., Cabra-Arias, C. and Carrillo, J. (2021). Intercropping with peppermint increases ground dwelling insect and pollinator abundance and decreases *Drosophila suzukii* in fruit, *Frontiers in Sustainable Food Systems* 5, 700842.

Greenleaf, S. S., Williams, N. M., Winfree, R. and Kremen, C. (2007). Bee foraging ranges and their relationship to body size, *Oecologia* 153(3), 589-596.

Guzman, A., Chase, M. and Kremen, C. (2019). On-farm diversification in an agriculturally-dominated landscape positively influences specialist pollinators, *Frontiers in Sustainable Food Systems* 3, 87. doi:10.3389/fsufs.2019.00087.

Hicks, D. M., Ouvrard, P., Baldock, K. C. R., Baude, M., Goddard, M. A., Kunin, W. E., Mitschunas, N., Memmott, J., Morse, H., Nikolitsi, M., Osgathorpe, L. M., Potts, S. G., Robertson, K. M., Scott, A. V., Sinclair, F., Westbury, D. B. and Stone, G. N. (2016). Food for pollinators: quantifying the nectar and pollen resources of urban flower meadows, *PLoS ONE* 11(6), e0158117. doi:10.1371/journal.pone.0158117.

Hodgkiss, D., Brown, M. J. F. and Fountain, M. T. (2019). The effect of within-crop floral resources on pollination, aphid control and fruit quality in commercial strawberry, *Agriculture, Ecosystems and Environment* 275, 112-122. doi:10.1016/j.agee.2019.02.006.

Housman, M., Tallman, S., Jones, C., Miller, P. and Zabinski, C. (2021). Soil biological response to multi-species cover crops in the Northern Great Plains, *Agriculture, Ecosystems and Environment* 313, 107373.

Johnson, G. A., Wells, M. S., Anderson, K., Gesch, R. W., Forcella, F. and Wyse, D. L. (2017). Yield tradeoffs and nitrogen between pennycress, camelina, and soybean in relay- and double-crop systems, *Agronomy Journal* 109, 2128-2135. doi:10.2134/agronj2017.02.0065.

Julier, H. E. and Roulston, T. H. (2009). Wild bee abundance and pollination service in cultivated pumpkins: farm management, nesting behavior and landscape effects, *Journal of Economic Entomology* 102(2), 563-573. doi:10.1603/029.102.0214.

Kearns, C. A. and Oliveras, D. M. (2009). Environmental factors affecting bee diversity in urban and remote grassland plots in Boulder, Colorado, *Journal of Insect Conservation* 13(6), 655-665. doi:10.1007/s10841-009-9215-4.

Kimoto, C., DeBano, S. J., Thorp, R. W., Taylor, R. V., Schmalz, H., DelCurto, T., Johnson, T., Kennedy, P. L. and Rao, S. (2012). Short-term responses of native bees to livestock and implications for managing ecosystem services in grasslands, *Ecosphere* 3(10), doi:10.1890/ES12-00118.1.

Klein, A. M., Vaissiere, B. E., Cane, J. H., Steffan-Dewenter, I., Cunningham, S. A., Kremen, C. and Tscharntke, T. (2007). Importance of pollinators in change landscapes for world crops, *Proceedings of the Royal Society. Series B,-Biological Sciences* 274, 303-313.

Klein Goldewijk, K. and Ramankutty, N. (2004). Land cover change over the last three centuries due to human activities: the availability of new global data sets, *GeoJournal* 61(4), 335-344.

Kordbacheh, F., Liebman, M. and Harris, M. (2020). Strips of prairie vegetation placed within row crops can sustain native bee communities, *PLoS ONE* 15(10), e0240354.

Kratschmer, S., Pachinger, B., Schwantzer, M., Paredes, D., Guzmán, G., Goméz, J. A., Entrenas, J. A., Guernion, M., Burel, F., Nicolai, A., Fertil, A., Popescu, D., Macavei, L., Hoble, A., Bunea, C., Kriechbaum, M., Zaller, J. G. and Winter, S. (2019). Response of wild bee diversity, abundance, and functional traits to vineyard inter-row management intensity and landscape diversity across Europe, *Ecology and Evolution* 9(7), 4103–4115. doi:10.1002/ece3.5039.

Kruess, A. and Tscharntke, T. (2002). Grazing intensity and the diversity of grasshoppers, butterflies, and trap-nesting bees and wasps, *Conservation Biology* 16(6), 1570-1580. doi:10.1046/j.1523-1739.2002.01334.x.

Larson, J. L., Redmond, C. T. and Potter, D. A. (2013). Assessing insecticide hazard to bumble bees foraging on flowering weeds in treated lawns, *PLoS ONE* 8(6), e66375. doi:10.1371/journal.pone.0066375.

Lázaro, A., Tscheulin, T., Devalez, J., Nakas, G. and Petanidou, T. (2016). Effects of grazing intensity on pollinator abundance and diversity, and on pollination services, *Ecological Entomology* 41(4), 400–412. doi:10.1111/een.12310.

Lehmkuhler, J. W., Ramos, M. H. and Albrecht, K. A. (2007). Cupplant silage as a replacement for corn silage in growing beef cattle diets, *Forage and Grazinglands* 5(1), 1–6.

Mallinger, R. E., Bradshaw, J., Varenhorst, A. J. and Prasifka, J. R. (2018). Native Solitary Bees Provide Economically Significant Pollination Services to Confection Sunflowers (Helianthus annuus L.) (Asterales: Asteraceae) Grown Across the Northern Great Plains, *Journal of Economic Entomology* 112(1), 40–48.

Mallinger, R. E., Franco, J. G., Prischmann-Voldseth, D. A. and Prasifka, J. R. (2019). Annual cover crops for managed and wild bees: optimal plant mixtures depend on pollinator enhancement goals, *Agriculture, Ecosystems and Environment* 273, 107–116.

Marja, R., Viik, E., Mänd, M., Phillips, J., Klein, A.-M. and Batáry, P. (2018). Crop rotation and agri-environment schemes determine bumblebee communities via flower resources, *Journal of Applied Ecology* 55(4), 1714–1724. doi:10.1111/1365-2664.13119.

Martin, A. R., Cadotte, M. W., Isaac, M. E., Milla, R., Vile, D. and Violle, C. (2019). Regional and global shifts in crop diversity through the Anthropocene, *PLoS ONE* 14(2), e0209788.

Mueller, A. L., Berger, C. A., Schittenhelm, S., Stever-Schoo, B. and Dauber, J. (2020). Water availability affects nectar sugar production and insect visitation on the cup plant *Silphium perfoliatum* L. (Asteraceae), *Journal of Agronomy and Crop Science* 206(5), 529–537.

Nicholls, C. I. and Altieri, M. A. (2013). Plant biodiversity enhances bees and other insect pollinators in agroecosystems. A review, *Agronomy for Sustainable Development* 33(2), 257–274.

Norfolk, O., Eichhorn, M. P. and Gilbert, F. (2016). Flowering ground vegetation benefits wild pollinators and fruit set of almond within arid smallholder orchards, *Insect Conservation and Diversity* 9(3), 236–243. doi:10.1111/icad.12162.

Norris, S. T., Blackshaw, R. P., Critchley, C. N. R., Dunn, R. M., Smith, K. E., Williams, J., Randall, N. P. and Murray, P. J. (2018). Intercropping flowering plants in maize systems increases pollinator diversity, *Agricultural and Forest Entomology* 20, 246-254. doi:10.1111/afe.12251.

Orford, K. A., Murray, P. J., Vaughan, I. P. and Memmott, J. (2016). Modest enhancements to conventional grassland diversity improve the provision of pollination services, *Journal of Applied Ecology* 53(3), 906–915. doi:10.1111/1365-2664.12608.

Ott, M. A., Eberle, C. A., Thom, M. D., Archer, D. W., Forcella, F., Gesch, R. W. and Wyse, D. L. (2019). Economics and agronomics of relay-cropping pennycress and Camelina with soybean in Minnesota, *Agronomy Journal* 111(3), 1281–1292.

Patel, S., Lenssen, A. W., Moore, K. J., Mohammed, Y. A., Gesch, R. W., Wells, M. S., Johnson, B. L., Berti, M.T. and Matthees, H. L. (2021). Interseeded pennycress and camelina yield and influence on row crops, *Agronomy Journal* 113, 2629–2647. doi:10.1002/agj2.20655.

Pecenka, J. R., Ingwell, L. L., Foster, R. E., Krupke, C. H. and Kaplan, I. (2021). IPM reduces insecticide applications by 95% while maintaining or enhancing crop yields through wild pollinator conservation, *Proceedings of the National Academy of Sciences of the United States of America* 118(44), e2108429118. doi:10.1073/pnas.2108429118.

Pereira, A. L. C., Taques, T. C., Valim, J. O. S., Madureira, A. P. and Campos, W. G. (2015). The management of bee communities by intercropping with flowering basil (Ocimum basilicum) enhances pollination and yield of bell pepper (Capsicum annuum), *Journal of Insect Conservation* 19(3), 479–486. doi:10.1007/s10841-015-9768-3.

Phalan, B., Onial, M., Balmford, A. and Green, R. E. (2011). Reconciling food production and biodiversity conservation: land sharing and land sparing compared, *Science* 333(6047), 1289–1291.

Pitts-Singer, T. L. and Cane, J. H. (2011). The alfalfa leafcutting bee, Megachile rotundata: the world's most intensively managed solitary bee, *Annual Review of Entomology* 56, 221–237. doi:10.1146/annurev-ento-120709-144836.

Requier, F., Odoux, J. F., Tamic, T., Moreau, N., Henry, M., Decourtye, A. and Bretagnolle, V. (2015). Honey bee diet in intensive farmland habitats reveals an unexpectedly high flower richness and a major role of weeds, *Ecological Applications : a Publication of the Ecological Society of America* 25(4), 881–890. doi:10.1890/14-1011.1.

Rering, C. C., Franco, J. G., Yeater, K. M. and Mallinger, R. E. (2020). Drought stress alters floral volatiles and reduces floral rewards, pollinator activity, and seed set in a global plant, *Ecosphere* 11(9), e03254.

Salzman, J., Bennett, G., Carroll, N., Goldstein, A. and Jenkins, M. (2018). The global status and trends of payments for ecosystem services, *Nature Sustainability* 1(3), 136–144.

Sampson, B., Noffsinger, S., Gupton, C. and Magee, J. (2001). Pollinator biology of the muscadine grape, *Hortscience* 36(1), 120–124.

Sanderson, M., Johnson, H. and Hendrickson, J. (2018). Cover crop mixtures grown for annual forage in a semi-arid environment, *Agronomy Journal* 110(2), 525–534.

Sardiñas, H. S., Tom, K., Ponisio, L. C., Rominger, A. and Kremen, C. (2016). Sunflower (Helianthus annuus) pollination in California's Central Valley is limited by native bee nest site location, *Ecological Applications : a Publication of the Ecological Society of America* 26(2), 438–447. doi:10.1890/15-0033.

Saunders, M. E., Luck, G. W. and Mayfield, M. M. (2013). Almond orchards with living ground cover host more wild insect pollinators, *Journal of Insect Conservation* 17(5), 1011–1025. doi:10.1007/s10841-013-9584-6.

Shuler, R. E., Roulston, T. H. and Farris, G. E. (2005). Farming practices influence wild pollinator populations on squash and pumpkin, *Journal of Economic Entomology* 98(3), 790–795.

Sidemo-Holm, W., Carrié, R., Ekroos, J., Lindström, S. A. M. and Smith, H. G. (2021). Reduced crop density increases floral resources to pollinators without affecting crop yield in organic and conventional fields, *Journal of Applied Ecology* 58(7), 1421-1430. doi:10.1111/1365-2664.13887.

Skidmore, A. R., Short, C. A., Dills, C., Goodell, K. and Bessin, R. T. (2019). Preference of Peponapis pruinosa (Hymenoptera: Apoidea) for tilled soils regardless of soil management system, *Environmental Entomology* 48(4), 961–967. doi:10.1093/ee/nvz052.

Stephen, W. P. (1960). Artificial bee beds for the propagation of the alkali bee, Nomia melanderi, *Journal of Economic Entomology* 53(6), 1025–1030. doi:10.1093/jee/53.6.1025.

Tepedino, V. J. (1981). The pollination efficiency of the squash bee (Peponapis pruinosa) and the honey bee (Apis mellifera) on summer squash (Cucurbita pepo), *Journal of the Kansas Entomological Society* 54(2), 359–377.

Tscharntke,T., Klein, A. M., Kruess, A., Steffan-Dewenter, I. and Thies, C. (2005). Landscape perspectives on agricultural intensification and biodiversity – ecosystem service management, *Ecology Letters* 8(8), 857–874.

Tscharntke, T., Clough, Y., Wanger, T. C., Jackson, L., Motzke, I., Perfecto, I., Vandermeer, J. and Whitbread, A. (2012). Global food security, biodiversity conservation and the future of agricultural intensification, *Biological Conservation* 151(1), 53–59.

Ullmann, K. S., Meisner, M. H. and Williams, N. M. (2016). Impact of tillage on the crop pollinating, ground-nesting bee, Peponapis pruinosa in California, *Agriculture, Ecosystems and Environment* 232, 240–246. doi:10.1016/j.agee.2016.08.002.

United State Department of Agriculture-Risk Management (2013). Agency (USDA-RMA), *Contract Price Addendum, a Risk Management Agency Fact Sheet*. Accessed May 2022, Available at: https://www.rma.usda.gov/pubs/rme/2013revisedorganicsfactsheet.pdf.

United States Department of Agriculture (2020). National Agricultural Statistics Service (USDA-NASS), *Quick Stats*. Accessed April 2022, Available at: https://quickstats.nass.usda.gov/.

United States Department of Agriculture (2021). National Agricultural Statistics Service (USDA-NASS), *Quick Stats*. Accessed April 2022, Available: https://quickstats.nass.usda.gov/.

van der Slujis, J. P., Simon-Delos, N., Goulson, D., Maxim, L., Bonmatin, J.-M. and Belzunces, L. P. (2013). Neonicotinoids, bee disorders and the sustainability of pollinator services, *Current Opinion in Environmental Sustainability* 5(3–4), 293–305.

Van Sambeek, J. W. (2017). Orchard management using cover crops to improve soil health and pollinator habitat in the midwestern United States, *The Nutshell* 71(3), 32–45.

Van Tassel, D. L., Albrecht, K. A., Bever, J. D., Boe, A. A., Brandvain, Y., Crews, T. E., Gansberger, M., Gerstberger, P., Gonzalez-Paleo, L., Hulke, B. S., Kane, N. C., Johnson, P. J., Pestsova, E. G., Picasso Risso, V.D., P., J.R., Ravetta, D. A., Schlautman, B., Sheaffer, C. C., Smith, K. P., Speranza, P. R., Turner, M. K., Vilela, A. E., von Gehren, P. and Wever, C. (2017). Accelerating Silphium domestication: an opportunity to develop new crop ideotypes and breeding strategies informed by multiple disciplines, *Crop Science* 57(3), 1274–1284.

Vulliamy, B., G. Potts, S. and G. Willmer, P. (2006). The effects of cattle grazing on plant-pollinator communities in a fragmented Mediterranean landscape, *Oikos* 114(3), 529–543.

Wallen, K. E. (2010). *Fire, Grazing, and Herbicide: Effects of Rangeland Management on Bee and Floral Communities in Oklahoma*. Oklahoma State University.

Wilson, H., Wong, J. S., Thorp, R. W., Miles, A. F., Daane, K. M. and Altieri, M. A. (2018). Summer flowering cover crops support wild bees in vineyards, *Environmental Entomology* 47(1), 63–69. doi:10.1093/ee/nvx197.

Witter, S., Nunes-Silva, P., Lisboa, B. B., Tirelli, F. P., Sattler, A., Hilgert-Moreira, S. B. and Blochtein, B. (2015). Stingless bees as alternative pollinators of canola, *Journal of Economic Entomology* 108(3), 880–886.

Xie, Z., Williams, P. H. and Tang, Y. (2008). The effect of grazing on bumblebees in the high rangelands of the eastern Tibetan Plateau of Sichuan, *Journal of Insect Conservation* 12(6), 695–703. doi:10.1007/s10841-008-9180-3.

Chapter 11

Ecological network approaches for promoting pollinators in agriculture

Darren M. Evans and Fredric M. Windsor, Newcastle University, UK

1 Introduction

We know much about biodiversity in many parts of the world, but still understand very little about how species interact with each other, both directly and indirectly. This is important as species interactions are as fundamental as species richness for understanding biodiversity dynamics and the response of ecosystems to global environmental change (Montoya, 2015). All organisms are embedded within complex networks of ecological interactions, ranging from antagonistic processes such as predation, through to mutualisms between plants and pollinators. What happens to one species or community can have knock-on effects for other species within an ecosystem. This in turn can affect ecological processes, such as pollination, that are a result of the complex ways in which species interact with each other through coevolution. These patterns can be studied through the lens of ecological networks, which are representations of the interactions that occur between species within a community, allowing properties of the network, such as structure and stability, to be examined.

Within an ecological network context, pollination can be studied as a mutualism between animal-pollinated plants and flower-visiting animals (i.e. a bipartite, mutualistic network). Although ecological networks can represent a

http://dx.doi.org/10.19103/AS.2022.0111.21

range of interaction types across multiple trophic levels (e.g. an ocean food web), it is the burgeoning study of plant-pollinator interactions in particular that has significantly contributed to our overall understanding of the structure and dynamics of mutualistic networks (Bascompte and Jordano, 2013). Indeed, many empirical plant-pollinator networks to date have been constructed in agro-ecosystems, even if not formally acknowledged, and offer a range of applications for understanding and enhancing pollination in agriculture.

Ecological networks can be used to assess: the effects of farm management intensity on plant-pollinator interactions, the importance of in-crop and off-crop plants as nectar and pollen sources; the contribution of non-crop habitats on farms; which species are 'important' (see later); the fragility of the network to species extinctions; and whether management interventions have been successful. Networks can also be used to plan management interventions and restoration schemes for promoting pollinators through the management of plants within the crop (e.g. weeds), in adjacent semi-natural habitats (e.g. pollinator-friendly strips via an agri-environment scheme) and/or in natural habitats that provide a source of pollinators within a wider farmed landscape.

In this chapter, we focus on pollination as a network-based phenomenon in agricultural systems. Starting with the first principles of plant-pollinator interactions, we describe how networks can be used to understand important ecological processes in agricultural environments. We demonstrate how concepts from the field of network ecology can be applied to develop strategies for sustainable management at different scales, from a single field to national landscapes. However, gaps in our understanding of pollination networks - for instance, the importance of nocturnal pollinators - limit the extent to which we understand the influence of pollinators in agricultural systems. We highlight areas that need further research and suggest ways to better understand plant-pollinator interactions and how they can be used to promote sustainable agriculture.

2 Plant-pollinator networks

2.1 Basics of pollination networks

Biotic pollination involves two groups of organisms, plants and animal pollinators, interacting with one another in a mutually beneficial manner. These interactions are generally considered to be trophic (i.e. organisms feeding on pollen and nectar), yet they have not often been included in food web studies (Hale et al., 2020). As such, it is common in community ecology to represent the interactions between plants and pollinators as bipartite networks (Fig. 1). A qualitative network simply shows which pollinator species (both vertebrates and invertebrates) are interacting with different plant species in flowers (Fig.

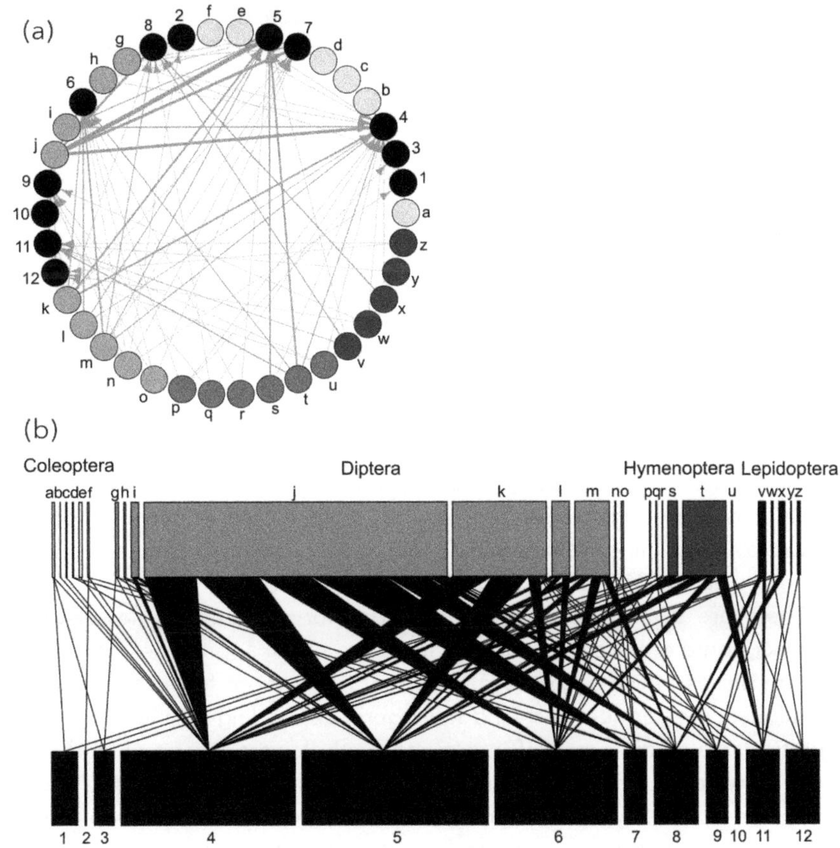

Figure 1 An example of a plant-pollinator network. (a) A qualitative representation of the network. (b) A quantitative representation of the network where the width of the bars represents the abundance of the insects and plants, and the width of the links represents the strength of the interactions (i.e. the frequency of visitation by each species of insect). Individual plant species are numbered, with pollinator species represented by a different letter. The latter are shaded differently to represent different insect orders.

1a), whereas a quantitative network shows the density of flowers, insect visitors and the frequency of the interactions between them, in common units (Fig. 1b). Seminal work by Memmott (1999) showed that the vast majority of plant-pollinator interactions are embedded in a complex ecological network, which can be studied in the same manner as conventional food webs. Focusing on an English meadow, she described how a quantitative plant visitation network was constructed for the plants and pollinators in the community (similar to the visualisation in Fig. 1b). Briefly, flower-visiting insects were sampled in a meadow plot (250 × 150 m in size) by walking transects lengthways and recording the identity of the species in each observed flower-insect interaction. Interactions were recorded in a swath 2 m wide along each transect, and an average of two transects per day were studied. Insects were either identified in the field or were collected and identified in the laboratory. In just a 1-month period, a total of 2722 interactions were recorded among 26 species of flowering plants and 79 species of pollinators, yet Memmott was quick to point out that this was only a small proportion of the expected total flower-insect interactions over a complete flowering season. Thus, we can see that although plant-pollinator networks can be relatively quick and straightforward to generate, important questions regarding sampling methodology, effort and other biases need to be considered before they can be formally analysed.

It is worth noting at this point that many studies of flower-visitor interactions refer to them as plant-pollinator networks, but this is technically incorrect unless pollination effectiveness is measured (e.g. direct contact with reproductive tissues of the flower and the transport and deposition of pollen), as flower visitation is generally considered to be a poor proxy for pollination (King et al., 2013). Thus, more informative and realistic plant-pollinator networks can be created by combining visitation data with pollen deposition measurements (Ballantyne et al., 2015), although most, to date, do not.

There are a range of different methods for constructing flower-visitor networks, some of which we cover in this chapter. All approaches have some inherent biases, which could have important implications for researchers and practitioners (i.e. farmers and land managers) depending on their questions. In particular, a whole range of interactions with flowers can be missed if the sampling window is too brief, or flowers and insects can be missed during surveys because they are too small to see and catch. Similarly, pollinating animals may operate at much larger scales than the focal plants within a defined study plot, so the full range of interactions are missed. To investigate the potential for variation in the structure of plant-pollinator interactions arising from different construction techniques, Gibson et al. (2011) compared plant-pollinator networks using two different methods (25 m transects versus timed observations that involved watching an approximately semi-circular area containing at least one individual of a single plant species for 15 min) at a single site and over three field seasons. Not surprisingly, these methods differed in the evenness of observation effort among plant species in the community, and this was shown to significantly affect the number of unique interactions observed in the network. The different sampling techniques ultimately altered some aspects of the network structure (see Section 2.2), which has implications for plant-pollinator studies. Perhaps not surprisingly, to obtain more complete and highly resolved species-interaction networks, studies are increasingly advocating a combination of sampling methods. Quintero et al. (2021) provide a good methodological overview of data-merging approaches for the construction of mutualistic networks pertinent to plant-pollinator interactions.

In addition to standardised transect surveys and timed observations of focal plants, there are a number of other methods and tools that can be implemented to collect data on plant-pollinator interactions. The most recent advances use DNA-based methods to decipher interactions by collecting samples and/or environmental DNA (eDNA) from insects and plants (Evans and Kitson, 2020). For example, Thomsen and Sigsgaard (2019) performed eDNA metabarcoding of flowers from several different plant species (i.e. a phytocentric approach) in Denmark using two sets of generic arthropod primers (targeting the 16S rRNA and COI mitochondrial genes). They found at least 135 arthropod species in 67 families and 14 orders, together representing diverse ecological groups including pollinators, parasitoids, gall inducers, predators and phytophagous species. Using an insect-focused (i.e. zoocentric) approach, Macgregor et al. (2019) constructed and compared moth pollen-transport networks using two methods: light microscopy and DNA metabarcoding. DNA metabarcoding detected pollen on more individual moths and detected multiple pollen types on more individuals than microscopy. After aggregating individuals of each species, metabarcoding detected more interactions per moth species. Importantly, they found that pollen-transport network metrics differed between methods because of variation in the ability of each to detect multiple pollen types per moth and to separate morphologically similar or related pollen. The DNA approach also detected unexpected but plausible moth-plant interactions, revealing new detail about nocturnal pollination systems.

Advances in computer vision and deep learning have a range of applications for studying plant-pollinator interactions. Cameras and other sensors can effectively, continuously and noninvasively perform entomological observations throughout diurnal and seasonal cycles (Høye et al., 2021). The physical appearance of specimens can also be captured by automated imaging in the laboratory and when trained on these data, deep learning models can provide estimates of insect abundance, biomass and diversity. Thus, there is considerable promise for not only counting and identifying insects in the field using this technology but capturing information on the plants they are interacting with (and going beyond simply observing which insects are visiting the flowers).

Before formal network analysis can begin, complete, highly resolved plant-pollinator networks are required. There are a number of ways in which this can be achieved for plant-pollinator networks. First, as already described, merging different sampling methods (e.g. plant and insect-focused approaches) is likely to produce the most complete networks and compensate for biases inherent to some of the methods described earlier (Quintero et al., 2021; Evans and Kitson, 2020; Wirta et al., 2014). Second, recent work has highlighted that most plant-pollinator networks to date are incomplete because they fail to include nocturnal pollinators (Macgregor et al., 2015). Combined diurnal and nocturnal networks may reveal the extent of redundancy and complementarity of pollination interactions in ecosystems. Indeed, it would appear that nocturnal moths are important but overlooked pollinators based on studies that have examined the pollen they carry (Devoto et al., 2011; Banza et al., 2015; Macgregor et al., 2017). Third, the quantitative analysis of ecological networks can be directly affected by the proportion of all interactions that have been observed. Sampling completeness of interaction data can be estimated using abundance-based and incidence-based species richness estimators (Macgregor et al., 2017; Traveset et al., 2015; Chacoff et al., 2012). The assessment of interaction datasets assists with the quantitative and comparative analyses and interpretation of network properties.

Transitioning towards highly detailed flower-visitor and pollination networks will allow for a greater level of predictive power and provides an important step towards operationalising network ecology for sustainable agricultural management. We address this challenge in Section 4.

2.2 Structure matters

Investigating the structure of plant–pollinator networks can tell us much about how they function, the importance of species or groups of species, their fragility to species loss and their potential ability to influence agricultural management strategies to benefit crop production. Basic information about which animals are interacting with which plants can be used to generate networks that shine a light on the structure, complexity and stability of ecosystems around the world. It is not our intention to provide a comprehensive overview of bipartite network topology, visualisation and analyses, which are fully described elsewhere (Dormann et al., 2009; Ings et al., 2009; Memmott, 2009; Tylianakis et al., 2010; Bascompte and Jordano, 2013). Rather, our aim is to introduce readers unfamiliar with network ecology to the basics in order to make informed decisions regarding agricultural management and/or pollinator conservation. This requires just a small step change in the way plants and pollinating insects are conventionally sampled. We start by looking at some simple network metrics/descriptors that are commonly used to describe the structure and complexity of plant-pollinator networks (Table 1) before showing how these can be used to answer a range of questions in ecology and evolution, with particular reference to agriculture.

Network ecology is rooted in the mathematical field of graph theory where a 'graph' is a representation of a set of elements, some of which are connected. In this context, a graph is made up of 'vertices' (singular: a 'vertex') or 'nodes' which are connected by 'edges' or 'links'. In essence, that is it: a network comprises pairs of nodes connected by links. For plant-pollinator networks, individual plants and pollinating animals are represented as nodes, with links representing the interactions between different animals and plants (Figs 1

Table 1 Some basic and commonly used network-level metrics to describe plant-pollinator networks. These are mostly qualitative examples, but a range of quantitative equivalents are often available.

Metric	Name	Definition
S	Species richness	Number of taxa (nodes) in the network
L	Trophic links	Number of interactions between the plants and animals in the network
L/S	Link density	Mean number of links per species
C	Connectance	The proportion of possible links between species that are realised. The most conventional algorithm is 'Directed Connectance' where $C = (L/S^2)$
G	Generality	Mean effective number of plant species per pollinator species
V	Vulnerability	Mean effective number of pollinator species per plant species
IE	Interaction evenness	Shannon's evenness of the network entries
H2'	Specialisation	A network-level measure of specialisation
NODF	Nestedness	One of a number of measures of the nestedness 'temperature' of the network
R	Robustness	A measure of the tolerance of the network to species extinctions

and 2). These are commonly represented as a matrix, with zeros indicating no interaction between a plant and pollinator and values of one indicating the presence of an interaction. More information, however, can be stored in these networks (and matrices). By incorporating additional information, such as abundances and frequency of interaction, there are a plethora of ways in which to describe the structure and complexity of these networks (both qualitatively and quantitatively) which are suited to particular ecological questions. Dormann et al. (2009) provide an introduction to metrics, graphs and null models that are produced in the 'bipartite' package in R (R Core Team, 2021) - perhaps the best and most used open access software for studying plant-pollinator and other bipartite networks.

Bersier et al. (2002) provide a good overview of qualitative food web metrics such as *Link Density* and *Connectance* (Table 1) before going on to show how to calculate quantitative metrics, that is, those incorporating more information on the strength of interactions (e.g. how many times a pollinator visits a flower). Although often couched in the language of trophic food webs, many of these metrics are suitable for the study of plant-pollinator interactions. For example, at the network level, we can examine consumer-resource asymmetries such as *Generality* and *Vulnerability*, which informs us about the mean number of plant species per pollinator (zoocentric) and the mean number of pollinator species per plant (phytocentric). In much the same way that species diversity

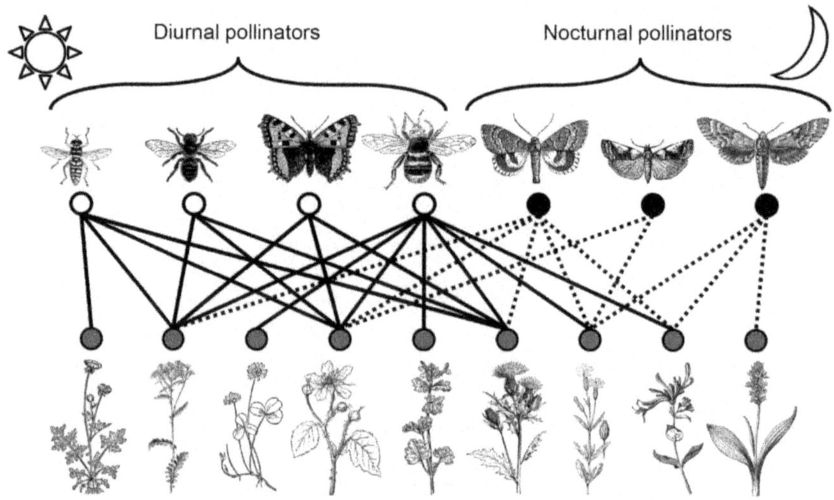

Figure 2 An illustrative example of a temperate grassland network combining diurnal and nocturnal pollination.Source: Adapted from Macgregor et al. (2015), used with permission. Nodes represent species: white = diurnal insects, black = nocturnal insects, grey = plants. Pollinators (from left): hoverfly (diptera), leaf-cutter bee (hymenoptera), butterfly (lepidoptera), bumblebee (hymenoptera), noctuid moth, pyralid moth, sphingid moth (all lepidoptera); plants (from left): *Ranunculus* sp. (ranunculaceae), *Jacobaea vulgaris* (asteraceae), *Trifolium* sp. (fabaceae), *Rubus* sp. (rosaceae), *Lamium* sp. (Lamiaceae), *Cirsium* sp. (asteraceae), *Silene latifolia* (caryophyllaceae), *Lonicera* sp. (caprifoliaceae), *Gymnadenia conopsea* (orchidaceae). Links represent hypothetical pollination interactions: solid = diurnal, dashed = nocturnal. Drawings of pollinators and plants are for illustration only and may not precisely represent the named plant or animal.

can be calculated in community ecology, the diversity of *interactions* can also be examined, for example, *Interaction Evenness*. Indeed, a benefit of studying plant–pollinator networks is that it provides information on biodiversity (at two trophic levels) and interactions (i.e. pollination), allowing a way for important questions regarding biodiversity and ecosystem functioning to be explored.

Of the many ecological network properties that can be described (and too numerous for this chapter), network *Robustness* has received particular attention, partly driven by advances in computational modelling (Kaiser-Bunbury et al., 2010; Staniczenko et al., 2010), but mostly by the desire to understand the real threat of biodiversity loss to ecosystem services and functioning (Pocock et al., 2012). This is a measure of the tolerance of the network to species extinctions (Dunne et al., 2002; Memmott et al., 2004). By simulating the sequential primary extinction of plants or pollinators within a network (either randomly or based on ecologically plausible scenarios), it is possible to plot the number of secondary extinctions that occur as a result (e.g. if the three plants that a pollinator only interacts with go extinct then so will the pollinator) and *Robustness* is calculated as the area under the curve (Figs 3a and b). Over the past decades, our understanding of network *Robustness* to species loss has advanced from studies of simple qualitative, bipartite mutualistic networks (Memmott et al., 2004), to investigations of patterns across ecosystems (Srinivasan et al., 2007; Evans et al., 2013) and to current quantitative approaches that take into account species abundance (Figs 3c and d) (Kaiser-Bunbury et al., 2010) or use a Bayesian approach (Young et al., 2021).

With multiple studies from around the world using such metrics, it is possible to start looking for universal patterns in plant-pollinator networks (with all of the caveats associated with standardisation). For example, empirical studies have found that the structure of plant-pollinator networks are typically more nested than expected by chance alone (Pawar, 2014) (note there are a range

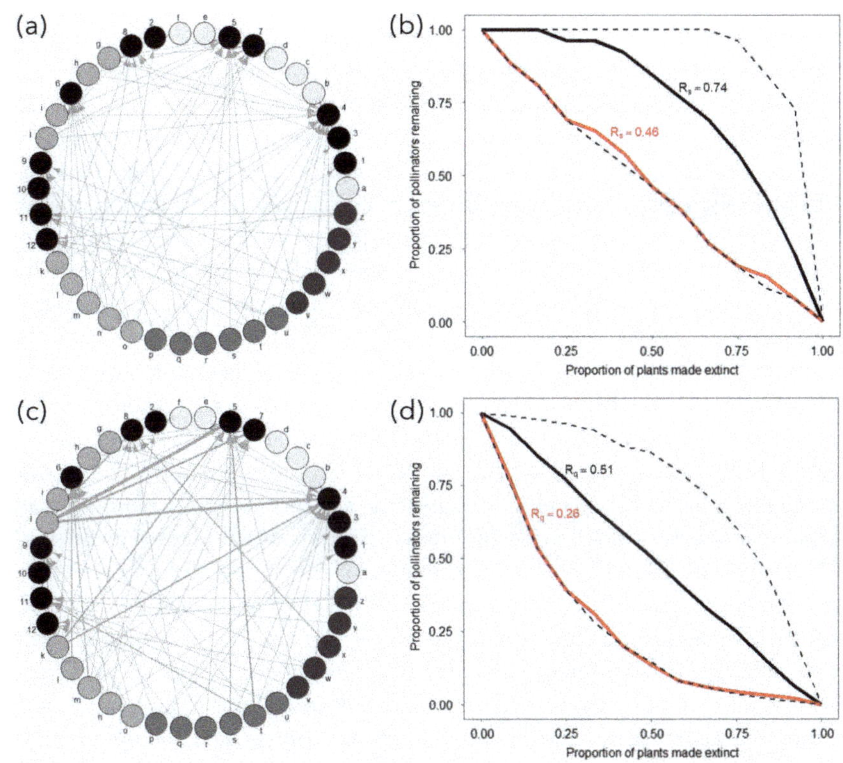

Figure 3 The robustness of a plant-pollinator network. (a-b) A qualitative network and robustness curve. (c-d) A quantitative network and the robustness curve. Black solid lines represent the median robustness curve, with black dotted lines representing the minimum and maximum robustness curves. All curves were derived from 1000 simulations, where plant species are randomly removed from the network. Red solid lines indicate scenarios where plant species are removed along a gradient from generalists to specialists (i.e. plants connected to the greatest number of pollinators through to those connected to the least number of pollinators), representing a potential worst-case scenario.

of methods to calculate *Nestedness*). Here, specialist pollinator species visit plant species that are subsets of those visited by more generalist pollinators. *Nestedness* has high ecological and evolutionary importance due to its role in enhancing species persistence and community stability. It should be noted, however, that *Nestedness* measures tend to be correlated with fundamental properties of networks, such as size and *Connectance*, and so *Nestedness* values (and indeed many other network metrics) must be normalised to enable fair comparisons between different ecological communities. Thankfully much work is going into providing statistical solutions and packages to overcome such issues (e.g. Hoeppke and Simmons, 2021), but these are nevertheless important considerations for anyone considering getting started in network ecology.

Over the past decades, studies have used metrics, such as *Connectance* and *Nestedness*, to characterise the structure of mutualistic and antagonistic networks and the roles of their constituent species. However, compressing a complex network into a single metric necessarily discards large amounts of information (especially regarding indirect interactions, see Section 2.3). Therefore it is common for studies to report a range of metrics pertinent to the ecological question being asked (see Tylianakis et al., 2007, for an example). Unfortunately, a few studies simply report 'commonly used' network metrics without necessarily considering their biological relevance, perhaps showing a disconnect between what is now computationally straightforward and fast using freely available software with a detailed understanding of the natural history of their study systems.

In addition to *network-level* metrics, a range of useful others can be calculated at the *species-level*, the most basic being *Degree* – the sum of links per species. Key plant species usually have high levels of *Centrality* – a suite of metrics that describe the contribution of individual species to network structure. For the study of plant–pollinator networks, the shannon *Diversity* of interactions per species or *Resource Range* (an index of unused resources (Poisot et al., 2012)) might be useful for answering particular questions, and others such as the pollination service/support index (PSI), which estimates the importance of a pollinator for all plant species (and vice versa), can also be determined. Again, by incorporating data on species *and* how they interact into an ecological network, we can move beyond conventional methods for describing ecological communities (such as species richness) to more sophisticated eco-evolutionary ways of studying biodiversity and ecosystem functioning, species 'importance' and extinction risks.

2.3 Direct and indirect interactions

Most of the interactions we have described so far refer to direct interactions, that is, a pollinator visiting a plant or two pollinators competing with one another. However, a range of indirect interactions can also occur when organisms interact with one another through an intermediate organism, and ecological networks lend themselves well to the study of them. We have already shown how studies use basic indices, such as *Connectance* and species *Degree*, to characterise the structure of ecological networks and the roles of their constituent species. But compressing a plant-pollinator network into a single metric necessarily discards large amounts of information about indirect interactions. This is significant as indirect effects in mutualistic networks are a major process governing coevolution (Guimarães et al., 2017) and a growing body of literature is showing the important and widespread indirect effects in nature (Wootton, 2002). Therefore, uncritical use of network and species-level indices as the sole method for characterising network structure risks missing all or part of this component (Simmons et al., 2019).

The concept of motifs (subgraphs representing patterns of interactions between a small number of species) is gaining momentum as it complements existing metric-based approaches described earlier by incorporating more explicit detail on realised direct interactions and thus potential indirect interactions. For example, Simmons et al. (2019) demonstrated their application by highlighting how indirect interactions can result in ecologically important differences between two pollinator species with similar direct interactions. In the study of a plant-pollinator community in the Seychelles, using the species-level metric *d'* (a specialisation index) they were able to show that both a native (*Lasioglossum mahense*) and a non-native (*Apis mellifera*) pollinator were super-generalists. However, by then using a motif approach they revealed that they were generalists in different ways, which would have been missed using conventional metrics. Their analysis showed that *L. mahense* and *A. mellifera* had significant roles over time and exploited different areas of interaction niche space. *L. mahense* was shown to be in greater competition with generalist pollinators than *A. mellifera* which visited more specialist plants and competed with more specialist pollinators.

In plant-pollinator networks, the potential for an indirect effect between co-flowering plant species may lead to facilitation, whereby the presence of one plant species increases pollinator visitation for another. Or it could lead to competition, whereby the presence of one plant species attracts effective pollinators away from another (Morales and Traveset, 2009). This can be a concern to farmers who fear that providing alternative foraging habitat will draw pollinators away from crops. An analysis of 750 quantitative plant-pollinator networks from 28 studies around the world (and representing diverse biomes) revealed that the potential for one plant species to influence another indirectly via shared pollinators was greater for plants whose resources were more abundant (higher number of flowers and nectar sugar content) and accessible (Carvalheiro et al., 2014). The study showed that potential indirect influence was also stronger between phylogenetically closer plant species and that there were important differences for bees compared to other groups.

While network and species-level metrics can be powerful tools for understanding plant-pollinator interactions, it is important to be aware of their limitations and that there are a range of ways of explicitly incorporating indirect effects into analyses. Nevertheless, an understanding of both direct and indirect effects, using a network approach, has considerable promise for sustainable agro-ecosystems in terms of in-crop and off-crop plant management, which we examine in the next sections.

2.4 Using networks to address agricultural challenges

While not explicitly stated, much of our general understanding of both mutualistic and antagonistic networks are thanks to empirical species-interaction data generated from a wide range of agro-ecosystem studies, in particular plant-pollinator networks. Tylianakis et al. (2007) set the standard for assessing the impact of agriculture on quantitative ecological network structures, and there remains huge scope for asking how different types of agriculture affect biodiversity and how biodiversity provides a range of ecosystem services (which are effectively the product of networks of interactions between species), such as pollination and pest control (see Miller et al., 2021, for the state of the art of host-parasitoid networks). More recently, network theory has advanced to include people in networks (i.e. social–ecological networks) for ecosystem service assessments (Dee et al., 2017) and we describe some of these recent advances in Section 5.

In terms of promoting pollinators for agriculture, network approaches have great potential, especially in the context of enhancing crop yield (although delivery of crop pollination services has been shown to be an insufficient argument for wild pollinator conservation (Kleijn et al., 2016)). This essentially involves linking network structure with ecosystem functioning, although studies measuring the impacts of pollinator-supporting practices on pollination service success - particularly in relation to long-term crop yield and economic profit - are still rare (Garibaldi et al., 2014). The key question is what in-crop and/or off-crop management techniques can enhance animal-pollinated crops?

Carvalheiro et al. (2010) provided one of the first network studies to show that pollination services to mango (*Mangifera indica* L.) fruit crops declined with distance from natural habitat even in biodiversity-rich areas, with plant-pollinator interaction data suggesting that maintaining a diversity of flower resources within farmland can help maintain pollinator communities. More general studies have also shown how agri-environment schemes can enhance pollinator diversity and plant reproductive success in nearby intensively managed farmland (Albrecht et al., 2007), with network approaches increasingly being used to inform agri-environment scheme design (Gresty et al., 2018).

Maia et al. (2019) calculated *Centrality* metrics and experimentally tested whether plant species' roles in pollination networks can be used to identify species with the most potential to recover plant-pollinator communities. Informed by network data, they found that the introduction of 'central' plant species attracted a higher richness and abundance of pollinators than the introduction of 'peripheral' species and that the introduced central plant species occupied the most important network roles. This approach could have implications for farmland management and restoration in terms of enhancing pollination. On the other hand, little is known about the pollinators of rare plants, such as arable weeds, in agro-ecosystems. Gibson et al. (2006) demonstrated how a network approach (including a pollinator importance index) can be used to determine the probable pollinators of three species of rare arable weeds among those suffering the greatest declines in the UK. They found that they were linked to other plant species in the community by shared pollinators, which in many cases constituted the primary food sources for the shared pollinators.

Pocock et al. (2012) advanced our understanding of the robustness of a farmland 'network of ecological networks' (see Section 3.2 for more details) by simulating the sequential loss of plants shared by a range of animal groups on an organic farm and calculating the number of secondary extinctions. They not only found that the flower-visitor network was the most fragile but identified which plants were disproportionately important to overall network integrity. This has implications for farmland restoration, especially the selection of plant mixes for agri-environmental schemes to enhance pollination and/or identifying weed species within crops that should be retained, although empirical testing is lacking (Windsor et al., 2021). A more general overview of network approaches (beyond just that of plant-pollinator interactions) for addressing contemporary challenges in agricultural research is presented by Bohan et al. (2013).

2.5 Summary

We have introduced the basics of network ecology for constructing flower-visitor and plant-pollinator networks and provided some examples of how they are currently being used to answer a range of questions in agricultural research. Next, we consider how this approach is being used to answer questions at different scales, again using examples of plant-pollinator studies within an agricultural context.

3 Networks across scales

Pollination networks can be captured at a range of scales, each of which offers an opportunity to understand different aspects of pollination ecology in agricultural systems. Previous studies have operated at several distinct scales, including within individual fields, across different farm habitats, among farms within an agricultural landscape and more broadly at the regional, national or continental scale. The following sections provide some examples of how ecological networks have been constructed for pollination networks across agricultural systems at all of these scales (Fig. 4).

3.1 Field

Early studies in agriculturally relevant environments, such as the pollinator networks in meadows we described earlier (Memmott, 1999), were completed at the field scale (i.e. a single habitat in a single location, Fig. 4). These original studies form the basis of our understanding of pollinator networks in agricultural systems, showing that interactions between plants and pollinators are common and that they can have important consequences for agriculturally important processes, such as crop production. At the field scale, it is possible to create networks with a far higher level of detail than at broader scales (including above- and belowground interactions), and therefore it is feasible to investigate processes important at a fine resolution (e.g. how changes to plant-microbe

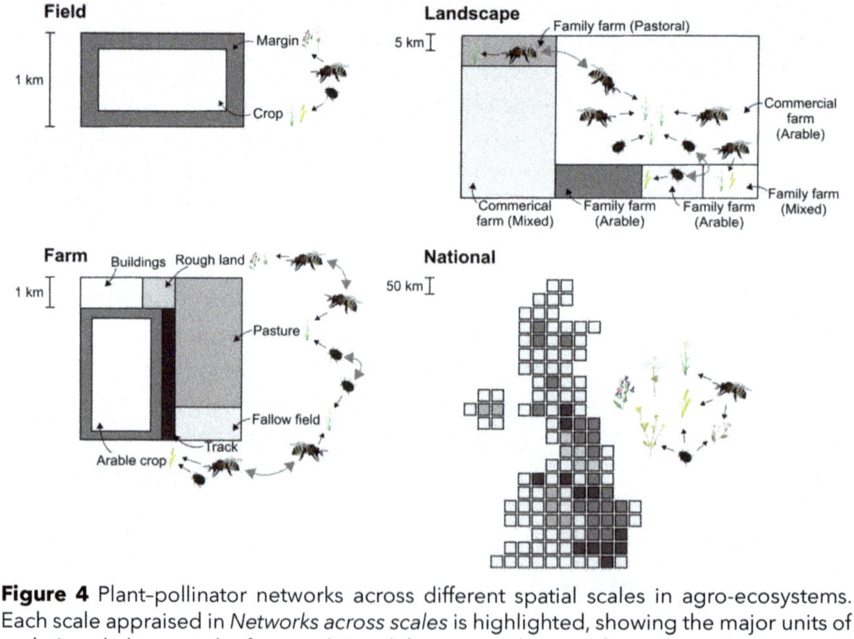

Figure 4 Plant-pollinator networks across different spatial scales in agro-ecosystems. Each scale appraised in *Networks across scales* is highlighted, showing the major units of scale (e.g. habitats at the farm scale) and the potential types of movement (grey arrows) and interactions (black arrows) between plants and pollinators.

interactions might affect individual pollinator behaviour and visitation (Bennett et al., 2019)). From this research, it has then been possible to branch out to broader scales with studies relying on the information gathered at field scales to support any findings generated across farms or landscapes (see Fig. 4).

Although more recent studies have tended to focus on assessing pollination across multiple habitats on the same farm or across multiple farms within an agricultural landscape, a range of contemporary studies continue to investigate processes at this relatively restricted spatial scale. This is because complex or detailed processes often require assessments at smaller scales. For example, to understand the extent to which spillover effects penetrate into cropped fields, highly replicated studies at the field scale are necessary. Examples of such research have shown that despite their perceived beneficial effects, flowering plants in the field margins only generate positive effects on pollination services (seed set, pollination deposition rate and flower visitation rates) <50 m into the cropped fields (Albrecht et al., 2020; Zamorano et al., 2020). To get these results, the study by Albrecht et al. (2020) collated data from 17 studies across 276 fields, investigating transects up to 100 m into the cropped fields, monitoring plant-pollinator interactions along the transect and comparing common pollinator species found visiting the field margin plants as well as the crops. This was a common phenomenon across different crops, including oilseed rape (*Brassica napus*), pumpkin (*Cucurbita maxima*), strawberry (*Fragaria ananassa*), watermelon (*Citrullus lanatus*), red clover (*Trifolium pratense*), highbush blueberry (*Vaccinium corymbosum*) and apple (*Malus domestica*). There is considerable scope for extending such studies into farm cropping systems across the world, providing much-needed information on direct and indirect plant-pollinator interactions and the contribution of natural and semi-natural habitats for ecosystem service provision more generally.

3.2 Farm

Few examples exist of farm-scale studies for pollinators, where a range of habitats across a single farm are investigated in detail. This is likely because a small number of proximal habitats (i.e. field margins) are the main focus of studies, but also that many studies instead monitor pollination across multiple sites with the aim of achieving a more representative landscape-scale appreciation of plant-pollinator interactions. Farm-scale assessments, however, are important as the data they produce are at the same scale at which most management decisions are made as well as the scale at which policy interventions are targeted, such as EU Common Agricultural Policy and the UK's Entry Level Stewardship (ELS) scheme.

One stand-out example of farm-scale monitoring, however, is the Hillesden experiment in the UK to understand the impacts of ELS on biodiversity and ecosystem function over a 5-year period (2006–2011) (Heard et al., 2012). The experiment consisted of a large-scale randomised block design, with three different management scenarios (conventional arable management, ELS and ELS extra) applied to 50–60 ha of land (five replicates). As part of a wider monitoring strategy, flower visitation by pollinators was assessed along 2 × 50 m transects in different treatment habitats (e.g. conventional crops, field edges, naturally regenerated habitats and actively planted habitats), as well as through pan trapping. Floral resources, measured as the abundances of flowers per hectare, for pollinators were markedly increased under both ELS and ELS X (1% and 5% area out of production, respectively), in comparison to conventional farming. This led to (i) a greater number of pollinators (butterflies, bees and flies), two and ten times greater in ELS and ELS X respectively when compared to conventional farming; and (ii) a significantly higher number of plant–pollinator interactions 7 and 60 times greater in ELS and ELS X. Many of these effects were generated through farm and landscape-scale processes, for example, the levels of non-crop plant species in the surrounding habitats. This work reinforces the need to gain a detailed understanding of plant–pollinator interactions at the farm scale, especially when looking to understand the benefits of biodiversity for farmers.

The most comprehensive study of species interactions in an agricultural context to date is the 'Norwood Farm Network' – a study of multiple plant–animal interactions on a single, 125 ha organic farm in Somerset, UK (sampled over 2 years during 2007 and 2008) (Pocock et al., 2012). This 'network of ecological networks' comprises of plants and 11 groups of animals; those feeding on plants (butterflies and other flower visitors, aphids, seed-feeding insects, and granivorous birds and mammals) and their dependents (primary and secondary aphid parasitoids, leaf-miner parasitoids, parasitoids of seed-feeding insects and rodent ectoparasites), with the specific aim of encompassing a wide taxonomic and functional range that included animals regarded as bioindicators and as ecosystem service providers (see Fig. 5). As mentioned earlier, the analysis revealed that the plant flower-visitor network exhibited lower robustness than other components, suggesting that perhaps this group should be the focus of biomonitoring and conservation. Evans et al. (2013) extended this work to examine the robustness of these networks to habitat loss (12 crop and non-crop types were measured), moving beyond individual species deletions in the analyses to the sequential loss of whole communities within a given habitat, while also accounting for spillover effects. This is a particularly powerful tool for assessing the impacts of, for example, converting woodland to arable crops or the impacts of removing hedgerows, while also identifying habitats that might be disproportionately important to network integrity for resilience planning and restoration purposes. The study identified two semi-natural habitats (waste ground and hedgerows together comprising <5% of the total area of the farm) as being disproportionately important to the integrity of the overall network

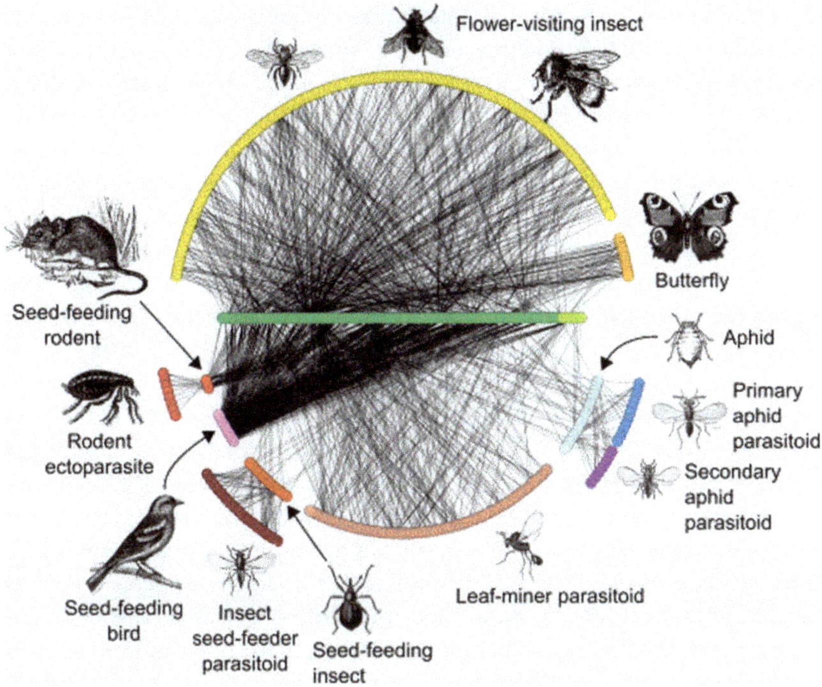

Figure 5 The Norwood Farm Network, UK. Originally presented in Pocock et al. (2012) and revised in Bohan et al. (2013), used with permission. Each species is represented by a circle and each interaction is represented by a line. Plants are shown by green circles in the centre of the network, with crops being shown in light green. The illustrations are used under licence from ClipartETC and the network was drawn using Pajek.

and demonstrated the utility of this approach at both the farm and landscape scale. We return to the Norwood Farm Network and discuss current work that merges plant-pollinator networks with other types of interactions (especially in the context of multiple ecosystem service provisions in agriculture) in Section 5.

3.3 Landscape

Studies assessing pollination networks across the landscape have become increasingly common as theoretical and technological advances have allowed for highly replicated and detailed plant-pollinator networks over large spatial scales. In agricultural systems, there are a number of studies that have investigated different aspects of plant-pollinator interaction networks across different crops, arable systems and regions of the globe. In the following section, we provide details on some particularly informative studies that have used landscape-scale assessments of pollination networks to answer applied questions in agriculture.

Restoration of non-crop habitats across the landscape has been a focus of studies investigating multiple farms. In particular, hedgerows have been investigated as they are generally regarded as beneficial habitats on farms (Evans et al., 2013). For example, the effects of hedge restoration in a variety of arable systems have been investigated across California, USA (Ponisio et al., 2017). A series of hedgerows were planted across sites and pollinators were caught after visiting the reproductive parts of flowers across hedgerows, crops and weedy field margins (which were used as controls to compare to hedgerows). The number of interactions between different plants and pollinators was recorded to produce quantitative networks across sites (1521 unique interactions). The study showed that over the 8 years of hedgerow development since their planting, the structure of plant-pollinator networks was highly variable and networks reorganised themselves several times (Ponisio et al., 2017). The levels of turnover between habitats (mature hedgerows, developing hedgerows and field margins) were, however, not significantly different, indicating that even rapidly developing habitats support diverse pollinator assemblages. Within these networks, it was also noted that opportunistic attachment was the dominant assembly process, which means that generalist species switch the plants they interact with to avoid high levels of competition over resources. This highlights the potential for non-crop habitats to support stable and diverse plant–pollinator interactions which in turn may enhance the levels of pollination in cropped fields by exporting pollinators (Morandin and Kremen, 2013; Ponisio et al., 2017). The combination of results indicates the important role of habitat management and restoration in agriculture and the potential benefits of such management activities for agricultural production.

Comparisons between different farming methods are also common in landscape-scale studies. Organic versus conventional farming is a common study design present in the literature, especially over the past several decades (e.g. Macfadyen et al., 2009). Most of these studies occurred in arable systems; however, a notable exception assessed the effects of pastoral farming regimes on plant–pollinator networks (Power and Stout, 2011). This study went further than most and investigated how changes in plant-pollinator networks affect the fruit set of the plants (proportion of pollinated flowers per site). Ten pairs of organic and conventional dairy farms were assessed across Ireland, and two fields from each farm were surveyed three times across 2009. Walking transects (100 × 2 m) in both the edge (hedgerows) and the centre (30 m from the edge) of the field were done to observe both obligate and facultative pollinators. The main finding of the study was that pollinator networks were larger on organic farms due to the higher floral abundances in organic fields, and there were greater overall levels of pollination. Network stability was not enhanced in organic farms and pollinators were still vulnerable due to their relatively small size compared to other pollinator networks studied in the wider agricultural literature (Power and Stout, 2011). A recommendation based on these findings was that strategic sowing of important plant species could be endorsed by agri-environment schemes to benefit pollinators across the wider agricultural landscape, not just in pastoral or arable fields, in concordance with some of the points we made earlier.

Extensive versus intensive agricultural practices can also be investigated using pollinator networks. A recent example by Guzman et al. (2019) aimed to understand whether on-farm diversification (polyculture) in an intensive agricultural landscape could positively influence specialist pollinator species. The study focused on a group of squash bees (*Peponapis* sp. and *Xenoglossa* sp.) which pollinate squash plants (*Cucurbita pepo*) in the San Joaquin Valley in California, USA. Both the abundance and number of flower visits were recorded for all pollinator species across five monoculture and five polyculture farms (>2 crop types) along two 30 m transects located 5 m from the edge of each squash crop. Contrary to expectations, the specialist 'squash bees' were not found in higher abundance in monocultures of squash, instead the polycultures supported a greater number of all types of generalist and specialist pollinators with the exception of honeybees (*Apis* sp.). Such findings offer the potential to develop farm-scale management, but with the added security that multiple farms were included in the study, thus incorporating potential variation in relation to many different environmental factors that can vary at the landscape scale.

3.4 National

Despite the potential for using networks for biomonitoring both biodiversity and ecosystem functioning (Bohan et al., 2017; Derocles et al., 2018), there are so far very few examples of national-scale monitoring of pollination networks. However, there are cases where a combination of non-target monitoring of pollinator species from biological records (citizen science databases) and secondary data on plant–pollinator interactions allows for the reconstruction of networks at coarse resolutions over large spatial scales. For Great Britain, one such example is presented by Redhead et al. (2018) and shows great promise. This study constructed plant–pollinator networks at 10 km resolution from species co-occurrence data and potential interactions (16 712 interactions between 485 pollinators and 499 plants). From these networks, it was possible to understand large-scale changes in pollinator networks, including positive relationships between agricultural land cover and pollinator generality and network robustness. The increase in robustness was due to the decreased prevalence of extinction-prone plant species and enhanced pollinator generality. It was therefore apparent that the changes in plant and pollinator communities in agricultural systems may make them more resilient to future environmental changes.

3.5 Summary

We have shown how ecological networks can be used to answer questions relevant to agricultural management across a range of spatial scales. Until now, agro-ecosystems have conceptually been regarded as crop monocultures, with a few associated plant and invertebrate species residing in a single field. However, in reality, these agro-ecosystems are far more complex: a network of ecological interactions occur between individuals in populations, between species (including humans), within and between communities and functional groups, and across the cropped fields and other natural and semi-natural habitats in the agricultural landscape. Network studies are beginning to quantify the ecosystem services provided by biodiversity, such as crop pollination, although their contributions are still poorly known and rarely acknowledged (see UN Sustainable Development Goal 15, 2015). There is clearly much more work to do in terms of quantifying plant–pollinator interactions in space, time and across biomes, and understanding how these are embedded in much bigger, complex networks. We consider both of these challenges in an agricultural context in the next sections.

4 Increasing the completeness of pollination networks

Although pollination networks have been widely studied in agro-ecosystems, we have already shown that some significant gaps remain in terms of both the coverage and level of detail surrounding plant–pollinator interactions. Here we provide some further important considerations.

Species coverage remains a fundamental issue, with several groups underrepresented in the vast majority of studies. The focus of much research has been on honey and bumblebees (mainly *Apis* sp. and *Bombus* sp.) as they are conspicuous, as well as sometimes managed, in agricultural environments. Yet it is widely appreciated that wild insect pollinators such as Diptera and Coleoptera play an important role in the pollination of many plants, including agricultural crops (Garibaldi et al., 2014). Furthermore, there are also a range of vertebrates (particularly in tropical systems) that contribute to overall pollination, including amphibians, bats and some larger mammals (Ratto et al., 2018). Better appreciating the role of these often cryptic or difficult to observe/identify species is an important stage in filling out our understanding of pollination in agro-ecosystems around the world.

As described earlier, it has become clear that nocturnal pollinators contribute significantly to plant pollination and reproductive success (Devoto et al., 2011; Banza et al., 2015; Macgregor et al., 2015), especially in agricultural environments (Walton et al., 2020). However, studies to date have almost all conducted pollinator surveys during daylight hours, and thus only incorporate diurnal pollinators.

By not including nocturnal pollinators, many studies may be misrepresenting the robustness and resilience of pollination networks to management and environmental change. This is important as drivers of environmental change, such as artificial light at night (ALAN), have been shown to affect nocturnal moth behaviour and physiology (Boyes et al., 2021), disrupt pollen transport by moths (Macgregor et al., 2017) and ultimately pollination (Knop et al., 2017). Indeed, recent work suggests ALAN can also impact diurnal plant-pollinator interactions (Giavi et al., 2021), although more research is required to confirm the impacts.

Many studies investigating agricultural pollinator networks have focused on qualitative networks, that is, the interactions are either present or absent. There are, however, other data that can be incorporated into a network-based framework. Most simply as we describe in *Basics of Pollination Networks* it is possible to quantify the network based on the number of interactions between a plant and pollinator. Further to this, traits of both the pollinators (e.g. proboscis length) and plants (e.g. corolla tube length) can be included, and this is especially important as certain traits strongly affect the nature and magnitude of interactions (Olito and Fox, 2015). For example, a pollinator with a short proboscis is unlikely to pollinate a plant with a long corolla tube length. This additional functional information is important for understanding how the structure of pollination networks translates to ecological processes and thus ecosystem services.

There is still significant work to be done in order to adequately describe the interactions between plant and vertebrate and invertebrate pollinators (both diurnal and nocturnal) in agro-ecosystems across the world. While this requires an increase in observation effort, we have demonstrated that network construction methods are relatively straightforward, and as other technological developments such as DNA-metabarcoding, cameras/sensors and citizen science become mainstream, important advances can be made in filling the gaps. However, the considerable advantages and applications of undertaking the task for promoting pollination in agriculture should now be evident.

5 Embedding pollination within wider agro-ecosystem networks

Pollination is just one of the many types of species interactions present within agro-ecosystems. Plants and pollinators are also involved in a complex web of antagonistic, mutualistic, commensalistic and amensalistic interactions, including plant-herbivore, plant-microbe, pollinator-predator, pollinator-parasite/parasitoid, as well as inter- and intra-specific competition between plants and pollinators. Through a series of direct and indirect interactions this wider ecological network (often termed a 'multiplex network') has the ability to influence plant-pollinator interactions. We have previously shown in the Norwood Farm Network (Fig. 5) how the flower-visitor network was the least robust to the loss of shared plants than the other animal groups within the network, and how robustness analysis can also be used to identify disproportionately important plants (and indeed habitats) for network integrity (see Section 3.2). Beyond the merits of this approach for the study of plant-pollinator interactions, it is possible to go much further analytically in terms of understanding the interdependencies of plants and animals in the wider network. For example, Pocock et al. (2012) examined whether robustness co-varied between the animal groups and found evidence that although some did, overall, this was less than expected. In terms of practical applications, this could mean that targeting particular groups (e.g. birds) for conservation management might not have cascading benefits across the networks. This could challenge the current conservation paradigm of species-focused interventions (especially large, charismatic animals), but clearly more empirical work is needed to examine the cascading effects of plant and animal management on the wider ecosystem.

Using the same Norwood Farm Network dataset, Windsor et al. (2021) evaluated the optimal mixes of plant species that could be used in field margin management to promote the delivery of ecosystem services (pollination and parasitism of pest insects) while minimising ecosystem disservices (herbivory). The premise here is that by only focusing on plant-pollinator networks to enhance crop yields (e.g. flower strips as part of an agri-environment scheme), farmers and land managers might inadvertently be providing a source for insect pests, while overlooking the potential for other types of ecosystem service provision, especially insect pest control. The study found that plant mixes designed solely for maximising pollinator species richness were not optimal for the provision of other ecosystem services and disservices. This example perfectly illustrates the importance of understanding both direct and indirect interactions within a farmed context, and why it is desirable to merge different types of ecological networks. Indeed, the ecological and evolutionary implications of merging different types of networks are a major focus in network ecology more generally (Fontaine et al., 2011).

In the examples previously described, it is assumed that the different interactions occur simultaneously (i.e. a pollinator can also be prey for a predator or a predator itself). Nevertheless, there is an additional element. Species may switch between different interaction networks depending on their life stage (e.g. through an ontogenetic niche shift). A relevant example is that some lepidopterans are herbivores of plants during their larval stages, yet important pollinators when adults (e.g. cabbage white butterflies *Pieris rapae* are herbivores of many crops but also act as pollinators). This is not often appreciated, especially in agricultural systems and it may be the case that the benefits of a pollinator may be so great as to offset the potential negative effects generated by its herbivorous early life stages. More research on these ontogenetic shifts in the role of common agricultural pests and beneficial insects, however, is required but can easily be incorporated into ecological network analyses.

It is not just other types of ecological interaction, however, that are important. As described earlier, expanding ecological networks to include spatial aspects of agricultural systems is also possible and desirable. In the case of pollination, it is important to understand how pollinators link together plants and other organisms in different habitats. As we have seen, linkages between field margins and cropped fields have been a focus of much current research in this area. But research has shown that pollinators link together multiple habitats and as such effects in one are able to translate into the other. In the case of field margins, increasing the floral diversity and selecting plant species that have the potential to enhance the species richness of pollinators and generate a spillover of pollinators into cropped fields (Zamorano et al., 2020). This also has implications for the resilience of these networks in response to either management or environmental change, as the effects of either changes need not occur in the habitat of interest (e.g. crops) to have substantial effects.

More broadly, agro-ecosystems are a combination of social and ecological networks, with top-down effects of management altering the nature of ecological systems, including the dynamics of plant–pollinator networks (Lescourret et al., 2015). Until recently, however, these two types of networks have been analysed in isolation. New studies have sought to combine social-ecological networks (SENs), providing insights into how management and biodiversity alter ecosystem functioning, including in agriculture (Hutchinson et al., 2019). Dee et al. (2017) provide a way of operationalising network theory for ecosystem service assessments, and Sayles et al. (2019) show how the degree of SEN articulation can be used to answer a range of questions. Essentially, within an SEN ecosystem services can be represented as nodes, links or even emergent properties. Although more research is necessary, a theoretical example shows the importance of combining social and ecological networks in agriculture. At the landscape scale, agricultural environments are composed of different types of farmers, ranging from family farmers through to commercial farm managers, who form nodes within a farmer–biota network. Farmers are linked through their effects on the wider biodiversity of the landscape, that is, agricultural management by one farmer may affect biota and in turn have impacts on other farmers. For example, the application of pesticides may reduce pollinator diversity and thus reduce pollination across other farms and crops – even those where pesticides are not applied. Recent work demonstrated how a network approach could be used to predict ecosystem service vulnerability to species losses within estuarine food webs, again by developing robustness analyses (Keyes et al., 2021). This approach could easily be extended to agro-ecosystems and is likely to be a rich area of future research.

6 Next steps for pollination networks

Building more comprehensive and detailed pollinator networks, particularly by including species traits and/or phylogenetic data, will allow for an increased understanding of how these systems can be managed to enhance the sustainability of agriculture. Such networks can now be constructed thanks to the development of novel techniques that provide a greater level of detail than the previous observation-based methods we have described throughout this chapter. In particular, recent advances in DNA-based network construction methods provide unprecedented opportunities to scale up the construction of highly resolved plant–pollinator networks in space and time (Derocles et al., 2018; Macgregor et al., 2019). Molecular techniques are particularly applicable to pollination network construction where interactions may be cryptic, infrequent, short-lived or difficult to observe (Vacher et al., 2016; Evans and Kitson, 2020). Furthermore, these techniques allow the inclusion of evolutionary information (e.g. phylogenetic data) that is particularly important in the context of using adaptive networks predictively for restoration purposes (Raimundo et al. 2018). For example, it may be possible to understand how both plant and pollinator populations respond to changing network structure (i.e. changing pollinator populations will in turn alter plant populations and vice versa).

Further to directly collecting more data on pollination, there are a number of existing data sources that can be mobilised to support applied research. Citizen science projects such as the UK Pollinator Monitoring Scheme (PoMS; https://ukpoms.org.uk/), but also routine monitoring data that previously has been difficult to access (e.g. Rothamsted Insect Survey; https://www.rothamsted.ac.uk/insect-survey), now provides a wealth of valuable data. An immediate advantage of such schemes is that only small changes to sampling are required in order to quickly construct networks (Petsopoulos et al., 2021), that can then be used to generate long-term network data and/or spatial coverage to answer a range of questions relevant to agriculture and/or insect conservation.

As identified in Section 4 there remain gaps in both data and understanding surrounding pollinator networks in agricultural systems. The above methods present options for filling some, but not all, of this missing information. But other options, such as inferring networks from co-occurrence data exist (but see Blanchet et al., 2020), which can be used to further improve our understanding of pollination and the role of species interactions in agro-ecosystems. This makes use of existing data and expert knowledge to construct networks. For example, based on recorded plant-pollinators from secondary literature it would be possible to construct a network from a simple list of plant and pollinator taxa at a farm, without having to complete the lengthy process of directly observing interactions across fields and other habitats.

Links between biodiversity and ecosystem functioning (BEF), for example, links between the community structure of pollinators and plants and the level of pollination, are currently poorly understood. Ecological networks present an opportunity to link together the structure and function of systems through mechanistic, process-based species interactions, for example, interactions between plants and pollinators directly describe the process of pollination. To further develop this aspect of research, however, there needs to be a greater focus on measuring functional endpoints alongside species interactions. For pollinators, measuring the implications of flower visits on the fruit or seed set of a given plant or plant population would help to link together the structure of ecological communities and the effects on ecosystem functioning.

The spatial and temporal dynamics of pollination networks are not included in most contemporary studies which tend to analyse the structure of plant–pollinator networks at a given site over a restricted time period (or a 'snapshot' of interactions in time). Yet those studies which have included a temporal dynamic show that pollination networks can vary drastically in space and time, across many different scales (Dupont et al., 2009). Adaptive network models, a novel suite of network analyses, present a way of dealing with the dynamic nature of ecological systems. These models offer an opportunity to incorporate additional information on the phylogeny and traits of the organisms that comprise a given ecological network (Raimundo et al., 2018), ultimately leading to the predictive use of networks. These adaptive network models incorporate potential functional or co-evolutionary changes that might manifest themselves in response to changes in the wider network, adding greater detail (Derocles et al., 2018). Developing adaptive network models within agricultural systems provides significant potential, especially in research investigating how human management may affect other aspects of the agricultural system. Examples of potential research include understanding the response of network structure and function to pesticide resistance, climate change and reductions in pollinator diversity, as well as developing precision fertiliser application and investigating the optimal combinations of non-crop plants in organic agricultural systems. By developing these methods of analysis, we can gain more information about pollination networks, and suitable management practices can be developed to make use of eco-evolutionary principles (Loeuille et al., 2013).

Automated monitoring of plant–pollinator networks provides promise for high-resolution construction and analysis of plant–pollinator networks while minimising the number of human hours required to collect data. Existing technologies lend themselves to this challenge, including video footage from camera traps (Høye et al., 2021) and radar methods used to detect flying insects (e.g. vertical-looking radar: Chapman et al., 2003). In the UK, vertical-looking radar is already used to detect flying insects such as ladybirds (Jeffries et al., 2013) and camera traps have been effectively used to determine plant-pollinator interactions (Balducci et al., 2020), and such methods could also be applied to detect other flying pollinators. Remotely monitoring the movement and interactions of pollinators, and then combining these data with novel computational techniques such as machine learning algorithms, also provides an opportunity to rapidly collect large amounts of data over multiple scales.

A final development is to understand how plant-pollinator networks respond to, and recover from, a series of environmental and anthropogenic pressures. Pollinator and plant populations are threatened by a range of potential stressors and a burgeoning number of studies are investigating multiple drivers of change in plant-pollinator networks. First, climate change is a substantial threat to pollination (Settele et al., 2016). In one study, ecological networks were used to show how phenological shifts as a result of a doubling of atmospheric CO_2 could theoretically reduce floral resources available for up to 50% of pollinator species, resulting in periods where no food plants were available (Memmott et al., 2007). The predicted result of these disruptions is the extinction of pollinators, plants and their crucial interactions. However, the potential functional consequences of the changes remain poorly understood. Second, ALAN, as explained in detail in Section 4, can disrupt both diurnal and nocturnal pollination networks and directly affect the levels of pollination. There are, however, a number of unanswered questions surrounding how plant and pollinator reproductive success is affected and how this translates to longer-term effects on pollination services (Grubisic and van Grunsven, 2021). Third, invasive species have the ability to threaten the stability of pollination networks, especially in agro-ecosystems where invasive plants and pollinating animals are able to outcompete native species and potentially reduce the overall levels of pollination. Indeed invasive species can affect the resilience of plant-pollinator networks to future shocks (Morales et al., 2017). Networks lend themselves well to studying how non-native plants can integrate into native plant-pollinator systems (Lopezaraiza-Mikel et al., 2007); for example, it may be possible to study the native networks of an invasive species to predict which plants and pollinators it will interact with in its non-native range but also its effect at a whole-community scale (Parra-Tabla and Arceo-Gómez, 2021). Fourth, a further climate-associated stressor which is becoming more frequent is wildfires. Although wildfires are natural phenomena, and thus have always presented a disturbance to plant-pollinator networks, human-induced climate change (and inappropriate management) has increased both the frequency and severity of these events (Duane and Brotons, 2021). Recent work on pollinator networks has suggested that if wildfires become more frequent under climate change, community resilience may be eroded (Banza et al., 2019), but that understanding the responses of such networks can inform management that promotes resilience and facilitates conservation at an ecosystem scale.

All of the above pressures, in combination with intensive agricultural management, can ultimately contribute to net losses of biodiversity, which in turn may have substantial effects on pollination (Grab et al., 2019). Most studies of plant-pollinator networks to date mainly focus on a single driver of change, but understanding how these may interact with farming practices, and developing management strategies that account for these alternative pressures on plant-pollinator networks in agro-ecosystems, is key to supporting pollination into the future.

7 Conclusion

In this chapter, we described some ways in which network theory can be used to answer important questions and address contemporary challenges surrounding pollination in agro-ecosystems. It is by no means a comprehensive review of the studies applying network theory to plant-pollination networks in agro-ecosystems. However, here we hope to have provided an introduction to the topic and demonstrated the benefits of applying network thinking to agricultural pollination research. We first described how to go about observing and constructing both flower-visitor and plant-pollinator networks, described the differences between the two, and the various metrics that can be used to summarise aspects of these networks of agricultural relevance (e.g. the number of pollinators visiting different plants). With examples we show how it is possible to use network metrics and analyses to answer important questions in agro-ecosystems, especially by linking network structure with ecosystem functioning to promote pollination. From this basis, we then described how plant-pollinator networks work at a range of scales from fields to national agricultural systems and provide examples of where contemporary studies have developed research at different scales to support decision-making by farmers and land managers. Finally, we showed how further work is required to create highly resolved plant-pollinator interaction networks and embed these networks within a wider network of networks involving other species interactions (e.g. predation, parasitism and competition) as well as human aspects such as decision-making and management practices. To enact these ideas, we also detailed the next stages for pollination network research in agro-ecosystems. The state of the art for this field is interdisciplinary research (see Dee et al., 2017, for a general review of such ideas in the context of ecosystem service assessment across ecosystems). Thus, to further develop understanding in these systems we need to bring together specialists across disciplines, including farmers, land managers, social scientists, ecologists and complexity scientists, among many others.

8 Where to look for further information

There is no single comprehensive text on plant-pollinator networks in agricultural systems, however, there are a few good introductions to the broader topic of plants, their pollinators and the interactions between the two groups:

- Ollerton, J. (2021). *Pollinators and Pollination: Nature and Society*. Pelagic Publishing Ltd., Exeter, UK.
- Waser, N. M., and Ollerton, J. (Eds.) (2006). Plant-pollinator interactions: from specialization to generalization. University of Chicago Press, Chicago, USA.

For a more general, theoretical discussion of mutualistic networks with examples we suggest:

- Bascompte, J. and Jordano, P. (2013). *Mutualistic Networks*. Princeton University Press, Princeton, USA.

Policies specifically targeting plant-pollinator interactions are uncommon, however, a number of strategies are beginning to use network thinking in a broader sense. Examples that refer specifically to plant-pollinator interactions and/or networks are:

- Intergovernmental Science-Policy Platform on Biodiversity and Ecosystem Services Assessment Report on Pollinators, Pollination and Food Production (https://ipbes.net/sites/default/files/downloads/pdf/individual_chapters_pollination_20170305.pdf).
- Defra National Pollinator Strategy 2014 to 2024 (https://www.gov.uk/government/publications/national-pollinator-strategy-2014-to-2024-implementation-plan).

9 Acknowledgements

The authors were supported by funding from the Royal Society Challenge-led Grant 'Developing network ecology to understand and secure resilient food systems' (grant number CHL\R1\180156). We thank Aifionn Evans for help with proofreading.

10 References

Albrecht, M., Duelli, P., Müller, C., Kleijn, D. and Schmid, B. (2007). The Swiss agri-environment scheme enhances pollinator diversity and plant reproductive success in nearby intensively managed farmland. *Journal of Applied Ecology* 44(4), 813–822.

Albrecht, M., Kleijn, D., Williams, N. M., Tschumi, M., Blaauw, B. R., Bommarco, R., Campbell, A. J., Dainese, M., Drummond, F. A., Entling, M. H., Ganser, D., Arjen de Groot, G., Goulson, D., Grab, H., Hamilton, H., Herzog, F., Isaacs, R., Jacot, K., Jeanneret, P., Jonsson, M., Knop, E., Kremen, C., Landis, D. A., Loeb, G. M., Marini, L., McKerchar, M., Morandin, L., Pfister, S. C., Potts, S. G., Rundlöf, M., Sardiñas, H., Sciligo, A., Thies, C., Tscharntke, T., Venturini, E., Veromann, E., Vollhardt, I. M. G., Wäckers, F., Ward, K., Westbury, D. B., Wilby, A., Woltz, M., Wratten, S. and Sutter, L. (2020). The effectiveness of flower strips and hedgerows on pest control, pollination services and crop yield: A quantitative synthesis. *Ecology Letters* 23(10), 1488–1498.

Balducci, M. G., Van der Niet, T. and Johnson, S. D. (2020). Diel scent and nectar rhythms of an African orchid in relation to bimodal activity patterns of hawkmoth pollinators. *Annals of Botany* 126(7), 1155–1164.

Ballantyne, G., Baldock, K. C. R. and Willmer, P. G. (2015). Constructing more informative plant-pollinator networks: Visitation and pollen deposition networks in a heathland plant community. *Proceedings of the Royal Society. Series B,: Biological Sciences* 282, 20151130.

Banza, P., Belo, A. D. F. and Evans, D. M. (2015). The structure and robustness of nocturnal Lepidopteran pollen-transfer networks in a Biodiversity Hotspot. *Insect Conservation and Diversity* 8(6), 538–546.

Banza, P., Macgregor, C. J., Belo, A. D. F., Fox, R., Pocock, M. J. O. and Evans, D. M. (2019). Wildfire alters the structure and seasonal dynamics of nocturnal pollen-transport networks. *Functional Ecology* 33(10), 1882–1892.

Bascompte, J. and Jordano, P. (2013). *Mutualistic Networks*. Princeton University Press, Princeton, NJ.

Bennett, A. E., Evans, D. M. and Powell, J. R. (2019). Potentials and pitfalls in the analysis of bipartite networks to understand plant-microbe interactions in changing environments. *Functional Ecology* 33(1), 107–117.

Bersier, L. F., Banasek-Richter, C. and Cattin, M. F. (2002). Quantitative descriptors of food-web matrices. *Ecology* 83(9), 2394–2407.

Blanchet, F. G., Cazelles, K. and Gravel, D. (2020). Co-occurrence is not evidence of ecological interactions. *Ecology Letters* 23(7), 1050–1063.

Bohan, D. A., Raybould, A., Mulder, C., Woodward, G., Tamaddoni-Nezhad, A., Bluthgen, N., Pocock, M. J. O., Muggleton, S., Evans, D. M., Astegiano, J., Massol, F., Loeuille, N., Petit, S. and Macfadyen, S. (2013). Networking agroecology. *Advances in Ecological Research* 49, 1–67.

Bohan, D. A., Vacher, C., Tamaddoni-Nezhad, A., Raybould, A., Dumbrell, A. J. and Woodward, G. (2017). Next-generation global biomonitoring: Large-scale, automated reconstruction of ecological networks. *Trends in Ecology and Evolution* 32(7), 477–487.

Boyes, D. H., Evans, D. M., Fox, R., Parsons, M. S. and Pocock, M. J. O. (2021). Is light pollution driving moth population declines? A review of causal mechanisms across the life cycle. *Insect Conservation and Diversity* 14(2), 167–187.

Carvalheiro, L. G., Biesmeijer, J. C., Benadi, G., Fründ, J., Stang, M., Bartomeus, I., Kaiser-Bunbury, C. N., Baude, M., Gomes, S. I. F., Merckx, V., Baldock, K. C. R., Bennett, A. T. D., Boada, R., Bommarco, R., Cartar, R., Chacoff, N., Dänhardt, J., Dicks, L. V., Dormann, C. F., Ekroos, J., Henson, K. S. E., Holzschuh, A., Junker, R. R., Lopezaraiza-Mikel, M., Memmott, J., Montero-Castaño, A., Nelson, I. L., Petanidou, T., Power, E. F., Rundlöf, M., Smith, H. G., Stout, J. C., Temitope, K., Tscharntke, T., Tscheulin, T., Vilà,

M. and Kunin, W. E. (2014). The potential for indirect effects between co-flowering plants via shared pollinators depends on resource abundance, accessibility and relatedness. *Ecology Letters* 17(11), 1389–1399.

Carvalheiro, L. G., Seymour, C. L., Veldtman, R. and Nicolson, S. W. (2010). Pollination services decline with distance from natural habitat even in biodiversity-rich areas. *Journal of Applied Ecology* 47(4), 810–820.

Chacoff, N. P., Vázquez, D. P., Lomáscolo, S. B., Stevani, E. L., Dorado, J. and Padrón, B. (2012). Evaluating sampling completeness in a desert plant-pollinator network. Journal of Animal Ecology 81(1), 190–200.

Chapman, J. W., Reynolds, D. R. and Smith, A. D. (2003). Vertical-looking radar: A new tool for monitoring high-altitude insect migration. *BioScience* 53(5), 503–511.

Dee, L. E., Allesina, S., Bonn, A., Eklöf, A., Gaines, S. D., Hines, J., Jacob, U., McDonald-Madden, E., Possingham, H., Schröter, M. and Thompson, R. M. (2017). Operationalizing network theory for ecosystem service assessments. *Trends in Ecology and Evolution* 32(2), 118–130.

Derocles, S. A. P., Bohan, D. A., Dumbrell, A. J., Kitson, J. J. N., Massol, F., Pauvert, C., Plantegenest, M., Vacher, C. and Evans, D. M. (2018). Biomonitoring for the 21st century: Integrating next-generation sequencing into ecological network analysis. *Advances in Ecological Research* 58, 1–62.

Devoto, M., Bailey, S. and Memmott, J. (2011). The "night shift": Nocturnal pollen-transport networks in a boreal pine forest. *Ecological Entomology* 36(1), 25–35.

Dormann, C. F., Fründ, J., Blüthgen, N. and Gruber, B. (2009). Indices, graphs and null models: Analysing bipartite ecological networks. *The Open Ecology Journal* 2(1), 7–24.

Duane, A. and Brotons, M. C. (2021). Towards a comprehensive look at global drivers of novel extreme wildfire events. *Climatic Change* 165(3–4), 43.

Dunne, J. A., Williams, R. J. and Martinez, N. D. (2002). Network structure and biodiversity loss in food webs: Robustness increases with connectance. *Ecology Letters* 5(4), 558–567.

Dupont, Y. L., Padrón, B., Olesen, J. M. and Petanidou, T. (2009). Spatio-temporal variation in the structure of pollination networks. *Oikos* 118(8), 1261–1269.

Evans, D. M. and Kitson, J. J. (2020). Molecular ecology as a tool for understanding pollination and other plant-insect interactions. *Current Opinion in Insect Science* 38, 26–33.

Evans, D. M., Pocock, M. J. O. and Memmott, J. (2013). The robustness of a network of ecological networks to habitat loss. *Ecology Letters* 16(7), 844–852.

Fontaine, C., Guimaraes, P. R., Jr., Kefi, S., Loeuille, N., Memmott, J., van der Putten, W. H., van Veen, F. J. F. and Thebault, E. (2011). The ecological and evolutionary implications of merging different types of networks. *Ecology Letters* 14(11), 1170–1181.

Garibaldi, L. A., Carvalheiro, L. G., Leonhardt, S. D., Aizen, M. A., Blaauw, B. R., Isaacs, R., Kuhlmann, M., Kleijn, D., Klein, A. M., Kremen, C., Morandin, L., Scheper, J. and Winfree, R. (2014). From research to action: Enhancing crop yield through wild pollinators. *Frontiers in Ecology and the Environment* 12(8), 439–447.

Giavi, S., Fontaine, C. and Knop, E. (2021). Impact of artificial light at night on diurnal plant-pollinator interactions. *Nature Communications* 12(1), 1690.

Gibson, R. H., Knott, B., Eberlein, T. and Memmott, J. (2011). Sampling method influences the structure of plant-pollinator networks. Oikos 120(6), 822–831.

Gibson, R. H., Nelson, I. L., Hopkins, G. W., Hamlett, B. J. and Memmott, J. (2006). Pollinator webs, plant communities and the conservation of rare plants: Arable weeds as a case study. *Journal of Applied Ecology* 43(2), 246–257.

Grab, H., Branstetter, M. G., Amon, N., Urban-Mead, K. R., Park, M. G., Gibbs, J., Blitzer, E. J., Poveda, K., Loeb, G. and Danforth, B. N. (2019). Agriculturally dominated landscapes reduce bee phylogenetic diversity and pollination services. *Science* 363(6424), 282–284.

Gresty, C. E. A., Clare, E., Devey, D. S., Cowan, R. S., Csiba, L., Malakasi, P., Lewis, O. T. and Willis, K. J. (2018). Flower preferences and pollen transport networks for cavity-nesting solitary bees: Implications for the design of agri-environment schemes. *Ecology and Evolution* 8(15), 7574–7587.

Grubisic, M. and van Grunsven, R. H. (2021). Artificial light at night disrupts species interactions and changes insect communities. *Current Opinion in Insect Science* 47, 136–141.

Guimarães, P. R., Jr., Pires, M. M., Jordano, P., Bascompte, J. and Thompson, J. N. (2017). Indirect effects drive coevolution in mutualistic networks. *Nature* 550(7677), 511–514.

Guzman, A., Chase, M. and Kremen, C. (2019). On-farm diversification in an agriculturally-dominated landscape positively influences specialist pollinators. *Frontiers in Sustainable Food Systems* 3, 87.

Hale, K. R. S., Valdovinos, F. S. and Martinez, N. D. (2020). Mutualism increases diversity, stability, and function of multiplex networks that integrate pollinators into food webs. *Nature Communications* 11(1), 2182.

Heard, M. S., Botham, M., Broughton, R., Carvell, C., Hinsley, S., Woodcock, B. and Pywell, R. F. (2012). *Quantifying the Effects of Entry Level Stewardship (ELS) on Biodiversity at the Farm Scale: The Hillesden Experiment*. UK Centre for Ecology & Hydrology, Lancaster, UK.

Hoeppke, C. and Simmons, B. I. (2021). maxnodf: An R package for fair and fast comparisons of nestedness between networks. *Methods in Ecology and Evolution* 12(4), 580–585.

Høye, T. T., Ärje, J., Bjerge, K., Hansen, O. L. P., Iosifidis, A., Leese, F., Mann, H. M. R., Meissner, K., Melvad, C. and Raitoharju, J. (2021). Deep learning and computer vision will transform entomology. *Proceedings of the National Academy of Sciences of the United States of America* 118(2), e2002545117.

Hutchinson, M. C., Mora, B. B., Pilosof, S., Barner, A. K., Kéfi, S., Thébault, E., Jordano, P. and Stouffer, D. B. (2019). Seeing the forest for the trees: Putting multilayer networks to work for community ecology. *Functional Ecology* 33(2), 206–217.

Ings, T. C., Montoya, J. M., Bascompte, J., Blüthgen, N., Brown, L., Dormann, C. F., Edwards, F., Figueroa, D., Jacob, U., Jones, J. I., Lauridsen, R. B., Ledger, M. E., Lewis, H. M., Olesen, J. M., Van Veen, F. J. F., Warren, P. H. and Woodward, G. (2009). Ecological networks – Beyond food webs. *Journal of Animal Ecology* 78(1), 253–269.

Jeffries, D. L., Chapman, J., Roy, H. E., Humphries, S., Harrington, R., Brown, P. M. J. and Handley, L. J. (2013). Characteristics and drivers of high-altitude ladybird flight: Insights from vertical-looking entomological radar. *PLoS ONE* 8(12), e82278.

Kaiser-Bunbury, C. N., Muff, S., Memmott, J., Müller, C. B. and Caflisch, A. (2010). The robustness of pollination networks to the loss of species and interactions: A quantitative approach incorporating pollinator behaviour. *Ecology Letters* 13(4), 442–452.

Keyes, A. A., McLaughlin, J. P., Barner, A. K. and Dee, L. E. (2021). An ecological network approach to predict ecosystem service vulnerability to species losses. *Nature Communications* 12(1), 1586.

King, C., Ballantyne, G. and Willmer, P. G. (2013). Why flower visitation is a poor proxy for pollination: Measuring single-visit pollen deposition, with implications for pollination networks and conservation. *Methods in Ecology and Evolution* 4(9), 811–818.

Kleijn, D., Winfree, R., Bartomeus, I., Carvalheiro, L. G., Henry, M., Isaacs, R., Klein, A. M., Kremen, C., M'Gonigle, L. K., Rader, R., Ricketts, T. H., Williams, N. M., Adamson, N. L., Ascher, J. S., Báldi, A., Batáry, P., Benjamin, F., Biesmeijer, J. C., Blitzer, E. J., Bommarco, R., Brand, M. R., Bretagnolle, V., Button, L., Cariveau, D. P., Chifflet, R., Colville, J. F., Danforth, B. N., Elle, E., Garratt, M. P. D., Herzog, F., Holzschuh, A., Howlett, B. G., Jauker, F., Jha, S., Knop, E., Krewenka, K. M., Le Féon, V., Mandelik, Y., May, E. A., Park, M. G., Pisanty, G., Reemer, M., Riedinger, V., Rollin, O., Rundlöf, M., Sardiñas, H. S., Scheper, J., Sciligo, A. R., Smith, H. G., Steffan-Dewenter, I., Thorp, R., Tscharntke, T., Verhulst, J., Viana, B. F., Vaissiére, B. E., Veldtman, R., Ward, K. L., Westphal, C. and Potts, S. G. (2016). Corrigendum: Delivery of crop pollination services is an insufficient argument for wild pollinator conservation. *Nature Communications* 7, 10841.

Knop, E., Zoller, L., Ryser, R., Gerpe, C., Hörler, M. and Fontaine, C. (2017). Artificial light at night as a new threat to pollination. *Nature* 548(7666), 206–209.

Lescourret, F., Magda, D., Richard, G., Adam-Blondon, A.-F., Bardy, M., Baudry, J., Doussan, I., Dumont, B., Lefèvre, F., Litrico, I., Martin-Clouaire, R., Montuelle, B., Pellerin, S., Plantegenest, M., Tancoigne, E., Thomas, A., Guyomard, H. and Soussana, J.-F. (2015). A social-ecological approach to managing multiple agro-ecosystem services. *Current Opinion in Environmental Sustainability* 14, 68–75.

Loeuille, N., Barot, S., Georgelin, E., Kylafis, G. and Lavigne, C. (2013). Eco-evolutionary dynamics of agricultural networks. *Advances in Ecological Research* 49, 339–435.

Lopezaraiza-Mikel, M. E., Hayes, R. B., Whalley, M. R. and Memmott, J. (2007). The impact of an alien plant on a native plant-pollinator network: An experimental approach. *Ecology Letters* 10(7), 539–550.

Macfadyen, S., Gibson, R., Polaszek, A., Morris, R. J., Craze, P. G., Planqué, R., Symondson, W. O. C. and Memmott, J. (2009). Do differences in food web structure between organic and conventional farms affect the ecosystem service of pest control? *Ecology Letters* 12(3), 229–238.

Macgregor, C. J., Evans, D. M., Fox, R. and Pocock, M. J. O. (2017). The dark side of street lighting: Impacts on moths and evidence for the disruption of nocturnal pollen transport. *Global Change Biology* 23(2), 697–707.

Macgregor, C. J., Kitson, J. J. N., Fox, R., Hahn, C., Lunt, D. H., Pocock, M. J. O. and Evans, D. M. (2019). Construction, validation, and application of nocturnal pollen transport networks in an agro-ecosystem: A comparison using light microscopy and DNA metabarcoding. *Ecological Entomology* 44(1), 17–29.

Macgregor, C. J., Pocock, M. J. O., Fox, R. and Evans, D. M. (2015). Pollination by nocturnal Lepidoptera, and the effects of light pollution: A review. *Ecological Entomology* 40(3), 187–198.

Maia, K. P., Vaughan, I. P. and Memmott, J. (2019). Plant species roles in pollination networks: An experimental approach. *Oikos* 128(10), 1446–1457.

Memmott, J. (1999). The structure of a plant-pollinator food web. *Ecology Letters* 2(5), 276-280.
Memmott, J. (2009). Food webs: A ladder for picking strawberries or a practical tool for practical problems? *Philosophical Transactions of the Royal Society of London. Series B, Biological Sciences* 364(1524), 1693-1699.
Memmott, J., Craze, P. G., Waser, N. M. and Price, M. V. (2007). Global warming and the disruption of plant-pollinator interactions. *Ecology Letters* 10(8), 710-717.
Memmott, J., Waser, N. M. and Price, M. V. (2004). Tolerance of pollination networks to species extinctions. *Proceedings. Biological Sciences* 271(1557), 2605-2611.
Miller, K. E., Polaszek, A. and Evans, D. M. (2021). A dearth of data: Fitting parasitoids into ecological networks. *Trends in Parasitology* 37(10), 863-874.
Montoya, J. M. (2015). Ecology: Dynamics of indirect extinction. *Current Biology* 25(23), R1129-R1131.
Morales, C. L., Sáez, A., Garibaldi, L. A. and Aizen, M. A. (2017). Disruption of pollination services by invasive pollinator species. In: *Impact of Biological Invasions on Ecosystem Services*, Vilà, M. and Hulme, P. E. (Eds), Springer, Cham, Switzerland, pp. 203-220.
Morales, C. L. and Traveset, A. (2009). A meta-analysis of impacts of alien vs. native plants on pollinator visitation and reproductive success of co-flowering native plants. *Ecology Letters* 12(7), 716-728.
Morandin, L. A. and Kremen, C. (2013). Hedgerow restoration promotes pollinator populations and exports native bees to adjacent fields. *Ecological Applications : A Publication of the Ecological Society of America* 23(4), 829-839.
Olito, C. and Fox, J. W. (2015). Species traits and abundances predict metrics of plant-pollinator network structure, but not pairwise interactions. *Oikos* 124(4), 428-436.
Parra-Tabla, V. and Arceo-Gómez, G. (2021). Impacts of plant invasions in native plant-pollinator networks. *New Phytologist* 230(6), 2117-2128.
Pawar, S. (2014). Ecology. Why are plant-pollinator networks nested? *Science* 345(6195), 383.
Petsopoulos, D., Lunt, D. H., Bell, J. R., Kitson, J. J. N., Collins, L., Boonham, N., Morales-Hojas, R. and Evans, D. M. (2021). Using network ecology to understand and mitigate long-term insect declines. *Ecological Entomology* 46(4), 693-698.
Pocock, M. J. O., Evans, D. M. and Memmott, J. (2012). The robustness and restoration of a network of ecological networks. *Science* 335(6071), 973-977.
Poisot, T., Canard, E., Mouillot, D., Mouquet, N. and Gravel, D. (2012). The dissimilarity of species interaction networks. *Ecology Letters* 15(12), 1353-1361.
Ponisio, L. C., Gaiarsa, M. P. and Kremen, C. (2017). Opportunistic attachment assembles plant-pollinator networks. *Ecology Letters* 20(10), 1261-1272.
Power, E. F. and Stout, J. C. (2011). Organic dairy farming: Impacts on insect-flower interaction networks and pollination. *Journal of Applied Ecology* 48(3), 561-569.
Quintero, E., Isla, J. and Jordano, P. (2021). Methodological overview and data-merging approaches in the study of plant-frugivore interactions. *Oikos* 2022(2), 1-18.
Raimundo, R. L. G., Guimarães, P. R., Jr. and Evans, D. M. (2018). Adaptive networks for restoration ecology. *Trends in Ecology and Evolution* 33(9), 664-675.
Ratto, F., Simmons, B. I., Spake, R., Zamora-Gutierrez, V., MacDonald, M. A., Merriman, J. C., Tremlett, C. J., Poppy, G. M., Peh, K. S.-H. and Dicks, L. V. (2018). Global importance of vertebrate pollinators for plant reproductive success: A meta-analysis. *Frontiers in Ecology and the Environment* 16(2), 82-90.

R Core Team (2021). *R: A Language and Environment for Statistical Computing*. R Foundation for Statistical Computing. https://www.R-project.org/.

Redhead, J. W., Woodcock, B. A., Pocock, M. J. O., Pywell, R. F., Vanbergen, A. J. and Oliver, T. H. (2018). Potential landscape-scale pollinator networks across Great Britain: Structure, stability and influence of agricultural land cover. *Ecology Letters* 21(12), 1821–1832.

Sayles, J. S., Mancilla Garcia, M., Hamilton, M., Alexander, S. M., Baggio, J. A., Fischer, A. P., Ingold, K., Meredith, G. R. and Pittman, J. (2019). Social-ecological network analysis for sustainability sciences: A systematic review and innovative research agenda for the future. *Environmental Research Letters: ERL [Web Site]* 14(9), 1–18, 35340667.

Settele, J., Bishop, J. and Potts, S. G. (2016). Climate change impacts on pollination. *Nature Plants* 2(7), 16092.

Simmons, B. I., Cirtwill, A. R., Baker, N. J., Wauchope, H. S., Dicks, L. V., Stouffer, D. B. and Sutherland, W. J. (2019). Motifs in bipartite ecological networks: Uncovering indirect interactions. *Oikos* 128(2), 154–170.

Srinivasan, U. T., Dunne, J. A., Harte, J. and Martinez, N. D. (2007). Response of complex food webs to realistic extinction sequences. *Ecology* 88(3), 671–682.

Staniczenko, P. P. A., Lewis, O. T., Jones, N. S. and Reed-Tsochas, F. (2010). Structural dynamics and robustness of food webs. *Ecology Letters* 13(7), 891–899.

Thomsen, P. F. and Sigsgaard, E. E. (2019). Environmental DNA metabarcoding of wild flowers reveals diverse communities of terrestrial arthropods. *Ecology and Evolution* 9(4), 1665–1679.

Traveset, A., Olesen, J. M., Nogales, M., Vargas, P., Jaramillo, P., Antolín, E., Trigo, M. M. and Heleno, R. (2015). Bird-flower visitation networks in the Galápagos unveil a widespread interaction release. Nature communications 6(1), 1–6.

Tylianakis, J. M., Laliberté, E., Nielsen, A. and Bascompte, J. (2010). Conservation of species interaction networks. *Biological Conservation* 143(10), 2270–2279.

Tylianakis, J. M., Tscharntke, T. and Lewis, O. T. (2007). Habitat modification alters the structure of tropical host-parasitoid food webs. *Nature* 445(7124), 202–205.

United Nations (2015). *Transforming Our World: The 2030 Agenda for Sustainable Development*. United Nations. https://sdgs.un.org/publications/transforming-our-world-2030-agenda-sustainable-development-17981.

Vacher, C., Tamaddoni-Nezhad, A., Kamenova, S., Peyrard, N., Moalic, Y., Sabbadin, R., Schwaller, L., Chiquet, J., Alex Smith, M. A., Vallance, J., Fievet, V., Jakuschkin, B. and Bohan, D. A. (2016). Learning ecological networks from next-generation sequencing data. *Advances in Ecological Research* 54, 1–39.

Walton, R. E., Sayer, C. D., Bennion, H. and Axmacher, J. C. (2020). Nocturnal pollinators strongly contribute to pollen transport of wild flowers in an agricultural landscape. *Biology Letters* 16(5), 20190877.

Windsor, F. M., Tavella, J., Rother, D. C., Raimundo, R. L. G., Devoto, M., Guimarães, P. R., Jr. and Evans, D. M. (2021). Identifying plant mixes for multiple ecosystem service provision in agricultural systems using ecological networks. *Journal of Applied Ecology* 58(12), 2770–2782.

Wirta, H. K., Hebert, P. D., Kaartinen, R., Prosser, S. W., Várkonyi, G. and Roslin, T. (2014). Complementary molecular information changes our perception of food web structure. Proceedings of the National Academy of Sciences 111(5), 1885–1890.

Wootton, J. T. (2002). Indirect effects in complex ecosystems: Recent progress and future challenges. *Journal of Sea Research* 48(2), 157–172.
Young, J. G., Valdovinos, F. S. and Newman, M. E. J. (2021). Reconstruction of plant-pollinator networks from observational data. *Nature Communications* 12(1), 3911.
Zamorano, J., Bartomeus, I., Grez, A. A. and Garibaldi, L. A. (2020). Field margin floral enhancements increase pollinator diversity at the field edge but show no consistent spillover into the crop field: A meta-analysis. *Insect Conservation and Diversity* 13(6), 519–531.

Chapter 12

Best management practices for pollinator protection in US apple production[1]

Julianna K. Wilson, Michigan State University, USA; David Biddinger, Penn State University, USA; David Epstein, Northwest Horticultural Council, USA; Anne Nielsen, Rutgers University, USA; Ngoc Phan, University of Arkansas, USA; Jim Walgenbach, North Carolina State University, USA; and Neelendra K. Joshi, University of Arkansas, USA

1 Introduction

Apples are one of the most valuable pollinator-dependent fruit crops grown in the United States. National apple production in 2020 was about 10.2 billion pounds, with a value of more than US$2.9 billion (USDA NASS, 2021). According to the Food and Agriculture Organization of the United Nations, the United States is the second-largest apple producer in the world behind China, followed by Poland, Turkey, and India (FAOSTAT, 2019). According to the United States Department of Agriculture, National Agricultural Statistics Service (USDA NASS), apple is grown on more than 295 000 acres in the US with 103 000 acres in the Eastern United States and another 192 000 acres in the Western United States (USDA NASS, 2020). In the east, Michigan, New York, Pennsylvania and Virginia are the largest producing states. Washington, California and Oregon

1 This chapter is adapted from *Pollination Best Management Practices for Apples*, a USApple publication developed in partnership with the Honey Bee Health Coalition. The chapter is published with the kind permission of USApple (www.usapple.org).

http://dx.doi.org/10.19103/AS.2022.0111.22

are the western states where most apples are grown. Where necessary, regional differences between eastern and western apple production are noted in this chapter.

Unlike many other crops, fresh market apple production is more dependent on qualitative characteristics such as size and shape for its highest value rather than on yield (Garratt et al., 2014). Apples are incapable of self-fertilization, even within a cultivar. This means they require cross-pollination by insects to set fruits and produce seeds. As an insect moves from one flower to another, it will transfer pollen from the anthers to the stigma to allow fertilization. Each apple has ten ovules. When the ovules are properly fertilized, the carpel will expand and protect the developing seeds. When full pollination occurs, large and symmetrically shaped apples are produced. If not, fruit distortion or dwarfing will lead to unmarketable apples (Brittain and Eidt, 1933; Partap et al., 2001; Ramírez and Davenport, 2013; Garratt et al., 2014; Sheffield et al., 2016). Fruit size is largely dependent on the number of fruit per unit area and it is common practice for growers to need to thin fruitlets down to the number of apples desired to achieve a particular size fruit at harvest. In practical terms this means that to achieve adequate crop pollination in apples, it is estimated that only 2-8% of apple blossoms require full pollination (Westwood, 1993).

Commercial apple growers in the United State typically rely on managed honey bees (*Apis mellifera*) as a reliable source of pollination services with a recommended stocking rate of at least two hives/acre. However, honey bees and other pollinator species have been threatened by multiple biotic and abiotic factors. These stresses include pathogens, parasites, agrochemical exposure, lack of forage and nesting habitat (Goulson et al., 2015; Brown et al., 2016; McMenamin et al., 2016), all of which may contribute to pollination failure risk and translate to increased pollinator rental costs (Breeze et al., 2011; Melathopoulos et al., 2015; Bauer and Wing, 2016). Due to increased demand for managed pollinators and the additional costs associated with managing Varroa mites in hives, growers of pollinator-dependent crops have seen an increase in rental costs from US$35/hive to over US$100/hive (Ferrier, 2018; USDA ARS, 2021; US EPA, 2021). Where and when demand is particularly high, commercial honey bee rental prices have quadrupled in the past 10 years (Caron, 2009; USDA NASS, 2017; FAO, 2018).

Increasing costs and problems with honey bee supply have led apple growers in some regions of the United States (e.g. in eastern states) to look for alternatives to supplement or even replace honey bees (Crassweller et al., 2017; Crassweller et al., 2020). For instance, some growers have employed other kinds of managed bees including bumble bees and mason bees (Park et al., 2012). While most bee species that visit orchards in bloom are native, some species such as the Japanese orchard bee (*Osmia cornifrons*) were introduced specifically for apple pollination by the USDA (Yamada et al., 1971; Batra, 1978, 1982; Kuhn and Ambrose, 1984; Sekita and Yamada, 1993; Bosch and Kemp, 2001, 2002; Sekita, 2001). Nests of mason bees such as *Osmia cornifrons* or *O. lignaria,* are typically stocked at densities that provide 250–500 females per acre based on rates used in almond pollination (Greer 1999; Kuhn and Ambrose 1984; Matsumoto *et al.* 2009).

Besides these managed bee species, over 50 other bee species have been recorded visiting apple orchards in North America, providing natural pollination services (Greer, 1999; Park et al., 2010; Moisset and Buchmann, 2011; Blitzer et al., 2016; Joshi et al. 2016, Hopwood *et al.* 2020). Surveys of New York and Pennsylvania apple growers in 2017 showed that wild bees can often provide all of the pollination needs of apple orchards when present in the surrounding landscape (Park et al., 2018). In the Western United States, honey bees continue to be the most important pollinator in apple production where pollination practices are driven by lack of non-crop floral resources and different wild bee communities.

This chapter presents a set of pollinator best management practices (BMPs) for US apple production. The primary goal is to provide guidance for growers, crop consultants and regulators on pollinator protection in apple orchards. This guidance considers practical production needs and supports outreach and education programs to foster adoption and implementation of identified BMPs. In the long term, the aim is to increase understanding within the US apple-growing community of the benefits and opportunities around protecting pollinators while also protecting the crop from pest damage.

2 Apple orchard pollinators

Apples produce flowers that many kinds of insects access including bees and hover flies. Among these, bees are the most important pollinators. Brought to North America more than 400 years ago, feral honey bees (*A. mellifera*) were once common across North America but are now largely absent because of the accidental introduction of the Varroa mite. Managed hives are maintained by beekeepers to control mites and are the most important crop pollinator because colonies can be rented and placed where they are needed and then moved out once crop bloom has ended. Where wild bees are insufficient in number, producers routinely contract with beekeepers for pollination services. There are other bee species (e.g. mason bees, bumble bees) that can be managed for orchard pollination, but availability is often limited.

More than 50 wild (non-*Apis*) bee species are known to visit apples during bloom in North America and can be grouped into one of the following categories based on nesting biology:

- Ground nesting bees that nest in orchard soils or in adjacent habitat – especially digger bees (*Andrena* spp.), but also cellophane (colletid) and sweat (halictid) bees;
- Mason bees (*Osmia* spp.) that nest above ground in preexisting holes in trees or in boxes supplied with paper straw, natural reeds or bamboo;
- Small carpenter bees (*Ceratina* spp.) that nest in woody plants with pithy stems (e.g. brambles); and
- Wild bumble bee queens collecting pollen and nectar in orchards while they build up their below-ground colonies for the season.

In some Eastern US orchards, wild bees may be abundant and in sufficient number to achieve adequate pollination without adding managed bees (Biddinger et al., 2013a).

3 Pests, diseases and their management in apple production

Apple production faces a number of arthropod pests and diseases which are still mainly managed by use of pesticides. It is important to remember that pollinator exposure to pesticides is not limited to insecticides. There is an increasing body of research focused on the impacts of fungicides, other pesticides and adjuvants on pollinator health (Hopwood et al., 2020).

3.1 Apple pests

North American apple orchards are subject to a diverse arthropod pest complex of up to 50 species that can cause from minor to severe damage. Economically, the most important arthropod pests are those that cause direct damage to fruit including codling moth (*Cydia pomonella*), oriental fruit moth (*Grapholita molesta*), plum curculio (*Conotrachelus nenuphar*), apple maggot (*Rhagoletis pomonella*), and a variety of plant bugs (Sutton et al. 2014). Indirect pests that attack trunks, limbs, leaves or buds, include borers, leafrollers, leafminers, leafhoppers, aphids, scale insects, and mites (Sutton et al. 2014). When they are abundant, scale insects and mites can also cause direct fruit damage.

Most insecticides are not labeled for use during bloom, but some key life stages of the pests listed above occur immediately before or at the end of bloom, a timing known as petal fall. Management of arthropod pests around bloom is a complicated task as the pesticides applied for the control of different pests during that time may later affect pollinator health. For instance, where sap-sucking insects such as the rosy apple aphid are abundant, the timing for management is at green tip, as soon as eggs hatch and young nymphs start to feed on leaf buds and fruit clusters, but before they can cause deformed fruit damage. A systemic insecticide is the best and most commonly applied option at this timing (Crassweller et al., 2020), but previous research has shown that pesticide residues used to control rosy apple aphid pre-bloom can be found later in the nectar and pollen of apple blossoms where they could potentially impact pollinators (Heller et al., 2020). Similarly, insecticides used against plum curculio when it emerges from overwintering around petal fall, present another timing that is tricky in terms of pollinator protection, especially when there are adjacent blocks of different cultivars at various stages of bloom. Toxicological data is used by the United States Environmental Protection Agency (EPA) to determine pesticide risk in the registration of new pesticides and in the review of older pesticides in light of new data.

3.2 Apple diseases

Fungicides have been used in North American apple production during bloom for decades to protect plants from major diseases such as apple scab (*Ventura inaequalis*) and powdery mildew (*Podosphaera leucotricha*). Failure to apply fungicides during the bloom infection period can lead to unmarketable fruit and/or greatly increased post-bloom applications when disease cycles are not disrupted at this critical time of the season (MacHardy, 1996; Turechek et al., 2004; Peter et al., 2018). While toxicological data on adult honey bees exposed to fungicides generally allowed fungicides to be considered safe for use during bloom, the study of sublethal effects from fungicide exposure in honey bees providing pollination services is an active area of research.

Apple scab is a fungal disease whose spores overwinter in the fallen leaves from the previous season. In the spring, spores will be mature at the green tip developmental stage of apple and disperse when spring rains occur. If primary infections of the leaves are not prevented early and through bloom, secondary infections of the fruit will occur throughout the rest of the season which can cause complete crop loss through deformed and unsaleable fruit as well as causing severe defoliation (MacHardy, 1996; Gauthier et al., 2018; Crassweller et al., 2020). Of the many fungicide options registered for use in apple for early season prevention of apple scab, the plant systemic products Syllit FL (dodine) and Rally (myclobutanil) have been found in apple nectar and pollen from pre-bloom applications (Heller et al., 2020). Widespread resistance by apple scab to various fungicides has meant these products are important rotational partners as part of a resistance management program (FRAC, 2018; Biggs and Peter, 2019).

Powdery mildew is a dry-weather disease that does not need free-standing water or a significant number of hours of leaf wetness for infection to be triggered. The spores overwinter in dormant flower or leaf buds and disperse via wind in the early spring up through apple bloom to develop in young leaf tissue. Primary infections occur from the tight cluster stage through the initiation of terminal shoots when temperatures range from 10°C to 25°C, and humidity is greater than 70%. These conditions are easily met in the micro-environment of the tree canopy in orchards (Turechek et al., 2004; Crassweller et al., 2020). Economic damage from powdery mildew is caused by reducing flower bud production and tree vigor. Myclobutanil is just one of the many options for early season management of powdery mildew with different modes of action alternated to prevent the formation of resistant fungal populations (Biggs and Peter, 2019).

4 Potential hazards to pollinators from apple pest management practices

Pollinators active in and around orchards, including managed bees hired for pollination services, have the potential to be negatively impacted by pesticide application (Heller et al., 2020; Belsky et al., 2022). However, there are some critical factors that may be overlooked which can minimize unintended exposures of bees to pesticides (Hooven et al., 2013). Within orchards, sources of exposure may include pesticide applications made in blocks not in bloom but which contain blooming weeds within the foraging range of hives or wild bee nests, and when systemic pesticides applied pre-bloom move through treated trees, ending up in nectar or pollen, later collected from those trees by foraging bees. Honey bees have a large foraging range, so pesticide exposure can also come from a wide variety of sources outside of the crop for which they were rented. Except for managing and mitigating drift, exposures from outside the crop are generally not something that either individual apple growers or beekeepers can control.

Bees also face potential hazards from pesticides within the hive. In honey bee hives, pesticides can build up in wax and stored pollen, including pesticides targeting control of Varroa mites. Residue exposure can be cumulative as hives are moved among various crops for their pollination services. Beekeepers who provide crop pollination services are advised to replace old frames with new frames annually to reduce this in-hive hazard.

Risks to bees from pesticides include those from insecticides, herbicides and fungicides, which are applied to crops via foliar sprays, soil applications or seed treatments (OPP, PMRA and CALDPR, 2014). There are two main routes of pesticide exposure for foraging bees: contact and ingestion. Foliar spray applications are commonly used in apple to control insects, mites, fungi, bacteria and other pests. The deposition of pesticides on plant surfaces (e.g. leaf, stem or flower) can directly contact foraging bees if they walk on these surfaces. Generally, growers do not spray contact insecticides during bloom to avoid contact exposure to bees, but fungicides, bactericides and plant growth regulators are often applied during bloom under the assumption that they are not harmful to pollinators.

Since insecticides are not sprayed during apple bloom, bee exposure is almost completely through ingestion of systemic products applied pre-boom as they concentrate in nectar and pollen. Exposure to fungicides, however, can be from direct contact exposure if the flowers are sprayed during bloom, as fungicides often are, or from plant systemic movement of pre-bloom and bloom sprays.

Most systemic pesticides also have contact activity, which may be short-lived depending on how quickly and to what extent they are absorbed into the crop tissues or degrade due to environmental exposure. In addition to their contact activity, systemic pesticides can be translocated to pollen or nectar via either xylem or phloem mobility (Sur and Stork, 2003; DeBoer and Satchivi, 2014; Bonmatin et al., 2015), providing exposure via the ingestion route. Systemic pesticides are also sheltered from environmental degradation such as weathering of surface residues to some extent by being in the plant tissues, and this can prolong their toxicity. During apple bloom, adult bees forage for and collect pollen and nectar as the main source of protein and energy.

Recently, it has been shown that fungicides can synergize insecticides if sprayed as mixtures leading to increased toxicity to pollinators, which results in a reduction of queen-rearing success and drastic decrease in forager survival for honey bees as well as disruption of nest recognition in solitary bees (Biddinger et al., 2013b; Johnson and Percel, 2013; Artz and Pitts-Singer, 2015; Fisher et al., 2017; Belsky et al., 2022). Fungicides alone can have adverse effects on the quality of bee colonies (Bernauer et al., 2015; Yoder et al., 2017), including higher incidence of diseases, reduction in worker productivity (in honey bees), smaller queens and less bee biomass (in bumble bees).

5 Best management practices to promote pollinators

Habitats containing flowers that are protected from pesticide drift benefit apple growers by providing additional forage (nectar and pollen) to pollinators. They can also provide nesting opportunities for wild pollinators. These habitats also provide food and refuge for agents of biological control – predators (e.g. hover flies, green lacewings, lady beetles) and parasitoids (tiny wasps that attack pests) that help suppress populations of apple insect pests. Depending on the region, different types of habitats can be planted or conserved to provide non-crop flowering plants on the farm. When the crop is not in bloom, these habitats can also help divert bees from fruit plantings.

Woodlands are a habitat that can be conserved and provide nesting and spring flowers for wild bees important to orchard pollination in the Eastern

United States. Wildflower meadows (a.k.a. prairie style pollinator plantings) can be established in some regions to provide summer and fall blooming resources but require planning, patience and maintenance to implement. Mass-flowering, summer-blooming cover crops like sunflowers, canola, buckwheat, etc. may be a good fit where more permanent plantings are impractical. Intercropping or polyculture farming may be the best way to provide additional bee forage in some parts of the United States where water conservation limits the planting of pollinator habitat outside of orchards.

In each region of the United States, there are a variety of organizations and programs ready to help with the planning and implementation of pollinator habitats. Some of these programs offer cost-sharing incentives. These programs include:

- USDA Natural Resource Conservation Service and Farm Service Agency (e.g. Conservation Reserve Program, Conservation Stewardship Program, Environmental Quality Incentives Program);
- US Fish & Wildlife Service (e.g. Partners for Fish and Wildlife);
- Various state Department of Natural Resource incentive programs; and
- Non-profit or private groups (e.g. The Bee & Butterfly Habitat Fund, Project Wingspan, Pheasants Forever, The Xerces Society).

Developing pollinator habitats promotes both commercial and wild bee populations and the pollination services they supply.

When contracting with a beekeeper for pollination services, it is important to communicate expectations. Growers and beekeepers are strongly encouraged to clarify expectations through a written contract that includes plans for record keeping, when and how many hives will be delivered, where hives will be placed on the farm, when they will be removed, planned pesticide use and other conditions to provide for the health of the bees, including provision of clean water. There should be a meeting between the grower and the beekeeper prior to the spring bloom period to discuss the upcoming bee season. If the grower is not also the pesticide applicator, open communication between the grower, whoever is applying the pesticides and the beekeeper will be essential. Ultimately, growers are responsible for communicating with their pesticide applicator.

Growers/applicators should provide sufficient time between pre-bloom sprays and placement of the hives to avoid direct exposure and to ensure that beekeepers can adhere to restricted-entry intervals on pesticide labels when placing or removing hives from the orchard. Orchardists should give beekeepers as much advance notice as possible on placing and removing hives. Ideally, 1 week's notice should be given for a prospective date of arrival and 2 days' notice when the exact night of arrival is decided upon.

It is also important to select a suitable location for hives on the farm. To minimize potential for pesticide drift reaching hives, select locations on the farm that are protected from wind, easy for the beekeeper to access and easy for the pesticide applicator to avoid when spraying. Honey bees are highly mobile, so hives do not need to be distributed throughout orchards. Set aside an area on the farm in a sunny location to enhance bee activity.

6 Best management practices: integrated pest and pollinator management

Integrated pest and pollinator management (IPPM) is a sustainable, science-based, decision-making process that combines scouting and multiple management tools (e.g. biological, cultural, physical, and chemical control) to identify, manage and reduce risk from pests in ways that minimize overall economic and environmental risk, while supporting human health. Pests are defined as any organism (microbes, plants or animals) that poses economic, health, aesthetic or environmental risk. Pests are context-specific, so an organism that is a pest in one environment may be benign or beneficial in others. Integrated pest management (IPM) practices are now an integral part of apple production systems, whether these production systems rely on synthetic or organically certified pesticides.

For many apple pests, weather-based or biologically predictive models are available and used to aid pathogen and insect pest management decisions. Successful implementation of these models requires IPM practitioners trained in their use and knowledgeable of economic or action thresholds specific to their commodities and production areas. Apple growers rely on a combination of information sources for pest management recommendations, including independent crop protection consultants and local extension specialists providing scouting reports and pest management guides. Examples of tools used in combination with predictive models include traps baited with semiochemicals, beat-tray samples, systematic scouting of plant tissues normally affected by target organisms, spore traps and remote sensing.

Pesticides are used in apple production to prevent, kill or suppress pest organisms from causing economic damage to the crop. Pesticides are selected according to many criteria, including their fit in a season-long IPPM program managing multiple co-occurring pests. Criteria include efficacy in suppressing different life stages of a pest (e.g. egg, larva and adult), cost, compatibility with biological control, regulatory restrictions on use related to target export markets, safety to farm workers, potential to harm non-target organisms and managing against pesticide resistance.

Growers also incorporate tactics to control pests such as biological control, pheromone-mediated mating disruption and other biorational practices as options that may help reduce the need for other pesticide applications. Currently, some key apple insect pests, such as codling moth and oriental fruit moth, are effectively controlled with pheromone-mediated mating disruption as part of an IPM approach, with greater than 90% of apple orchards in the Western United States implementing this technology. The practice of mating disruption has reduced the need for companion insecticide sprays by 40–60% for the targeted insect.

Pesticides still play a crucial role in the management of apple pests. In the recent past, different strategies have been developed to mitigate potential non-target impacts of orchard pesticides. For instance, Biddinger and Rajotte (2015) proposed an IPPM approach that, with slight shifts in pesticide types and application timings, pests can be controlled without harming pollinators (Biddinger and Rajotte, 2015; Biddinger et al., 2018). Heller et al. (2020) analyzed residue levels of commonly used orchard pesticides in apple nectar and pollen after pre-bloom applications using IPPM principles by using the QuEChERS procedure. The result showed that pesticide residue levels were 1000 to 10000 times lower compared with conventional pesticide concentrations applied to the plant surface and much lower than the lethal concentration of pesticide needed to kill 50% of a population (LC_{50}) of honey bees (Hopwood et al., 2016; Shugrue, 2016). This knowledge provides the flexibility to modify pest management activities to favor pollinator health in an IPPM framework.

7 Best management practices for pesticide application

7.1 Regulations, labels and record keeping

Pesticide applicators are required by law to follow current product labels when applying pesticides. Pesticide labels can be found in a variety of places online, including the CDMS.net database (www.cdms.net). All new and recently updated pesticide labels are likely to contain mandatory and advisory language related to bee safety. This language can appear in multiple places within a single pesticide label, and may be accompanied by an icon of a bee to alert applicators of use restrictions related to pollinators. In addition, many state departments of agriculture maintain their own pesticide use requirements.

Growers and crop protection consultants should be knowledgeable of state and federal rules and regulations guiding pesticide applications. If not required by law in a particular state, keeping good records of all pesticide applications provides the applicator with the ability to demonstrate that applications were made according to the label if an issue arises. This can also provide a record to evaluate the efficacy or impacts of pesticide applications retroactively.

7.2 Pesticide selection and application timing

There are several regional guides available for advising growers and crop protection consultants on the selection of pesticides that are least toxic to bees and other beneficial insects. Important considerations include the following:

- Consider whether the insecticide or fungicide must be applied during bloom. Can it wait until bee colonies are removed from the orchard?
- When selecting from among pesticides permitted for use during crop bloom, give due consideration to pesticides that are least toxic to bees and other beneficial organisms. For example, for codling moth, granulosis virus products can be effective against the pest while having no effect on pollinators.
- Time applications to coincide with when bees are least active; most bees are less active in cool temperatures and low light, so spraying pesticides after sunset can greatly reduce the risk of direct exposure.
- Daytime applications when temperatures are expected to remain at 50°F or below will generally reduce risk of exposure. However, colony strength, food resource levels within hives and recent weather events that may have inhibited foraging all combine to influence bee flight activity, even under low temperature conditions.
- For night applications it is recommended to wait until bees have quit foraging before starting a spray application, with a stop time for spraying before 3:00 am to allow adequate time for the application to dry before bees begin to forage in the morning. The drying time for pesticides will differ depending on active ingredient and formulation. Check regional guides for more information.
- Above all, always apply the product according to the label.

7.3 Drift prevention and mitigation

Applicators should always think about spray drift prevention and mitigation. This starts with selection of the sprayer, orchard architecture, understanding how tractor application speed can impact deposition and knowing how to calibrate sprayers to maximize coverage and minimize loss to drift. When purchasing new sprayers, look for advances in sprayer technology that help make proper calibration or drift prevention easier to achieve, including consideration of electrostatic sprayers and tower sprayers that easily convert to accommodate different growing systems. Other considerations include the following:

- Applicators should be well-versed in how to maintain, modify and otherwise calibrate sprayers to maximize spray deposition to the target site.

- Good calibration tools are available (e.g. https://sprayers101. com/ orchardmax) as are calibration training workshops provided by land-grant university extension services and/ or state departments of agriculture in most regions.
- Wherever consistent with orchard architecture and topography, shields or diverters can be added to modify existing sprayers to maximize coverage and minimize drift.
- Orchard architecture, canopy density and physical and chemical properties of the pesticides and adjuvants will dictate the best combination of nozzles, pressures, gallonage and driving speed to maximize deposition and minimize drift.
- When actively spraying in an orchard, turn off the sprayer near hives and pollinator habitat; turn off nozzles directed out of the orchard in the outer row and end of rows.
- Avoid spraying under windy conditions and during temperature inversions. Many pesticide labels require that growers check for inversion layers in their area before spraying and some states (e.g. Washington) require recording such conditions in their spray records.
- Plant wind breaks where possible and maintain buffers between treated areas and hives or pollinator habitats.

7.4 Weed management

Non-crop flowering plants that invade orchard grass row middles and are attractive to bees can become sources of contaminated pollen and nectar. Some growers attach a mowing implement to their sprayer to cut flowering weeds as they apply pesticides to the crop. Selective herbicides may be used to suppress flowering forbs in orchards, but the best timing will depend on the chemical selected. If flowering habitat has been deliberately incorporated into orchard row middles, it will be important to apply pesticides that are least harmful to bees and when bees are not actively foraging.

7.5 Preventing contamination of open water

Honey bees consume large amounts of water for regulating in-hive temperatures. Contamination of open water sources may result in unintended exposures leading to bee poisoning. After applying pesticides, always clean equipment and dispose of pesticide products safely - do not leave contaminated water where bees can access it. Consider using/draining all waste water from pesticide fill sites daily to prevent bees from using waste water as a drinking source.

In addition, consider providing a source of clean water for bees within an orchard under a pollination contract. Test irrigation systems well before pollinators are scheduled to arrive to allow time to make any needed repairs and to provide dry conditions within the orchard, thereby reducing the chance of pollinators drinking compromised (contaminated) water.

Some states have voluntary programs to assist with siting and building pesticide storage facilities to prevent surface and groundwater contamination. One example is the Michigan Agriculture Environmental Assurance Program. Check for similar resources with your state department of agriculture.

7.6 Pest management at different crop stages

In this section we provide additional guidance on mitigating hazards to bees with respect to pest management at different crop stages.

- **Pre-bloom to pink stage**: when systemic pesticides are selected for use, consider restricting applications to the half-inch green developmental stage or selecting a non-systemic alternative, if an application is needed within 14 days before bloom. The pesticide label may include additional restrictions on use around bloom.
- **During bloom**: when selecting from among pesticides permitted for use during crop bloom, consider use of pesticides that are least toxic to bees and other beneficial organisms. Consider using biorational alternatives where effective and available (e.g. pheromone mating disruption).
- **Petal fall**: petal fall is the most challenging timing for insecticide use with respect to pollinators in orchards. It is common for blocks to be planted with mixed varieties for cross pollination, or for adjacent blocks to contain later blooming varieties. This can make it difficult to assess when petal fall has been achieved. Issues include the following:
    - Individual pesticide labels may contain definitions of 'petal fall'.
    - Growers should work closely with beekeepers to coordinate hive removal.
    - Drift prevention and mitigation will be critical with respect to adjacent blocks still in bloom or where bees are actively foraging.
    - Consider use of pheromone mating disruption for key moth pests where available.
    - Eliminating flowering broad-leaf weeds from orchards will both remove a source of potential pesticide exposure to pollinators and will reduce populations of disease vectoring insects active at this time (e.g. leafhoppers).
    - When insecticide applications are used against key pests active at this time, consider insecticides with a short residual, and spray after sunset to minimize exposure to pollinators.

- **Fruit thinning**: for materials that are labeled for use during petal fall for fruit thinning, be aware of bee activity in orchards and consider applying thinners in the evening once the bees have stopped foraging for the night.
- **Remainder of the season through harvest**: continue practices that prevent or mitigate spray drift on flowering plants in orchard row middles and adjacent habitats.

8 Acknowledgements

This chapter is adapted from a USApple publication developed in partnership with the Honey Bee Health Coalition. Two commercial apple growers, Harold Austin of Zircle Fruit Company and Mike Van Agtmael of Van Agtmael Orchards, and two commercial beekeepers, Tim Hiatt of Hiatt Honey LLC and Eric Olson of Olson's Honey, served as members of the advisory panel involved in the publication for USApple. The authors and publisher also wish to acknowledge the contribution of the stakeholder reviewers involved in reviewing the original version of BMPs guidance: Amy Irish-Brown, Michigan State University; Ana Heck, Michigan State University; Steve Hiatt, Hiatt Honey; Chris Hiatt, Hiatt Honey; Rufus Isaacs, Michigan State University; Jeff Leonardini, Washington Fruit & Produce Co.; and Aaron Riggs, Cache Orchards. Additional thanks are due to the Honey Bee Coalition for facilitating the development of the BMPs and the inclusion of data and diverse perspectives.

9 References

Artz, D. R. and Pitts-Singer, T. L. (2015). Effects of fungicide and adjuvant sprays on nesting behavior in two managed solitary bees, *Osmia lignaria* and *Megachile rotundata*, *PLoS ONE* 10(8), e0135688. https://doi.org/10.1371/journal.pone.0135688.

Batra, S. W. T. (1978). *Osmia cornifrons* and *Pithitis smaragdula*, two Asian bees introduced into the United States for crop pollination. In: *Proceedings of the IVth International Symposium on Pollination: Increasing Production of Agriculture Through Increased Insect Pollination*, Maryland Agricultural Experiment Station, pp. 307–312.

Batra, S. W. T. (1982). The hornfaced bee for efficient pollination of small farm orchards. In: Kerr, H. W. and Knutson, L. V. (Eds) *Research for Small Farms* (Vol. 1422), USDA Miscellaneous Publication, pp. 116–120.

Bauer, D. M. and Wing, I. S. (2016). The macroeconomic cost of catastrophic pollinator declines, *Ecological Economics* 126, 1–13. https://doi.org/10.1016/j.ecolecon.2016.01.011.

Belsky, J. E., Biddinger, D. J., Seiter, N. and Joshi, N. K. (2022). Various routes of formulated insecticide mixture whole-body acute contact toxicity to honey bees (*Apis mellifera*), *Environmental Challenges* 6, 100408. https://doi.org/10.1016/j.envc.2021.100408.

Bernauer, O. M., Gaines-Day, H. R. and Steffan, S. A. (2015). Colonies of bumble bees (*Bombus impatiens*) produce fewer workers, less bee biomass, and have smaller mother queens following fungicide exposure, *Insects* 6(2), 478–488. https://doi.org/10.3390/insects6020478.

Biddinger, D. J., Joshi, N. K., Rajotte, E. G., Halbrendt, N. O., Pulig, C., Naithani, K. J. and Vaughan, M. (2013a). An immunomarking method to determine the foraging patterns of *Osmia cornifrons* and resulting fruit set in a cherry orchard, *Apidologie* 44(6), 738–749. https://doi.org/10.1007/s13592-013-0221-x.

Biddinger, D. J., Robertson, J. L., Mullin, C., Frazier, J., Ashcraft, S. A., Rajotte, E. G., Joshi, N. K. and Vaughn, M. (2013b). Comparative toxicities and synergism of apple orchard pesticides to *Apis mellifera* (L.) and *Osmia cornifrons* (Radoszkowski), *PLoS ONE* 8(9), e72587. https://doi.org/10.1371/journal.pone.0072587.

Biddinger, D. J. and Rajotte, E. G. (2015). Integrated pest and pollinator management - Adding a new dimension to an accepted paradigm, *Current Opinion in Insect Science* 10, 204–209. https://doi.org/10.1016/j.cois.2015.05.012.

Biddinger, D. J., Rajotte, E. G. and Joshi, N. K. (2018). Integrating pollinator health into tree fruit IPM: A case study of Pennsylvania apple production. In: Roubik, D. W. (Ed.) *The Pollination of Cultivated Plants: A Compendium for Practitioners* (2nd edn.). Rome: Food and Agriculture Organization of the United Nations, pp. 67–80.

Biggs, A. R. and Peter, K. A. (2019). *2019 Guideline Chart for Fungicide Resistance in Apples*.

Blitzer, E. J., Gibbs, J., Park, M. G. and Danforth, B. N. (2016). Pollination services for apple are dependent on diverse wild bee communities, *Agriculture, Ecosystems and Environment* 221, 1–7. https://doi.org/10.1016/j.agee.2016.01.004.

Bonmatin, J. M., Giorio, C., Girolami, V., Goulson, D., Kreutzweiser, D. P., Krupke, C., Liess, M., Long, E., Marzaro, M., Mitchell, E. A., Noome, D. A., Simon-Delso, N. and Tapparo, A. (2015). Environmental fate and exposure; neonicotinoids and fipronil, *Environmental Science and Pollution Research International* 22(1), 35–67. https://doi.org/10.1007/s11356-014-3332-7.

Bosch, J. and Kemp, W. P. (2001). *How to Manage the Blue Orchard Bee as an Orchard Pollinator, Sustainable Agricultural Network*.

Bosch, J. and Kemp, W. P. (2002). Developing and establishing bee species as crop pollinators: the example of *Osmia* spp. (Hymenoptera: Megachilidae) and fruit trees, *Bulletin of Entomological Research* 92(1), 3–16. https://doi.org/10.1079/BER2001139.

Breeze, T. D., Bailey, A. P., Balcombe, K. G. and Potts, S. G. (2011). Pollination services in the UK: how important are honey bees?, *Agriculture, Ecosystems and Environment* 142(3–4), 137–143. https://doi.org/10.1016/J.AGEE.2011.03.020.

Brittain, W. H. and Eidt, C. C. (1933). Seed content, seedling production and fruitfulness in apples, *Canadian Journal of Research* 9(4), 307–333. https://doi.org/10.1139/cjr33-088.

Brown, M. J. F., Dicks, L. V., Paxton, R. J., Baldock, K. C., Barron, A. B., Chauzat, M. P., Freitas, B. M., Goulson, D., Jepsen, S., Kremen, C., Li, J., Neumann, P., Pattemore, D. E., Potts, S. G., Schweiger, O., Seymour, C. L. and Stout, J. C. (2016). A horizon scan of future threats and opportunities for pollinators and pollination, *PeerJ* 4, e2249. https://doi.org/10.7717/peerj.2249.

Caron, D. M. (2009). *Bee Colony Pollination Rental Prices, Eastern US with Comparison to West Coast*.

Crassweller, R. M., Kime, L. F. and Harper, J. K. (2017). Apple production, Penn State Extension. Available at: https://extension.psu.edu/apple-production (Accessed: 17 September 2021).

Crassweller, R. M., Krawczyk, G. and Peter, K. A. (2020). *Penn State Tree Fruit Production Guide 2020-2021*. University Park, PA: Penn State Extension.

DeBoer, G. J. and Satchivi, N. (2014). Comparison of translocation properties of insecticides versus herbicides that leads to efficacious control of pests as specifically illustrated by Isoclast™ Active, a new insecticide, and Arylex™ Active, a new herbicide. In: *ACS Symposium Series*. Indianapolis: American Chemical Society, pp. 75-93. https://doi.org/10.1021/bk-2014-1171.ch004.

FAO. (2018). *The Pollination of Cultivated Plants: A Compendium for Practitioners* (2nd edn., vol. 1) Roubik, D. (Ed.). Food and Agriculture Organization of the United Nations. Ancon, Republic of Panama: Smithsonian Tropical Research Institute. Available at: https://www.researchgate.net/publication/326158454_The_Pollination_of_cultivated_Plants_A_Compendium_for_Practitioners_Volume_1 (Accessed: 8 September 2022).

FAO and STAT. (2019). World apple production. Available at: http://www.fao.org/faostat/en/#data (Accessed: 16 September 2021).

Ferrier, P. (2018). Pollination service fees, Amber Waves - The economics of food, farming, natural resources, and rural America. Available at: https://www.ers.usda.gov/amber-waves/2018/march/pollination-service-fees/ (Accessed: 17 September 2021).

Fisher, A., Coleman, C., Hoffmann, C., Fritz, B. and Rangel, J. (2017). The synergistic effects of almond protection fungicides on honey bee (Hymenoptera: Apidae) forager survival, *Journal of Economic Entomology* 110(3), 802-808. https://doi.org/10.1093/jee/tox031.

FRAC. (2018). *FRAC Code List 2018: Fungicides Sorted by Mode of Action*.

Garratt, M. P. D., Breeze, T. D., Jenner, N., Polce, C., Biesmeijer, J. C. and Potts, S. G. (2014). Avoiding a bad apple: insect pollination enhances fruit quality and economic value, *Agriculture, Ecosystems and Environment* 184(100), 34-40. https://doi.org/10.1016/J.AGEE.2013.10.032.

Gauthier, N., Vaillancourt, L. J. and Hartman, J. R. (2018). 'Apple scab', *The Plant Health Instructor* [Preprint]. Available at: https://doi.org/10.1094/phi-iPHI-I-2000-1005-01.

Goulson, D., Nicholls, E., Botías, C. and Rotheray, E. L. (2015). Bee declines driven by combined stress from parasites, pesticides, and lack of flowers, *Science* 347(6229), 1255957. https://doi.org/10.1126/science.1255957.

Greer, L. (1999). Alternative pollinators: Native bees, *Appropriate Technology Transfer for Rural Area*, pp. 1-14. Available at: https://doi.org/http://www; jswcd.org/Files/ATTRA%20native%20pollinators.pdf.

Heller, S., Joshi, N. K., Chen, J., Rajotte, E. G., Mullin, C. A. and Biddinger, D. J. (2020). Pollinator exposure to systemic insecticides and fungicides applied in the previous fall and pre-bloom period in apple orchards, *Environmental Pollution* [Preprint] 265(A), 114589. https://doi.org/10.1016/j.envpol.2020.114589.

Hooven, L., Sagili, R. and Johansen, E. (2013). How to reduce bee poisoning from pesticides. Available at: https://catalog.extension.oregonstate.edu/sites/catalog/files/project/pdf/pnw591.pdf (Accessed: 12 September 2019).

Hopwood, J., Rajotte, E., Gill, K., Vaughan, M., Lee-Mader, E., May, E., Code, A., Joshi, N. K. and Biddinger, D. (2020). *Managing Eastern Apple Orchards for Pollinators and Other Beneficial Insects: Integrated Pest Management, Habitat Enhancement and Managed Bees*. University Park, PA: Penn State University Press, p. 98.

Hopwood, J., Vaughan, M., Shepherd, M., Biddinger, D., Mader, E., Black, S. H. and Mazzacano, C. (2016). *Are Neonicotinoids Killing Bees? A Review of Research into the Effects of Neonicotinoid Insecticides on Bees* (2nd edn.) Morris, S. (Ed.). Portland, OR: The Xerces Society for Invertebrate Conservation.

Johnson, R. M. and Percel, E. G. (2013). Effect of a fungicide and spray adjuvant on queen-rearing success in honey bees (Hymenoptera: Apidae), *Journal of Economic Entomology* 106(5), 1952–1957.

Joshi, N. K., Otieno, M., Rajotte, E. G., Fleischer, S. J. and Biddinger, D. J. (2016). Proximity to woodland and landscape structure drives pollinator visitation in apple orchard ecosystem. Frontiers in *Ecology* and *Evolution* 4, 38.

Kuhn, E. D. and Ambrose, J. T. (1984). Pollination of "Delicious" apple by Megachilid bees of the genus *Osmia* (Hymenoptera), *Journal of the Kansas Entomology Society* 57(2), 169–180.

MacHardy, W. E. (1996). *Apple Scab: Biology, Epidemiology, and Management*. St. Paul, MN: American Phytopathological Society (APS Press).

Matsumoto, S., Abe, A. and Maejima, T. (2009). Foraging behavior of Osmia cornifrons in an apple orchard, *Scientia Horticulturae* 121(1), 73–79.

McMenamin, A. J., Brutscher, L. M., Glenny, W. and Flenniken, M. L. (2016). Abiotic and biotic factors affecting the replication and pathogenicity of bee viruses, *Current Opinion in Insect Science* 16, 14–21. https://doi.org/10.1016/J.COIS.2016.04.009.

Melathopoulos, A. P., Cutler, G. C. and Tyedmers, P. (2015). Where is the value in valuing pollination ecosystem services to agriculture?, *Ecological Economics* 109, 59–70. https://doi.org/10.1016/j.ecolecon.2014.11.007.

Moisset, B. B. and Buchmann, S. (2011). Bee basics: An introduction to our native bees. pp: 40. Published by the USDA, US Forest Service and Pollinator Partnership. Available on-line at: http://www. pollinator.org.

OPP, PMRA and CALDPR. (2014). *Guidance for Assessing Pesticide Risks to Bees*. Washington, D.C.; Ottawa, ON; and Sacramento, CA.

Park, M. G., Danforth, B., Losey, J., Biddinger, D., Vaughan, M., Dollar, J., Rajotte, E. and Angello, A. (2012). Wild pollinators of Eastern apple orchards and how to conserve them. Cornell University, Penn. State University, and the Xerces Society. Available at: http://www.northeastipm.org/park2012.

Park, M. G., Joshi, N. K., Rajotte, E. G., Biddinger, D. J., Losey, J. E. and Danforth, B. N. (2018). Apple grower pollination practices and perceptions of alternative pollinators in New York and Pennsylvania, *Renewable Agriculture and Food Systems* 35(1), 1–14.

Park, M. G., Orr, M. C. and Danforth, B. N. (2010). The role of native bees in apple pollination, *New York Fruit Quarterly* 18(1), 21–25.

Partap, U. M. A., Partap, T. E. J. and Yonghua, H. E. (2001). 'Pollination failure in apple crop and farmers' management strategies in Hengduan Mountains, China, *Acta Horticulturae* 561(561), 225–230. https://doi.org/10.17660/ActaHortic.2001.561.32.

Peter, K. A. (2017a). Apple scab, Penn State extension - Apple disease. Available at: https://extension.psu.edu/apple-disease-apple-scab (Accessed: 17 September 2021).

Peter, K. A. (2017b). Powdery mildew, Penn State extension - Apple disease. Available at: https://extension.psu.edu/apple-disease-powdery-mildew (Accessed: 17 September 2021).

Peter, K. A., Crassweller, R. M. and Krawczyk, G. (2018). Penn State tree fruit production Guide 2018-2019. PennState Extension.

Ramírez, F. and Davenport, T. L. (2013). Apple pollination: A review, *Scientia Horticulturae* 162, 188-203. https://doi.org/10.1016/J.SCIENTA.2013.08.007.

Sekita, N. (2001). Managing *Osmia cornifrons* to pollinate apples in Aomori Prefecture, Japan, *Acta Horticulturae* 561(561), 303-307. https://doi.org/10.17660/ActaHortic.2001.561.46.

Sekita, N. and Yamada, M. (1993). Use of *Osmia cornifrons* to pollinate apples in Aomori Prefecture, Japan, *Japan Agricultural Research Quarterly* 26, 264-270.

Sheffield, C. S., Ngo, H. T. and Azzu, N. (2016). *A Manual on Apple Pollination*. Rome: Food and Agriculture Organization.

Shugrue, S. E. (2016). *Pesticide use and habitat manipulation and management changes as factors in pollinator sustainability in Pennsylvania apple orchards*. Thesis. The Pennsylvania State University.

Sur, R. and Stork, A. (2003). Uptake, translocation and metabolism of Imidacloprid in plants, *Bulletin of Insectology* 56(1), 35-40.

Sutton, T. B., Aldwinckle, H. S., Agnello, A. M. and Walgenbach, J. F. (eds). (2014). Compendium of *Apple* and *Pear Diseases* and *Pests (pp. 20-21)*. St. Paul, MN: American Phytopathological Society.

Turechek, W. W., Carroll, J. E. and Rosenberger, D. A. (2004). Powdery mildew of apple, *Tree Fruit IPM*. New York: Cornell University & the New York State IPM Program, p. 3.

US EPA. (2021). Colony collapse disorder. Available at: https://www.epa.gov/pollinator-protection/colony-collapse-disorder (Accessed: 17 September 2021).

USDA ARS. (2021). Colony collapse disorder, ARS honey bee health. Available at: https://www.ars.usda.gov/oc/br/ccd/index/ (Accessed: 17 September 2021).

USDA NASS. (2017). *Cost of Pollination*.

USDA NASS. (2020). 2020 state agriculture overview for Pennsylvania. Available at: https://www.nass.usda.gov/Quick_Stats/Ag_Overview/stateOverview.php?state=PENNSYLVANIA (Accessed: 16 September 2021).

USDA NASS. (2021). National statistics for apples. Available at: https://quickstats.nass.usda.gov/results/C4D9BA73-243D-3330-9D22-AC09511CE3AD?pivot=short_desc (Accessed: 16 September 2021).

Westwood, N. W. (1993). *Temperate Zone Pomology: Physiology and Culture* (3rd edn.). Portland, OR: Timber Press, Inc., p. 523.

Yamada, M., Oyama, N., Sekita, N., Shirasaki, S. and Tsugawa, C. (1971). The ecology of the Megachilid bee *Osmia cornifrons* and its utilization for apple pollination, *Bulletin of Aomori Experiment Station* 15, 1-80 (in Japanese).

Yoder, J. A., Nelson, B. W., Jajack, A. J. and Sammataro, D. (2017). Fungi and the effects of fungicides on the honey bee colony. In: Vreeland, R. H. and Sammataro, D. (Eds) *Beekeeping - From Science to Practice*. Cham: Springer International Publishing, pp. 73-90. https://doi.org/10.1007/978-3-319-60637-8_5.

Chapter 13

Entomovectoring: using pollinators to spread biocontrol agents

Jari Temmermans, Ghent University and Antwerp University, Belgium; Marie Legein and Sarah Leeber, Antwerp University, Belgium; and Guy Smagghe, Ghent University, Belgium

1 Introduction

Plants and insects have co-evolved in a mutually beneficial relationship for pollination. Ollerton et al. (2011) consider plant-pollinator relationships to be one of the most ecologically and economically important relationships on Earth as 75% of the world's major food crops show increased fruit and/or seed set as a result of animal pollination. Important pollinators include butterflies, beetles, flies, bats, and moths, but the vast majority of pollination is done by bees (Halder et al., 2019). In the greenhouse, pollination by wild pollinators is generally not feasible. Commercially available pollinators are then used to obtain a reasonable fruit set in sheltered cultivation (Guerra-Sanz, 2008). For the strawberry plant, for example, the most important commercially used pollinators in sheltered cultivation are the honeybee (*Apis mellifera*) and the bumblebee (*Bombus* spp.) (Abrol et al., 2019; Paydas et al., 1998; Zaitoun et al., 2006).

Insects such as the honeybee and the bumblebee are known pollinators and have the capacity to carry small particles (such as pollen) from plant to plant (Mommaerts et al., 2010; Wael et al., 1990). These pollinators can also transport

http://dx.doi.org/10.19103/AS.2022.0111.23

bacterial cells and fungal spores from flower to flower (Mommaerts et al., 2010). It is the combination of a pollination service and the use of the pollinators as vectors for the transport of beneficial organisms such as biocontrol agents (BCAs) to combat plant pathogens, which gave rise to the entomovectoring system. In the system, the pollinators fulfill an important dual role of pollination and crop protection (Mommaerts and Smagghe, 2011).

In this chapter, we review the most important factors for successful entomovectoring. Several previous applications of the entomovectoring technology using two commercially used bumblebee species, *Bombus terrestris* and *Bombus impatiens,* will be discussed. Finally, we will present a case study focusing on vectoring bacterial BCAs using bumblebees to combat *Botrytis cinerea*, the causal agent of gray mold, a commonly feared plant pathogen on strawberry.

2 Factors important for successful entomovectoring

The term 'entomovectoring technology' was introduced by Hokkanen and Menzler-Hokkanen (2007) indicating the use of pollinators as disseminators of BCAs against plant pathogens (Kevan et al., 2020). However, other terms such as apivectoring or bee vectoring are used as well (Coates et al., 2019). Nowadays there are companies such as Biobest (Belgium), Bee Vectoring Technology (Canada), and Assatek (Finland) which have developed business models based on entomovectoring technology. Entomovectoring has several advantages over conventional applications of BCAs, such as spraying. Among others, it guarantees that the BCAs will be applied to the flowers from the moment they open (Karise et al., 2016). This is important as, in many cases, the flowers are the main target of infection (Bristow et al., 1986; Mommaerts and Smagghe, 2011). This transport to the target location is also much more efficient than spraying the BCA because biocontrol organisms will not be lost to the surrounding plant tissue and soil to such a large extent (Smagghe et al., 2012). However, when using an entomovectoring system, not only the flowers but also the leaves can be protected with BCAs. For example, Kapongo et al. (2008a) reported that 90% of the leaves of the tomato plant and 76% of the leaves of the bell pepper plant contained the fungal biocontrol organism *Clonostachys rosea* after entomovectoring with *Bombus impatiens* inside a greenhouse. In addition, each flower is subjected to several visits, which implies the continuous application of the biocontrol organisms directly onto the flower (Macedo et al., 2020). Furthermore, entomovectoring requires less inoculum, decreases the need for external inputs such as energy, chemicals, and water, reduces the cost and labor, and minimizes non-target organisms exposure (Macedo et al., 2020). This also fits within the framework of the European Green Deal in which the European Commission aims to reduce net greenhouse gas emissions by at least 55% by 2030 and to overcome climate change and environmental degradation (https://ec.europa.eu/info/strategy/priorities-2019-2024/european-green-deal_en).

The success of entomovectoring depends on its various components and their interactions. A suitable vector, BCA, formulation, dispenser, and the guarantee of safety for the environment, vector, and human health are all important factors (Fig. 1) (Mommaerts and Smagghe, 2011).

2.1 The vector

The number of insect species that have been used as vectors in entomovectoring systems is quite limited because most insects are only commercially available for a limited part of the year. As a consequence, the entomovectoring system cannot always be optimally synchronized with the flowering period of the plant needing protection (Mommaerts and Smagghe, 2011). Honeybees and bumblebees that are commercially available throughout the year are

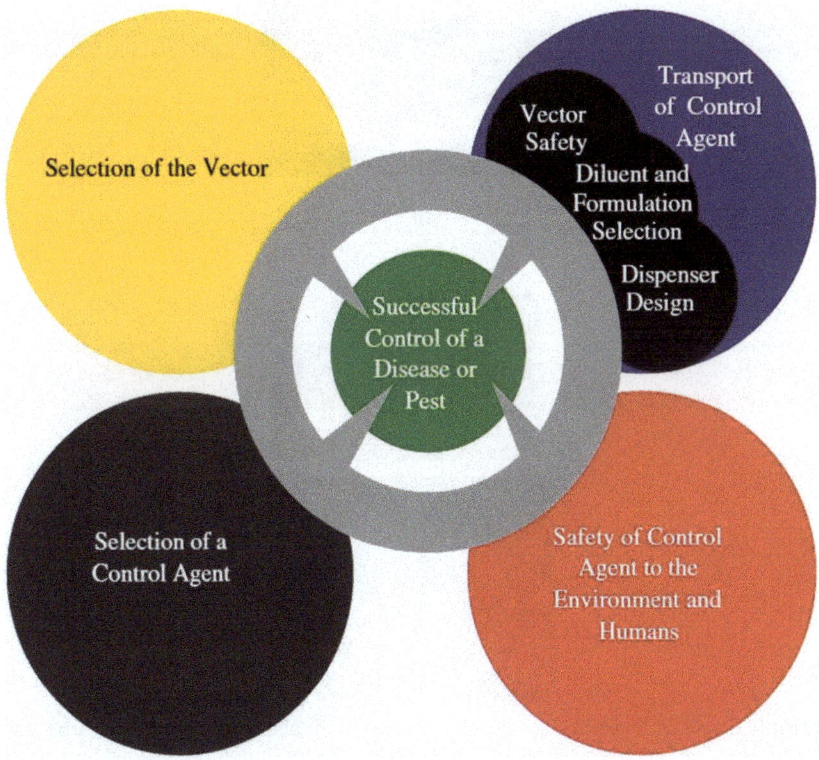

Figure 1 Schematic overview of the factors important for successful entomovectoring and their interactions. Adapted from Mommaerts and Smagghe (2011).

therefore a very popular choice for use and have been the vectors in many scientific investigations of entomovectoring (Kevan et al., 2020). Commercially, honeybees (*Apis mellifera*) (Fig. 2c) have been used in North America, Europe, South America, and Asia (Kevan et al., 2020; Kovach et al., 2000; Kapongo et al., 2008b). Bumblebees have been successfully deployed in North America (*Bombus impatiens*) (Fig. 2b), Europe (*Bombus terrestris*) (Fig. 2a) and Korea (*Bombus terrestris*) (Karise et al., 2016; Kevan et al., 2020; Kovach et al., 2000). Other important reasons for using bumblebees, aside from their commercial availability, are their morphological and behavioral characteristics. Bumblebees, whose bodies are covered with branched hairs in large part, can trap and transport pollen grains and thus also BCAs easily (Mänd et al., 2020).

When choosing between honeybees and bumblebees, the conditions under which entomovectoring is occurring have to be considered. Greenhouses without temperature control are prone to temperature fluctuations as a result of

Figure 2 Detail of important pollinators: (a) the buff-tailed bumblebee or large earth bumblebee *Bombus terrestris* (Sadd et al., 2015), (b) the common eastern bumblebee *Bombus impatiens* (Sadd et al., 2015), and (c) the western honeybee or European honeybee *Apis mellifera* (De la Rúa et al., 2014).

refracted sunlight on the glass of the greenhouse. Honeybees are less suitable for flying under greenhouse conditions and bumblebees are often the vector of choice because they can tolerate temperature extremes better and are more resistant to temperature fluctuations than are honeybees (Goulson, 2010; Mommaerts et al., 2011). Bumblebees also have a better vision than honeybees, which gives them a navigational advantage in a greenhouse (Guerra-Sanz, 2008). Another important advantage that bumblebees have over honeybees is that they perform buzz pollination, a very efficient pollination technique in which the flower stamens are shaken intensively by the vibrating wings of the insect, resulting in the formation of a cloud of pollen grains (Corbet et al., 1988). Buzz pollination is critical for the pollination success of some crops such as tomatoes and blueberries (Goulson, 2010; Guerra-Sanz, 2008).

The choice of a particular vector and more specifically its health play a role in the choice of the BCA. It is important that the BCA does not exert any adverse effects on the vector or on surrounding non-target organisms for which the biocontrol is not intended (Kevan et al., 2008). Examples of commercially available products that are already considered safe for *Bombus terrestris* are Binab-T-vector, a 50/50 mixture of *Trichoderma atroviride* and *Hypocrea parapilulifera*, and Prestop-Mix with *Gliocladium catenulatum* as its active component (Mommaerts et al., 2010).

2.2 The dispenser

A crucial step in the vectoring of BCAs by pollinators is to load the vector in an efficient manner to ensure sufficient dispersal of the BCA (Maccagnani et al., 2020). To achieve this, designing a suitable dispenser, a device loaded with the BCA in powder form through which the vectors have to walk in order to leave the hive, is essential (Mänd et al., 2020). A good dispenser should apply sufficient powdery BCA to the vector, be safe for the insect, have no or as little effect as possible on the foraging of the vector in question, and the refilling interval should be as long as possible (preferably greater than 1 day) (Smagghe et al., 2012). The two basic models available are the one-way dispenser where the entrance and exit of the dispenser are in the same space and the two-way dispenser where the exit and entrance are separated. The latter has proven to be the most suitable dispenser in terms of applying BCA to both bumblebees and honeybees, its effect on the foraging behavior of the bee vectors, and the time interval between dispenser recharging (Smagghe et al., 2012). Furthermore, from a practical point of view, complexity, cost of construction, and ease of use should be considered as well (Mommaerts et al., 2010).

The data as far available on the compatibility of dispensers with bumblebees are rather limited (see review by Maccagnani et al. (2020)). Yu and Sutton (1997) conducted the first dispenser study with bumblebees (*Bombus impatiens*), showing that *Bombus impatiens* could be loaded with BCAs to the same extent as *Apis mellifera*. Dispenser designs include the Houle dispenser (Albano et al., 2009), the one-way side-by-side passageway (Maccagnani et al., 2005) and the two-way overlapping passageway developed for entomovectoring with *Bombus terrestris* by Maccagnani et al. (2005). Dispenser designs generally lack information about their effect on induced grooming by the vectors upon contact with the powdery BCA product or their effect on foraging activity, although both of these might affect the ability of the vectors to disseminate the inoculum to the target crops (Maccagnani et al., 2020).

Mommaerts et al. (2010) developed another dispenser specifically for *Bombus terrestris* (Fig. 3a). This system achieved a load of Prestop-Mix (with the fungal BCA *Gliocladium catenulatum*) of 10^4 CFU per bumblebee worker. Unlike the one-way side-by-side passageway dispenser, it did not have any effect on the insect's foraging, with no significant effect on the number of bumblebees leaving the hives. In greenhouse experiments with strawberry plants, Mommaerts et al. (2010) found that, after the bumblebees had fledged, approximately 100 CFU per flower of the strawberry plant was detected after a single flight. Karise et al. (2016) used this dispenser developed by Mommaerts et al. (2010) in which *Bombus terrestris* loaded with Prestop-Mix successfully

(a)

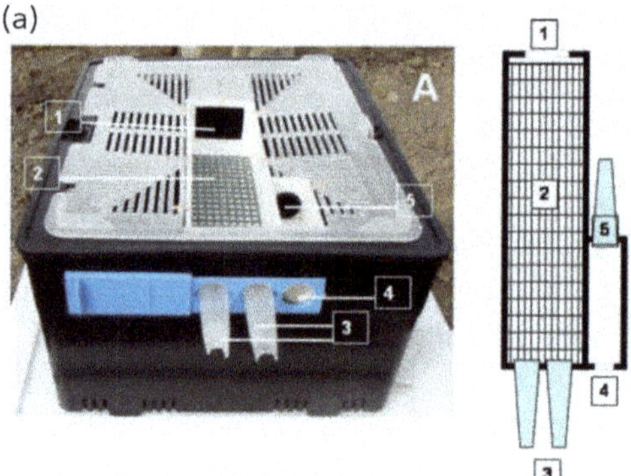

(b)

Figure 3 (a) Mommaerts dispenser developed for bumblebees (Mommaerts et al., 2010). (b) Flying Doctors® hive from Biobest (www.biobestgroup.com; Westerlo, Belgium).

suppressed the plant pathogen *Botrytis cinerea* on the strawberry plant. In good weather conditions, the number of infected flowers was reduced 2.9 times in the first year and 1.7 times in the second year. Furthermore, Mommaerts et al. (2011) have successfully used this dispenser in the greenhouse to combat *Botrytis cinerea* in strawberry plants with *Bombus terrestris* and Prestop-Mix. Here, the total yield (pre- and post-harvest) increased 2–2.5 fold compared to the control.

Finally, the Flying Doctors® hive (Fig. 3b) is a patented and commercially available nest for vectoring using *Bombus terrestris* and *Bombus impatiens* with an integrated dispenser based on the previously described dispenser developed by Mommaerts et al. (2010). The workers can leave the hive by walking through a tray that can be loaded with the powdery BCA. After disseminating the BCA on the target crop, the bees can return to the hive through a separate entrance.

Compared to dispensers for bumblebees, the first dispensers for honeybees were already developed in the 1950s (Maccagnani et al., 2020). These dispensers were used to load the vector with pollen in order to obtain cross-pollination. However, these devices were unfit for biological control due to an insufficient loading as reviewed by Maccagnani et al. (2020). Peng et al. (1992) were the first to develop a honeybee dispenser fit for biological control. The authors used the dispenser to transport the fungus *Gliocladium catenulatum* to combat *Botrytis cinerea* on the strawberry plant. The Gross-dispenser (Gross et al., 1994), the Triwaks dispenser (Bilu et al., 2004), and the Houle dispenser (Albano et al., 2009) followed. Furthermore, one commercially available honeybee dispenser, BeeTreat, has been developed and was described by Smagghe et al. (2012). This dispenser was shown to successfully load honeybees with *Gliocladium catenulatum* for the control of *Botrytis cinerea* on strawberry (Hokkanen et al., 2015).

Unfortunately, no thorough scientific comparison between these different dispensers has been made. In order to further optimize entomovectoring by honeybees, a scientific comparative study between all available dispensers would be necessary and very useful (Maccagnani et al., 2020).

Solitary bees, unlike social bees, build their own nest and raise their own progeny, and can be used for the dissemination of BCAs as well (Maccagnani and Sgolastra, 2020). However, because the dispenser has to be integrated in the shelter, this is limited to those species that nest in preexisting cavities and exhibit some properties of social behavior (Maccagnani et al., 2020). Maccagnani et al. (2006) were first to develop a dispenser for *Osmia cornuta* to vector *Bacillus subtilis* BS-F4 for the control of fire blight on pear (*Pyrus communis*; cultivar Abbé Fétel). The number of BCAs per vector of *Osmia cornuta* ranged between 10^4 and 10^7 CFU/bee (Maccagnani et al., 2006). However, the dispenser was not ideal as after a time, bees learned to exit the dispenser without passing through the powdery BCA formulation. Maccagnani et al. (2008) later successfully increased the dispenser's efficiency by, among other modifications, relocating the dispenser towards a higher part of the shelter.

2.3 The biocontrol agent product

Various microorganisms including viruses, bacteria, and fungi have been suggested as possible microbial BCAs, and at least a dozen have already been tested (reviewed by Kevan et al., 2020). Typically, the products used in entomovectoring studies are powdery BCA formulations available as commercial products (Mommaerts and Smagghe, 2011). In order to obtain a cost-effective, maximal dispersibility and dosage onto the plant, BCAs are often diluted (Kevan et al., 2008). The main purpose of these formulation products or carriers is to reduce the loss of BCA powder in flight (Mommaerts and Smagghe, 2011). The carriers used can be divided into mineral (talc and clays) and vegetable (flours and starch) materials (Kevan et al., 2020). The vegetable carriers have proven to work especially well under laboratory and experimental conditions (Kevan et al., 2008).

For example, Mommaerts et al. (2011) used a mixture of Prestop-Mix and corn starch (Maizena) together with *Bombus terrestris* in strawberry plants to combat *Botrytis cinerea*. The authors observed that the use of Maizena led to an equally efficient dissemination of the BCA into the flowers with only half the initial concentration of Prestop-Mix (Mommaerts et al., 2011). Furthermore, this mixture had no significant effect on the foraging behavior of the bumblebees (Mommaerts et al., 2011).

Commercial formulations have to meet several criteria. First of all, they have to adhere effectively to the vector's body and dislodge when pollination of the target flower occurs. Second, they should have no effect on the behavioral characteristics (e.g. foraging or grooming) of the vector or cause any health problems whatsoever (Kevan et al., 2020). Furthermore, it is important that they are food grade, easily available, inexpensive, and safe (Mommaerts and Smagghe, 2011). Finally, they should remain as dry and accessible as possible inside the dispensers and be readily applicable by beekeepers and growers.

2.4 Biosafety

In order to assure that entomovectoring technology is safe for use in the human food supply chain, safe for growers to use, safe for the cultivated crops, and has no or minimal environmental effects, environmental safety and registration must be assessed (Kevan et al., 2020). This implies that an environmental and human risk assessment is required before BCAs can be used with this technology

(Kevan et al., 2008). Often, the registration period for microbial pesticides is shorter than the one for conventional pesticides (McVey and Wassenberg, 2020).

Besides human health, the health of non-target organisms such as the vector transporting the BCAs needs to be safeguarded as well. According to Mommaerts and Smagghe (2011), there are three levels at which side effects of the vector need to be assessed including acute and chronic mortality, sublethal effects on worker and nest reproduction, and effects on pollinator foraging behavior. As mentioned above, a wide variety of BCAs have been suggested; however, tests on safety of those agents to pollinators vary from very thorough to completely absent (Kevan et al., 2020).

Mommaerts et al. (2012) developed a laboratory miniature-dispenser-based bioassay to screen for lethal and sublethal effects. Using this bioassay system, the authors found that the carrier kaolin was toxic for *Bombus terrestris*. Using a one-way dispenser, mortality of over 70% was observed after 5 weeks of exposing the bee to the carrier. In addition, it was shown that kaolin had a significant negative effect on the vector's reproductive abilities when nests were exposed to the carrier (Mommaerts et al., 2012). During the same experiments, cellulose and wheat flour had no significant effects on the viability of the workers (Mommaerts et al., 2012).

Thus far, very few BCAs have been tested on their effect on pollen viability and germination rate, embryogenesis, and seed set (Kevan et al., 2020). However, Ngugi et al. (2005) attempted to investigate the adverse effects of applying Serenade® with *Bacillus subtilis* as active bacteria BCA to suppress floral infection by *Monilinia vaccinii-corymbosi* on blueberry. Application of Serenade on detached blueberry flowers had no effect on the number of pollen tubes or their growth rates. When honeybees were utilized to vector Serenade to open flowers, the fruit weight was not significantly altered. However, the fruit set was significantly reduced from 49.1% to 38.1% and the number of seeds was halved (Ngugi et al., 2005). Nonetheless, even in treatments receiving no Serenade, the fruit weight and number of seeds were already low, indicative of poor pollination overall (Ngugi et al., 2005). The authors concluded that the use of Serenade® had no inherently adverse effects but that caution should be applied when using this product in circumstances otherwise unfavorable for pollination.

3 Examples of bumblebees as vectors for entomovectoring: successes and failures

Honeybees or bumblebees have been used as vectors for biocontrol for more than 25 years. One of the first comprehensive studies was one by Peng et al. (1992) in which the BCA *Gliocladium roseum* was delivered by honeybees to the flowers of the strawberry plant to control *Botrytis cinerea*. Since then, several studies using bumblebees have been conducted (Table 1).

Kapongo et al. (2008b) demonstrated how a mix of the fungi *Clonostachys rosea* and *Beauveria bassiana* transported by *Bombus impatiens* reduced *Botrytis cinerea* infection by 57% and 46%, respectively, in the flowers and leaves of the tomato plant and by 59% and 47%, respectively, in the flowers and leaves of the sweet pepper. Kovach et al. (2000) investigated the use of both honeybees and bumblebees compared to spraying for the transmission of the fungus *Trichoderma harzianum* on the strawberry plants in open fields. The authors found that despite the lower concentration of *Trichoderma harzianum* on the flowers using the entomovectoring system, entomovectoring still resulted in a better biocontrol than the more conventional spraying. The authors credited this to the more targeted application of the BCA at the appropriate time and place where needed, (i.e. when the flowers open). The spraying of *Trichoderma harzianum* was done before all flowers had opened, whereas the vectors flew continuously with an almost immediate protection of all flowers at opening as a result. Furthermore, when making use of the insects as a vector, the strawberries contained on average 22% more seeds resulting in a yield increase of 26–40% by weight (Kovach et al., 2000). When using the entomovectoring system with *Trichoderma harzianum*, both the level of control and the yield were at least as good or better compared to the use of commercial fungicides plus pollination (Kovach et al., 2000). The authors believed that the chemical fungicides (including captan, iprodione, benomyl, and vinclozolin) had a negative effect on pollination, resulting in fewer seeds and therefore a lower yield. Mommaerts et al. (2011) performed a study, using the previously described Mommaerts dispenser, Prestop-Mix, and the vector *Bombus terrestris* to combat *Botrytis cinerea* under greenhouse conditions. The authors observed that the use of this BCA and vector (in combination with Maizena as a formulation product) could provide a significantly higher yield compared to the control group without biocontrol organisms, even under conditions that were considered ideal for the fungus. The authors also found that the powder had no effect on the vector's foraging. Karise et al. (2016) observed that the weather conditions clearly influenced the proper functioning of Prestop-Mix transported by *Bombus terrestris* in the open field. The incidence of *Botrytis cinerea* was strongly reduced in good weather conditions, but the biocontrol was substandard during very rainy periods. Another important conclusion from these different experiments was that the use of *Bombus terrestris* resulted in a similar biocontrol of *Botrytis cinerea* compared to the honeybee *Apis mellifera*. Albano et al. (2009) tested the disseminating of the biofungicide Rootshield® with active component *Trichoderma harzianum* to strawberry plants in the field, using *Apis mellifera*, and in the greenhouse, using *Bombus impatiens*. The authors examined

Table 1 Examples of biological control agents vectored for crop protection by managed pollinating bumblebees

BCA	Crop	Pathogen	Vector	Effect	Reference
Clonostachys rosea + *Beauveria bassiana*	Field tomato Sweet pepper	*Botrytis cinerea*	*Bombus impatiens*	57% incidence 59% incidence	Kapongo et al. (2008b)
Trichoderma harzianum	Field strawberry	*Botrytis cinerea*	*Bombus impatiens*	Yield: +26–40%	Kovach et al. (2000)
Gliocladium catenulatum	Greenhouse strawberry	*Botrytis cinerea*	*Bombus terrestris*	Yield: ×2–2.5%	Mommaerts et al. (2011)
Gliocladium catenulatum	Field strawberry	*Botrytis cinerea*	*Bombus terrestris*	Weather dependent	Karise et al. (2016)
Clonostachys rosea	Field sunflower	*Sclerotinia sclerotiorum*	*Bombus terrestris*	Yield: +25%	Terzić et al. (2020)

the transmission of the powder from the Houle dispenser to the vector and by extension from the vector to the strawberry crops. They found that both honeybees and bumblebees were capable of dispersing the powder. However, no data were reported on the level of disease control (Albano et al., 2009).

Terzić et al. (2020) investigated entomovectoring with *Bombus terrestris* and *Clonostachys rosea* for the protection of sunflowers (*Helianthus annuus*), an important crop cultivated for oil, bird seed, and improved seed stock, against the pathogen *Sclerotinia sclerotiorum*, the causal agent of head rot. During the experiments, the authors observed that the BCA *Clonostachys rosea* affected the behavior of the bumblebees who actively avoided going through the biocontrol powder (Terzić et al., 2020). Decreased foraging of *Bombus terrestris* in the presence of *Clonostachys rosea* was observed as well. Despite these difficulties, the entomovectoring system managed to reduce the disease incidence (on average 2%, 13%, and 57% compared to 6%, 26%, and 68% in the controls) (Terzić et al., 2020). Although the presence of *Sclerotinia* was not significantly lower, the negative effects on the plant itself were significantly reduced by the entomovectoring system as the yield of the treated plants was on average 25% higher than in the control (Terzić et al., 2020).

4 Case study: entomovectoring of bacteria by *Bombus terrestris* against *Botrytis cinerea* in strawberry

An important goal of our own ongoing research is to find BCAs as alternatives for the spraying of chemical fungicides to combat *Botrytis cinerea* on the strawberry plant. For this, we have used *Bombus terrestris* as the vector for the transfer of different bacteria selected based on their *in vitro* inhibitory effect against *Botrytis cinerea*.

4.1 Strawberry production and the importance of Botrytis cinerea

According to Debode et al. (2013), strawberry (*Fragaria × ananassa*) can be considered one of the most widely cultivated small fruit crops on a global scale with a worldwide cultivation of more than 350,000 ha and fruit production estimated at 8.3 million tons (FAO, 2018).

Fungal diseases in strawberry may be responsible for heavy economic losses in this crop (Debode et al., 2013). For example, the fungus *Botrytis cinerea*, the causal agent of gray mold, is the most common reason for fruit loss causing crop losses of 20–25% among European strawberry growers (European Commission, 2013) and further losses during shipping, storage, and retail, and for end consumers (Abbey et al., 2018; Dean et al., 2012). *Botrytis* can cause both quantitative losses (reduced yield) and qualitative damage, such as an inferior taste or smell (Fillinger and Elad, 2015).

Botrytis cinerea is capable of infecting all above-ground parts of the plant including, for example, flowers, leaves, and shoots (Cota et al., 2008). However, the most damaging infections occur during the flowering of the host plant where infection often starts on the petals, stamens, and pistils and can subsequently colonize the fruit via the ovary (Cota et al., 2008). Therefore, to prevent infection, it is of utmost importance that treatment methods are timed to coincide with new flower budding (Departement Landbouw and Visserij, 2015; Smagghe, 2020). Although infection of the fruit occurs in the field, often symptoms will not be noticed until the strawberries ripen and the fungus begins to form spores (Bristow et al., 1986).

4.2 Control of **Botrytis cinerea**

At present, *Botrytis cinerea* is still mainly controlled by chemical fungicides applied to the plants both before and after harvest (Petrasch et al., 2019; Wedge et al., 2007). There is no question that fungicides have improved disease control, crop quality, and economic opportunities (Ragsdale and Alford, 1991). However, recently consumers are demanding a reduction in the amounts of fungicides present on the fruit and used during its production (Debode et al., 2013; Sams, 2018). There has also been an increasing legislative pressure to shift towards more environmentally friendly alternatives (https://ec.europa.eu/info/strategy/priorities-2019-2024/european-green-deal_en) by reducing the number of registered authorized active components (European Parliament and the Council, 2021; FOD, 2020). Several active components are no longer sold or have lost their registration for use on the strawberry plant in North America (https://www.epa.gov/) (https://www.canada.ca/en/health-canada/services/consumer-product-safety/pesticides-pest-management.html). In Europe, a similar trend has been taking place (FOD, 2020; Wedge et al., 2007), with iprodione and thiram having lost their registration in Europe for use on strawberry for combating *Botrytis cinerea* in 2017 and 2018, respectively.

Limiting the use of chemical fungicides is important to prevent deleterious and often lethal effects on beneficial insects and fungi (Budi et al., 2000). For example, chitosan may cause modifications in the cell walls of *Fusarium oxysporum*, a ubiquitous soil-bound fungus known for the degradation of lignin (Sutherland et al., 1983). Cyproconazole, in turn, can affect the cell viability of *Athelia rolfsii*, a saprotrophic but facultatively pathogenic soil fungus (Budi et al., 2000). Besides effects on beneficial fungi, the fungicides captan, thiophanate methyl, and benomyl can have a negative effect on the ability of pollen to germinate, resulting in a lower strawberry yield (Kovach et al., 2000).

Reducing chemical fungicide use can also mitigate fungal resistance to available fungicides and reduce fungicidal residues on the fruit, thus protecting human health (Fillinger and Elad, 2015).

The increase in resistance to chemical fungicides displayed by pathogens is another recurrent problem associated with fungicide use (Dean et al., 2012; Fillinger and Elad, 2015; Mommaerts et al., 2011; Petrasch et al., 2019; Wedge et al., 2007). *Botrytis cinerea* is designated as a high-risk pathogen by the Fungicide Resistance Action Committee because of its wide genetic variability, adaptation skills, and resistance. By modifying the target molecules on which the fungicides act and by actively transporting fungicides outwardly, *Botrytis cinerea* can develop resistance to several antimicrobial agents simultaneously (Fillinger and Elad, 2015). An increasing number of both *in vitro* and *in situ* tests indicate an increased resistance in *Botrytis* spp. for, among others, methyl benzimidazole carbamate-generating components (benomyl, carbendazim, and thiophanate-methyl) and the dicarboximides (vinclozolin and iprodione) (Wedge et al., 2007). As early as 1984, Barak and Edgington (1984) demonstrated significant resistance to the active components captan, folpet, captafol, etem, thiram, and chlorothalonil.

Fungal resistance combined with increased environmental concerns, continued consumer demand, and more stringent policies in many countries have increased the need for alternatives to fungicides. To this end, the European Commission has issued a directive to encourage integrated pest management (European Commission, 2021). This implies the integration of measures that combat pathogenic organisms and keeps the use of chemical protection products and other interventions to an economically and ecologically justified level. Part of this strategy is to preferentially use sustainable biological methods over chemical methods if they can sufficiently protect the plant (European Commission, 2021). Within an Integrated Pest Management (IPM) framework, bacterial BCAs are a promising alternative.

4.3 Measuring acquisition of viable bacterial biocontrol agents on workers of Bombus terrestris

To evaluate the viability, loading, and dispersal of potential bacterial BCAs onto bumblebees (*Bombus terrestris*; Biobest, Westerlo, Belgium) to control *Botrytis cinerea*, we used a dry bacterial formulation placed into a dispenser (Fig. 4a,b). To load the bumblebees, the bacterial formulation was mixed with corn starch (Maizena) (Carrefour, Belgium) in a 1:10 ratio by weight. Because it mimics pollen, Maizena can facilitate the loading of the bacteria onto the bee vector to reduce the loss of bacterial spores during flight towards the strawberry plant (Smagghe et al., 2012). Diluting and mixing the bacteria with Maizena minimally affected the viability of the BCA and bacterial loading occurred on 73% of the bees coming

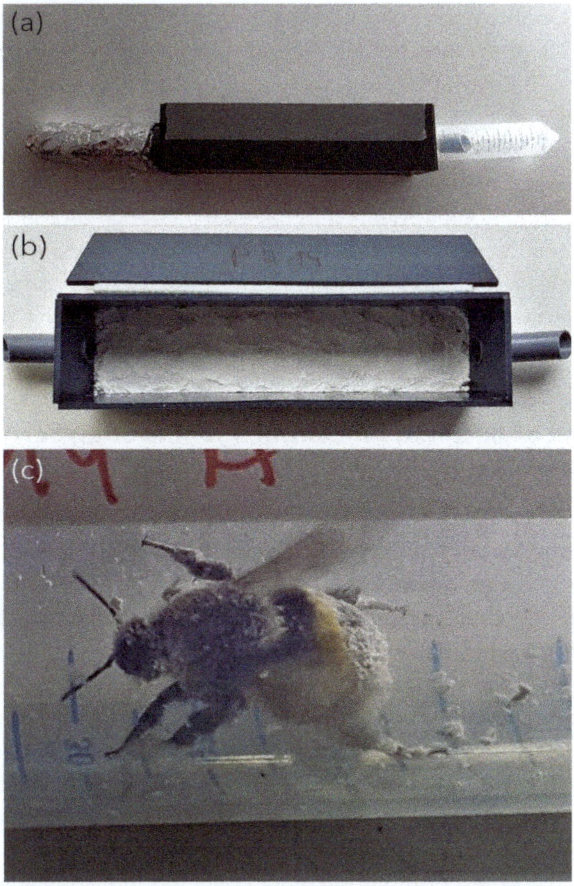

Figure 4 (a) Overview experimental setup with miniature dispenser (length = 20 cm). *Bombus terrestris* was collected in a dark 50 mL tube (left side of the dispenser) and, attracted by a light source, captured on the opposite side. (b) Detail inside dispenser filled with 1:10 bacteria with Maizena. (c) Bumblebee worker (*Bombus terrestris*) exiting the dispenser, loaded with 1:10 bacterial formulation. After passage through the dispenser, the worker was captured, washed using leaf washing buffer (LWB), and subsequently plated out. Plating method: 10 µL of a 1 mg/mL suspension of the 1:10 bacterial formulation was plated in triplicate on selective agar plates containing cycloheximide (0.1 g/L) in a dilution series. CFUs were counted in the countable dilution.

in contact with the BCA-Maizena mixture in the dispenser. As the bumblebees passed through the dispenser, their hairy legs, thorax, and abdomen were loaded with the bacterial powder (Fig. 4c). This is in agreement with the findings of Mommaerts et al. (2010) for *Bombus terrestris* who identified the thorax, abdomen, and legs as the main loading sites with *Gliocladium catenulatum*.

However, the bacterial load on the bumblebee workers was quite variable. This can partly be explained by the imperfect mixing of the bacterial formulation, but also by the walking path of the bees through the powder formulation inside the dispenser. Furthermore, not all bumblebees stayed in the dispenser for the same amount of time, which could result in higher loading with the bacterial mix for those individuals that remained in the dispenser longer. The bumblebees also differed in size; thus, the potential surface area available for acquisition with the bacterial powder varied among individuals. Finally, storage at 4°C of the bacterial formulation appeared to have an influence on the acquisition by the bumblebees because the powder was hygroscopic and became clumpy. The change in texture could not only lead to a negative effect on the acquisition by the bumblebees (Fig. 4c) but also lead to a decrease in the viability of the BCA.

4.4 Transport of bacterial biocontrol agent to strawberry plants in a single visit

After observing that bacteria could consistently be loaded on *Bombus terrestris*, we evaluated the amount of bacterial transfer from the dispenser to the flowers of strawberry in a single visit using a mass balance approach that measured the amount of BCA that was picked up by the bee (*Bombus terrestris*) from the dispenser, the amount of BCA that was lost in flight, and the amount of BCA that was deposited onto the flower

After releasing the bumblebees but before adding the bacterial formulation to the dispenser, three random flowers, the blanks, were cut off and transferred into a sterile 50 mL tube. Next, a dispenser containing 1:10 bacterial mixture with Maizena was attached to the hive. Subsequently, a bumblebee worker passed through, flew, and landed on the plants (Fig. 5c). Both the pollinated flower (Fig. 6d) and the bumblebee were collected individually in a sterile 50 mL tube and kept refrigerated for transportation to the lab (Fig. 5e).

In total, four different bumblebees were captured using tweezers and subsequently passed through the dispenser filled with freshly mixed bacterial formulation. Subsequently, all bumblebees flew towards a first and single flower of one of the three strawberry plants (Fig. 5a). Bacterial powder was present between the hairs of the bumblebee after passing through the dispenser (Fig. 5c). Following the capture of the bumblebee, in some cases, traces of the white powder could also be observed on the flower (Fig. 5d).

By plating out on a selective medium, no loading with bacteria could be detected on the three blank flowers cut before the start of the experiment, showing that no background bacteria were present. Thus, it can be assumed that all bacteria encountered on the flowers were in fact the result of entomovectoring with *Bombus terrestris*. The four flowers that were visited first by one of the bumblebees were all significantly loaded with bacteria. After visiting a flower, the bumblebees were recaptured and evaluated showing that

Figure 5 Greenhouse experiment with strawberry plants: transport of bacteria during single first flight (a) Overview of experimental greenhouse (4 × 3 m) with strawberry plants. (b) One representative plot of six strawberry plants. (c) Worker of *Bombus terrestris* with powder on her body during a first, single flower visit. (d) Presence of powder in flower after first, single flower visit. The washing and plating out of all flowers and bumblebees were completely analogous to the previous dispenser experiment as was explained in Figure 4.

the number of CFU on these bumblebees varied relatively greatly. The reasons for this variability can be of various origins. First, as previously described, there is variation in bacterial acquisition on the bumblebees exiting the dispenser. Some bumblebees flew longer or even displayed grooming behavior before flying towards the plant, which likely resulted in a lower load. It should also be noted that catching the bumblebees with tweezers is a stressor for these insects, making them less likely to prioritize visiting a plant. Furthermore, it was observed that for two bumblebees, after capture, bacterial loads were lower than the loads on the flower they visited, implying that the bumblebees in question had left most of their initial payload on the plant and/or had lost it in flight.

4.5 Transport of biocontrol agent by vectors over 4 days

After having determined the loading of a strawberry flower as a result of a single flight, we mapped the loading and hence the potential protection of the

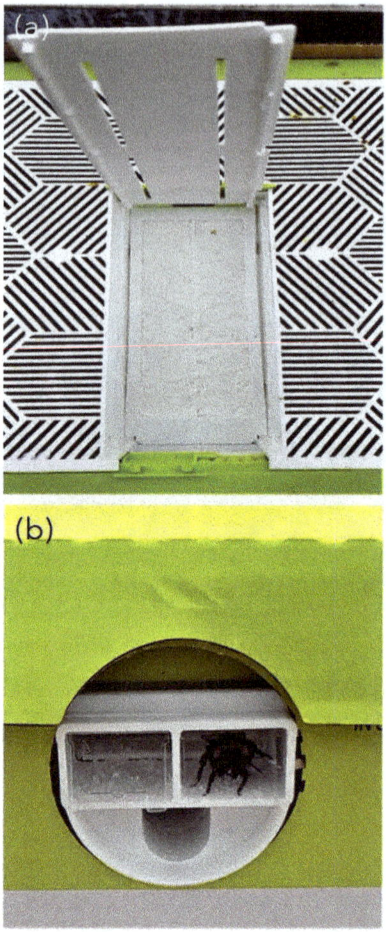

Figure 6 Set-up of the greenhouse experiment (4 × 3 m) at Ghent University with strawberry plants as in Figure 5 to study the transport of bacteria during 4 days of entomovectoring. (a) Loading of powder product (bacteria + Maizena, 1:10, w:w) in Flying Doctors® system. (b) A worker bee (*Bombus terrestris*) leaving the exit of the Flying Doctors nest. Washing and plating of the flowers were done analogous to the previously described experiments (Figures 4 and 5).

plant from the plant's point of view over a 4-day entomovectoring period. We measured whether all flowers were loaded and to what extent the loading and thus the potential protection was homogeneously distributed.

Figure 6a provides an overview of the experimental design: the grower, cultivar, and location were the same as described during the previous experiment as presented in Fig. 5. For 4 days, bumblebees were able to pollinate the flowers at any time (Fig. 6c). The dispenser of a Flying Doctors hive was daily replenished with 1:10 bacterial mixture with Maizena (Fig. 6b).

At 24 h, 48 h, 72 h, and 96 h after filling the platform for the first time, 6–16 of the open flowers were collected for a total of 37 collected flowers representing 30–50% of the entire flower population present (determined on a visual basis). Before filling the platform for the first time, three flowers were collected as blanks.

Bacteria were detectable in all collected flowers at every time point, and all flowers sampled had significantly higher loadings than the three blanks. However, it should be noted that the current experiment was carried out on a small scale, so it is uncertain how these results would translate to a larger scale.

In contrast to the results of the experiment after a single flight (see above), the loaded flowers in this experiment may have been visited repeatedly by more than one bee and/or by bees visiting consecutive flowers during a single flight. Furthermore, the survival of the BCA is accounted for better over the longer term. The average load on strawberries after 4 days of entomovectoring was approximately 1.14% of the initial load after a single flight. One explanation for this decrease over time is a decrease in the viability of bacteria on the phyllosphere as well as inside the dispenser due to external stressors such as drought stress. Another important potential cause is redistribution. This implies that during the time period of the experiment, the bacteria were redistributed from flower to flower by bumblebees visiting different flowers sequentially. This is a very important fact, especially on a commercial scale, as on that scale not all flowers would be first in line to be visited by a bumblebee right after leaving the dispenser. This immediately shows the power of the entomovectoring system to distribute a homogeneous load of BCA to flowers in an efficient manner.

The obtained results show the extent and consistency with which bacteria are deposited on the flowers of *Fragaria × ananassa* by *Bombus terrestris* are promising. These results prompt us to think about the next step, that of carrying out the experiments on a larger scale to ensure that results apply to the commercial production context, either in the greenhouse or in the field.

5 Conclusion and future trends

Entomovectoring has shown much promise (Karise et al., 2016; Kovach et al., 2000; Mommaerts et al., 2011; Peng et al., 1992, case study). Nevertheless, there is room for improvement of the systems' efficiency under adverse environmental conditions such as bad weather days (Espinosa et al., 2020; Karise et al., 2016). In addition, new products with suitable formulations need to be developed in order to improve the acquisition of the formulation by the vectors and provide alternatives for growers.

The bacteria discussed above have displayed the potential to perform as BCA. During the dispenser experiments, 11 out of 15 bumblebee workers (~72%) were significantly loaded with bacteria. Furthermore, after a first and single flight, there was consistent transfer of the bacteria from the bumblebee to the flower.

This consistency was confirmed as 100% of the flowers sampled (i.e. at least 30–50% of the total number of flowers) were found to be loaded with bacteria and redistribution from flower to flower took place via the activity of the bumblebees.

To move one step closer to the commercialization of bacteria, their inhibitory activity in the field or in commercial greenhouses must also be determined. To assess whether bacteria can also effectively protect the strawberry plant on a commercial scale, more large-scale experiments need to be performed and the duration of the experiment should be increased.

Several other optimization possibilities should be considered. For example, the hygroscopic and clumpy texture of the bacterial formulation with Maizena should be addressed with a possible higher acquisition by *Bombus terrestris* as a result.

It is also important to investigate the effect of the BCA bacteria on bumblebees, the environment, and humans. After all, the strawberry plant is a food crop and will therefore have to meet strict conditions. Hence, it is important that the BCA used does not exert any harmful effects on the strawberries, the consumer, or the vector.

More testing is required not only for new formulations but also for other bee species. In future studies, native species should be considered as well as their pollination efficiency might be higher due to closer co-evolutionary relationships with the native crops under protection or improved adaptation (Espinosa et al., 2020). The level of attraction between the bees and plants under protection should be investigated as well (Espinosa et al., 2020).

Another important factor that should be considered for further improvements to entomovectoring systems is the dispenser. By tracking and using the results that commercial growers obtain by making use of commercially available dispensers, interesting new information should surface for further improvement (Maccagnani et al., 2020). By investigating the separated traits of the dispenser as was done by Mommaerts et al. (2010) for their developed dispenser, optimization of the dispenser could be achieved (Maccagnani et al., 2020). To ease further testing, standard protocols should be developed.

A final important aspect is the evaluation of socioeconomic factors. After all, producers often display resistance to change during commercialization (Espinosa et al., 2020). In some cases, the traditional agricultural methods limit the implementation of alternate technologies such as entomovectoring technology, for example, the use of products that are harmful to bees (Espinosa et al., 2020). In this regard, work has to be done to make producers see the entomovectoring technology as part of a strategy with a more sustainable and integrated management approach to protect crops against damage by pests and diseases. It may be possible to integrate entomovectoring into a larger crop protection system in which chemical fungicides are applied to plants before flowering and after harvest to reduce fungal loads and BCAs are applied during flowering to reduce exposure to fungicide residues for bees and reduce fungicide residues on fruit consumed by humans. This option may offer the best of both worlds.

6 Where to look for further information

Smagghe et al. (2020) describes research and development of entomovectoring for use both in pollination and in the delivery of biocontrol agents against crop and pollinator diseases, pests, and parasites. The book describes studies on use of entomovectoring in both open fields and greenhouse settings. Mommaerts and Smagghe (2011) provides a nice overview of the different components of the entomovectoring system and how it works. Temmermans and Smagghe (2022) shares a critical view on the (dis)advantages of entomovectoring and on transport of pollinators for the purpose of entomovector systems. Professional organizations already commercializing the entomovectoring system are Biobest (https://www.biobestgroup.com/) and Bee Vectoring Technology (https://www.beevt.com/).

7 References

Abbey, J. A., Percival, D., Abbey, L., Asiedu, S. K., Prithiviraj, B. and Schilder, A. (2018). Biofungicides as alternative to synthetic fungicide control of grey mould (*Botrytis cinerea*)-prospects and challenges. *Biocontrol Science and Technology* 29(3), 241-262.

Abrol, D. P., Gorka, A. K., Ansari, M. J., Al-Ghamdi, A. and Al-Kahtani, S. (2019). Impact of insect pollinators on yield and fruit quality of strawberry. *Saudi Journal of Biological Sciences* 26(3), 524-530.

Albano, S., Chagnon, M., de Oliveira, D., Houle, É., Thibodeau, P. O. and Mexia, A. (2009). Effectiveness of *Apis mellifera* and *Bombus impatiens* as dispersers of the Rootshield® biofungicide (*Trichoderma harzianum*, strain T-22) in a strawberry crop. *Hellenic Plant Protection Journal* 2, 57-66.

Barak, E. and Edgington, L. V. (1984). Cross-resistance of *Botrytis cinerea* to captan, thiram, chlorothalonil, and related fungicides. *Canadian Journal of Plant Pathology* 6(4), 318-320.

Bilu, A., Dag, A., Elad, Y. and Shafir, S. (2004). Honey bee dispersal of biocontrol agents: an evaluation of dispensing devices. *Biocontrol Science and Technology* 14(9), 607-617.

Bristow, P. R., Mcnicol, R. J. and Williamson, B. (1986). Infection of strawberry flowers by *Botrytis cinerea* and its relevance to grey mould development. *Annals of Applied Biology* 109(3), 545-554.

Budi, S. W., van Tuinen, D., Arnould, C., Dumas-Gaudot, E., Gianinazzi-Pearson, V. and Gianinazzi, S. (2000). Hydrolytic enzyme activity of *Paenibacillus* sp. strain B2 and effects of the antagonistic bacterium on cell integrity of two soil-borne pathogenic fungi. *Applied Soil Ecology* 15(2), 191–199.

Coates, C., Willis Chan, S., Kevan, P., Stanisavljević, L. and Smagghe, G. (2019). First international advanced course on using managed pollinators for dissemination of biological control agents for suppression of insect, fungal & other pests of crops. *Bulletin of the Entomological Society of Canada* 51(3), 159–161.

Corbet, S. A., Chapman, H. and Saville, N. (1988). Vibratory pollen collection and flower form: bumble bees on Actinidia, Symphytum, Borago and Polygonatum. *Functional Ecology* 2(2), 147–155.

Cota, L. V., Maffia, L. A., Mizubuti, E. S. G., Macedo, P. E. F. and Antunes, R. F. (2008). Biological control of strawberry gray mold by *Clonostachys rosea* under field conditions. *Biological Control* 46(3), 515–522.

Dean, R., Van Kan, J. A., Pretorius, Z. A., Hammond-Kosack, K. E., Di Pietro, A., Spanu, P. D., Rudd, J. J., Dickman, M., Kahmann, R., Ellis, J. and Foster, G. D. (2012). The top 10 fungal pathogens in molecular plant pathology. *Molecular Plant Pathology* 13(4), 414–430.

De la Rúa, P., Paxton, R. J., Moritz, R. F. A., Roberts, S., Allen, D. J., Pinto, M. A., Cauia, E., Fontana, P., Kryger, P., Bouga, M., Buechler, R., Costa, C., Crailsheim, K., Meixner, M., Siceanu, A. and Kemp, J. R. (2014). *Apis mellifera. The IUCN Red List of Threatened Species 2014*. Accessed 01/08/2021.

Debode, J., Van Hemelrijck, W., Creemers, P. and Maes, M. (2013). Effect of fungicides on epiphytic yeasts associated with strawberry. *Microbiology Open* 2(3), 482–491.

Departement Landbouw en Visserij (2015). *Ziekten en Plagen in Aardbeien*. Technical Report. Brussel: Department Landbouw & Visserij.

Espinosa, S., Figueroa, J., Kevan, P. G., Baéz, C., Solarte, V., Smagghe, G. and Sánchez, A. (2020). Advances in the implementation of apivectoring technology in Colombia: strawberry case (Fragaria x ananassa). In: Smagghe, G., Boecking, O., Maccagnani, B., Mänd, M. and Kevan, P. G. (Eds.), *Entomovectoring for Precision Biocontrol and Enhanced Pollination of Crops*. Switzerland: Springer Nature, pp. 201–237.

European Commission (2013). Biosensor-based instrumentations to be used in vineyards and wineries for fast and sensitive detection of *Botrytis cinerea*, grey rot. In: *Grapes*. Accessed 30/12/2020. Available at: https://cordis.europa.eu/project/id/232453/reporting/de.

European Commission (2021). Integrated pest management. Accessed 21/03/2021. Available at: https://ec.europa.eu/food/plant/pesticides/sustainable_use_pesticides/ipm_en.

European Parliament and the Council (2021). Uitvoeringsverordening (EU) 2017/2091 van de Commissie. Accessed 21/03/2021. Available at: https://eur-lex.europa.eu/eli/reg_impl/2017/2091/oj.

FAO (2018). FAOSTAT. Accessed 04/12/2020. Available at: http://www.fao.org/faostat/en/{#}data/QC.

Fillinger, S. and Elad, Y. (2015). *Botrytis - The Fungus, The Pathogen and Its Management in Agricultural Systems*. Springer International Publishing.

FOD (2020). Fytoweb: Gewasbeschermingsmiddelen en Bemestingsproducten. Accessed 29/12/2020. Available at: https://fytoweb.be/nl.

Goulson, D. (2010). *Bumblebees: Behaviour, Ecology, and Conservation*. Oxford: Oxford University Press.

Gross, H. R., Hamm, J. J. and Carpenter, J. E. (1994). Design and application of a hive-mounted device that uses honey bees (Hymenoptera: Apidae) to disseminate *Heliothis* nuclear polyhedrosis virus. *Environmental Entomology* 23(2), 492–501.

Guerra-Sanz, J. M. (2008). Chapter 3: crop pollination in greenhouses. In: Pitts-Singer, T. L. (Ed.), *Research Review James Bee Pollination in Agricultural Ecosystems*. New York: Oxford University Press, pp. 27–47.

Halder, S., Ghosh, S., Khan, R., Khan, A. A., Perween, T. and Hasan, M. A. (2019). Role of pollination in fruit crops: a review. *The Pharma Innovation Journal* 8, 695–705.

Hokkanen, H. and Menzler-Hokkanen, I. (2007). Use of honeybees in the biological control of plant diseases. *Entomological Research* 37, A62–A63.

Hokkanen, H. M. T., Menzler-Hokkanen, I. and Lahdenpera, M.-L. (2015). Managing bees for delivering biological control agents and improved pollination in berry and fruit cultivation. *Sustainable Agriculture Research* 4(3), 89.

Kapongo, J. P., Shipp, L., Kevan, P. and Broadbent, B. (2008a). Optimal concentration of *Beauveria bassiana* vectored by bumble bees in relation to pest and bee mortality in greenhouse tomato and sweet pepper. *BioControl* 53(5), 797–812.

Kapongo, J. P., Shipp, L., Kevan, P. and Sutton, J. C. (2008b). Co-vectoring of *Beauveria bassiana* and *Clonostachys rosea* by bumble bees (*Bombus impatiens*) for control of insect pests and suppression of grey mould in greenhouse tomato and sweet pepper. *Biological Control* 46(3), 508–514.

Karise, R., Dreyersdorff, G., Jahani, M., Veromann, E., Runno-Paurson, E., Kaart, T., Smagghe, G. and Mänd, M. (2016). Reliability of the entomovector technology using Prestop-Mix and *Bombus terrestris* L. as a fungal disease biocontrol method in open field. *Scientific Reports* 6, 31650.

Kevan, P. G., Kapongo, J. P. and Al Mazra'awi, M. (2008). Chapter 5: Honey bees, bumble bees and biocontrol: new alliances between old friends. In: Pitts-Singer, T. L. (Ed.), *Research Review James Bee Pollination in Agricultural Ecosystems*. New York: Oxford University Press, pp. 65–82.

Kevan, P. G., Shipp, L. and Smagghe, G. (2020). Ecological intensification: managing biocomplexity and biodiversity in agriculture through pollinators, pollination and deploying biocontrol agents against crop and pollinator diseases, pests and parasites. In: Smagghe, G., Boecking, O., Maccagnani, B., Mänd, M. and Kevan, P. G. (Eds.), *Entomovectoring for Precision Biocontrol and Enhanced Pollination of Crops*. Switzerland: Springer Nature, pp. 19–51.

Kovach, J., Petzoldt, R. and Harman, G. E. (2000). Use of honey bees and bumble bees to disseminate *Trichoderma harzianum* 1295-22 to strawberries for *Botrytis* control. *Biological Control* 18(3), 235–242.

Maccagnani, B., Giacomello, F., Fanti, M., Gobbin, D., Maini, S. and Angeli, G. (2008). *Apis mellifera* and *Osmia cornuta* as carriers for the secondary spread of *Bacillus subtilis* on apple flowers. *BioControl* 54(1), 123–133.

Maccagnani, B., Mocioni, M., Ladurner, E., Gullino, M. L. and Maini, S. (2005). Investigation of hive-mounted devices for the dissemination of microbiological preparations by *Bombus terrestris*. *Bulletin of Insectology* 58(1), 3–8.

Maccagnani, B., Pisman, M. and Smagghe, G. (2020). Dispensers for entomovectoring: for every bee a different type? In: Smagghe, G., Boecking, O., Maccagnani, B., Mänd,

M. and Kevan, P. G. (Eds.), *Entomovectoring for Precision Biocontrol and Enhanced Pollination of Crops*. Switzerland: Springer Nature, pp. 95–122.

Maccagnani, B., Bazzi, C., Biondi, E., Tesoriero, D. and Maini, S. (2006). Potential of Osmia cornuta as a carrier of antagonist bacteria in biological control of fire blight: A comparison with Apis mellifera. *Acta Horticulturae* 704, 379–386. https://doi.org/10.17660/ActaHortic.2006.704.59

Maccagnani, B. and Sgolastra, F. (2020). Solitary bees as pollinators. In: Smagghe, G., Boecking, O., Maccagnani, B., Mänd, M. and Kevan, P. G. (Eds.), *Entomovectoring for Precision Biocontrol and Enhanced Pollination of Crops*. Switzerland: Springer Nature, pp. 63–79.

Macedo, J., Viana, B., Freitas, B., Medeiros, A., Kevan, P. G. and Vergara, C. H. (2020). The potential of bee vectoring on coffee in Brazil. In: Smagghe, G., Boecking, O., Maccagnani, B., Mänd, M. and Kevan, P. G. (Eds.), *Entomovectoring for Precision Biocontrol and Enhanced Pollination of Crops*. Switzerland: Springer Nature, pp. 165–181.

Mänd, M., Karise, R. and Smagghe, G. (2020). Bumble bees and entomovectoring in open field conditions. In: Smagghe, G., Boecking, O., Maccagnani, B., Mänd, M. and Kevan, P. G. (Eds.), *Entomovectoring for Precision Biocontrol and Enhanced Pollination of Crops*. Switzerland: Springer Nature, pp. 81–93.

McVey, E. A. and Wassenberg, J. (2020). Regulatory processes surrounding the risk assessment of microbial pesticides for pollinators. In: Smagghe, G., Boecking, O., Maccagnani, B., Mänd, M. and Kevan, P. G. (Eds)., *Entomovectoring for Precision Biocontrol and Enhanced Pollination of Crops*. Switzerland: Springer Nature, pp. 251–261.

Mommaerts, V., Put, K. and Smagghe, G. (2011). *Bombus terrestris* as pollinator-and-vector to suppress *Botrytis cinerea* in greenhouse strawberry. *Pest Management Science* 67(9), 1069–1075.

Mommaerts, V., Put, K., Vandeven, J., Jans, K., Sterk, G., Hoffmann, L. and Smagghe, G. (2010). Development of a new dispenser for microbiological control agents and evaluation of dissemination by bumblebees in greenhouse strawberries. *Pest Management Science* 66(11), 1199–1207.

Mommaerts, V., Put, K., Vandeven, J. and Smagghe, G. (2012). Miniature dispenser-based bioassay to evaluate the compatibility of powder formulations used in an entomovectoring approach. *Pest Management Science* 68(6), 922–927.

Mommaerts, V. and Smagghe, G. (2011). Entomovectoring in plant protection. *Arthropod-Plant Interactions* 5(2), 81–95.

Ngugi, H. K., Dedej, S., Delaplane, K. S., Savelle, A. T. and Scherm, H. (2005). Effect of flower-applied Serenade biofungicide (*Bacillus subtilis*) on pollination-related variables in rabbiteye blueberry. *Biological Control* 33(1), 32–38.

Ollerton, J., Winfree, R. and Tarrant, S. (2011). How many flowering plants are pollinated by animals? *Oikos* 120(3), 321–326.

Paydas, S., Eti, S., Kaftanoglu, O., Yasa, E. and Derin, K. (1998). Effects of pollination of strawberries grown in plastic greenhouses by honeybees and bumblebees on the yield and quality of the fruits. *Acta Horticulturae* 513(513), 443–452.

Peng, G., Sutton, J. C. and Kevan, P. G. (1992). Effectiveness of honey bees for applying the biocontrol agent *Gliocladium roseum* to strawberry flowers to suppress *Botrytis cinerea*. *Canadian Journal of Plant Pathology* 14(2), 117–129.

Petrasch, S., Knapp, S. J., van Kan, J. A. L. and Blanco-Ulate, B. (2019). Grey mould of strawberry, a devastating disease caused by the ubiquitous necrotrophic fungal pathogen *Botrytis cinerea*. *Molecular Plant Pathology* 20(6), 877–892.

Ragsdale, N. and Alford, H. (1991). *Health and Environmental Factors Associated With Agricultural Use of Fungicides, National Agricultural Pesticide Impact Assessment Program*.

Sadd, B., Barribeau, S., Bloch, G., de Graaf, D., Dearden, P., Elsik, C., Gadau, J., Grimmelikhuijzen, C., Hasselmann, M., Lozier, J., Robertson, H., Smagghe, G., Stolle, E., VanVaerenbergh, M., Waterhouse, R., Bornberg-Bauer, E., Klasberg, S., Childers, A., Camara, F. and Worley, K. (2015). The genomes of two key bumblebee species with primitive eusocial organization. *Genome Biology* 16, 76.

Sams, C. (2018). Better safe than sorry on chemicals used in agriculture. *Guardian Letters*.

Smagghe, G. (2020). Case studies on entomovectoring in the greenhouse and open field. In: Smagghe, G., Boecking, O., Maccagnani, B., Mänd, M. and Kevan, P. G. (Eds.), *Entomovectoring for Precision Biocontrol and Enhanced Pollination of Crops*. Switzerland: Springer Nature, pp. 123–136.

Smagghe, G., Boecking, O., Maccagnani, B., Mänd, M. and Kevan, P. G. (2020). Entomovectoring for precision biocontrol and enhanced pollination of crops. https://doi.org/10.1007/978-3-030-18917-4

Smagghe, G., Mommaerts, V., Hokkanen, H. and Menzler-Hokkanen, I. (2012). Chapter 5: multitrophic interactions: the entomovector technology. In: Smagghe, G. and Diaz, I. (Eds.), *Arthropod-Plant Interactions: Novel Insights and Approaches for IPM*. Dordrecht: Springer Netherlands, pp. 127–157.

Sutherland, J. B., Pometto, A. L., III and Crawford, D. L. (1983). Lignocellulose degradation by Fusarium species. *Canadian Journal of Botany* 61(4), 1194–1198.

Temmermans, J. and Smagghe, G. (2022). Different bees as vectors for entomovectoring with enhanced pollination and crop protection control: current practices, use cases and critical view on transport. *Revue Scientifique et Technique (International Office of Epizootics)* 41(1), 107–116. https://doi.org/10.20506/rst.41.1.3308

Terzić, S., Dedić, B., Tančić Živanov, S., Milovac, Ž., Franeta, F., Zorić, M., Stanisavljević, L. and Kevan, P. G. (2020). Using bumblebees (*Bombus terrestris*) as bioagent vectors to control Sclerotinia head rot on sunflower in Serbia. In: Smagghe, G., Boecking, O., Maccagnani, B., Mänd, M. and Kevan, P. G. (Eds.), *Entomovectoring for Precision Biocontrol and Enhanced Pollination of Crops*. Switzerland: Springer Nature, pp. 183–199.

Wael, L., Greef, M. and Laere, O. (1990). The honey bee as a possible vector of *Erwinia amylovora* (Burr.). *Acta Horticulturae* 273, 107–114.

Wedge, D. E., Smith, B. J., Quebedeaux, J. P. and Constantin, R. J. (2007). Fungicide management strategies for control of strawberry fruit rot diseases in Louisiana and Mississippi. *Crop Protection* 26(9), 1449–1458.

Yu, H. and Sutton, J. C. (1997). Effectiveness of bumblebees and honeybees for delivering inoculum of *Gliocladium roseum* to raspberry flowers to control *Botrytis cinerea*. *Biological Control* 10(2), 113–122.

Zaitoun, S. T., Al Ghzawi, A. A. M., Shannag, H. K. and Al-Tawaha, A. R. M. (2006). Comparative study on the pollination of strawberry by bumble bees and honeybees under plastic house conditions in Jordan Valley. *Journal of Food, Agriculture and Environment* 4(2), 237–240.

Index

www.ingramcontent.com/pod-product-compliance
Ingram Content Group UK Ltd.
Pitfield, Milton Keynes, MK11 3LW, UK
UKHW021825100926
8556IPUK00003B/165

9 781801 460989